Spectrum and Power Allocation in Cognitive Radio Systems

Kannadhasan Suriyan
Study World College of Engineering, India

Dhaya R.
King Khalid University, Saudi Arabia

R. Nagarajan
Gnanamani College of Technology, India

Alagar Karthick
KPR Institute of Engineering and Technology, India

IGI Global
PUBLISHER of TIMELY KNOWLEDGE

A volume in the Advances in Wireless Technologies and Telecommunication (AWTT) Book Series

Published in the United States of America by
IGI Global
Information Science Reference (an imprint of IGI Global)
701 E. Chocolate Avenue
Hershey PA, USA 17033
Tel: 717-533-8845
Fax: 717-533-8661
E-mail: cust@igi-global.com
Web site: http://www.igi-global.com

Library of Congress Cataloging-in-Publication Data

CIP DATA PROCESSING

2024 Information Science Reference
ISBN(hc): 9798369328934
ISBN(sc): 9798369350652
eISBN: 9798369328941

British Cataloguing in Publication Data
A Cataloguing in Publication record for this book is available from the British Library.

For electronic access to this publication, please contact: eresources@igi-global.com.

Advances in Wireless Technologies and Telecommunication (AWTT) Book Series

Xiaoge Xu
University of Nottingham Ningbo China, China

ISSN:2327-3305
EISSN:2327-3313

Mission

The wireless computing industry is constantly evolving, redesigning the ways in which individuals share information. Wireless technology and telecommunication remain one of the most important technologies in business organizations. The utilization of these technologies has enhanced business efficiency by enabling dynamic resources in all aspects of society.

The **Advances in Wireless Technologies and Telecommunication Book Series** aims to provide researchers and academic communities with quality research on the concepts and developments in the wireless technology fields. Developers, engineers, students, research strategists, and IT managers will find this series useful to gain insight into next generation wireless technologies and telecommunication.

Coverage

- Mobile Technology
- Radio Communication
- Telecommunications
- Wireless Broadband
- Wireless Technologies

IGI Global is currently accepting manuscripts for publication within this series. To submit a proposal for a volume in this series, please contact our Acquisition Editors at Acquisitions@igi-global.com or visit: http://www.igi-global.com/publish/.

The Advances in Wireless Technologies and Telecommunication (AWTT) Book Series (ISSN 2327-3305) is published by IGI Global, 701 E. Chocolate Avenue, Hershey, PA 17033-1240, USA, www.igi-global.com. This series is composed of titles available for purchase individually; each title is edited to be contextually exclusive from any other title within the series. For pricing and ordering information please visit http://www.igi-global.com/book-series/advances-wireless-technologies-telecommunication/73684. Postmaster: Send all address changes to above address.

Titles in this Series

For a list of additional titles in this series, please visit: www.igi-global.com/book-series

Security Issues and Solutions in 6G Communications and Beyond
Digvijay Pandey (Department of Technical Education, Government of Uttar Pradesh, India) Binay Kumar Pandey (Department of Information Technology, Govind Ballabh Pant University of Agriculture and Technology, India) and Tanveer Ahmad (Tata Consultancy Services, UK)
Information Science Reference • copyright 2024 • 339pp • H/C (ISBN: 9798369329313) • US $295.00 (our price)

Radar and RF Front End System Designs for Wireless Systems
Shilpa Mehta (Auckland University of Technology, New Zealand) and Rupesh Kumar (SRM University, India)
Information Science Reference • copyright 2024 • 354pp • H/C (ISBN: 9798369309162) • US $255.00 (our price)

Metamaterial Technology and Intelligent Metasurfaces for Wireless Communication Systems
Shilpa Mehta (Auckland University of Technology, New Zealand) and Arij Naser Abougreen (University of Tripoli, Libya)
Information Science Reference • copyright 2023 • 364pp • H/C (ISBN: 9781668482872) • US $225.00 (our price)

Opportunities and Challenges of Industrial IoT in 5G and 6G Networks
Poshan Yu (Soochow University, China & Australian Studies Centre, Shanghai University, China) Xiaohan Hu (Shanghai University, China) Ajai Prakash (University of Lucknow, India) Nyaribo Wycliffe Misuko (KCA University, Kenya) and Gu Haiyue (Shanghai University, China)
Information Science Reference • copyright 2023 • 304pp • H/C (ISBN: 9781799892663) • US $250.00 (our price)

Applications of Artificial Intelligence in Wireless Communication Systems
Karan Kumar (Maharishi Markandeshwar University (Deemed), Mullana, India)
Information Science Reference • copyright 2023 • 284pp • H/C (ISBN: 9781668473481) • US $250.00 (our price)

Designing and Developing Innovative Mobile Applications
Debabrata Samanta (Rochester Institute of Technology, Kosovo)
Information Science Reference • copyright 2023 • 426pp • H/C (ISBN: 9781668485828) • US $225.00 (our price)

IGI Global
PUBLISHER of TIMELY KNOWLEDGE

701 East Chocolate Avenue, Hershey, PA 17033, USA
Tel: 717-533-8845 x100 • Fax: 717-533-8661
E-Mail: cust@igi-global.com • www.igi-global.com

Table of Contents

Detailed Table of Contents

B. Yamini, SRM Institute of Science and Technology, India
P. J. Beslin Pajila, Vel Tech Rangarajan Dr. Sagunthala R&D Institute of Science and Technology, India
T. P Anish, R.M.K. College of Engineering and Technology, India
B. Gracelin Sheena, Sathyabama Institute of Science and Technology, India
M. G. Dinesh, EASA College of Engineering and Technology, India
M. Nalini, S.A. Engineering College, India
Siva Subramanian R., R.M.K. College of Engineering and Technology, India

Cognitive radio networks (CRNs) revolutionise wireless communications by providing dynamic spectrum access and management to overcome static spectrum allocation. This overview explores CRNs from fundamentals to cutting-edge protocols and applications. It addresses cognitive radio theory, hardware, networking, and difficulties for academics and practitioners. Starting with CRNs' history and motivation, it describes research scope and qualities. The research examines dynamic spectrum access, waveform design, and resource allocation. Detailed networking protocol and standard analysis covers medium access control and interference management. Additionally, cooperative communications and database-driven CRNs are considered. To solve spectrum sharing economics and regulations, next-gen technologies, energy-efficient designs, creative applications, and business models are explored. Examining flexible CRN feature machine learning methods. The analysis concludes with future trends and guides cognitive radio network research and innovation.

Sivasankari Jothiraj, Ultra College of Engineering and Technology, Madurai
M. Brindha, Velammal College of Engineering and Technology, India
V Persis Jeffrey, Velammal College of Engineering and Technology, India

Cognitive radio networks (CRNs) represent a dynamic and intelligent paradigm for efficient spectrum utilization by allowing unlicensed users, known as secondary users, to opportunistically access underutilized spectrum bands. However, the open and shared nature of the spectrum in CRNs exposes them to various security threats and attacks. In this context, the need for robust intrusion detection mechanisms becomes paramount to safeguard the integrity of the network. This work proposes a novel approach for detecting attackers in cognitive radio networks using a combination of self-attention mechanism and long short-term memory (LSTM) networks. The self-attention LSTM model is designed to capture complex temporal dependencies and spatial correlations in the dynamic spectrum access patterns, making it well-suited

heat sensors, and sophisticated algorithms, these systems swiftly identify potential fire outbreaks. Real-time monitoring facilitates prompt responses, thereby reducing the likelihood of extensive damage. As global efforts intensify towards bolstering safety protocols, the ongoing evolution of fire detection and suppression systems emerges as a crucial factor in mitigating the catastrophic consequences of fire incidents. The proposed project entails an IoT-based fire management system that integrates computer vision technology to revolutionize conventional firefighting methodologies. This comprehensive system prioritizes precise fire detection, accurate object identification, and optimized extinguishing strategies based on temperature analysis.

In cognitive radio networks, secondary users face spectrum scarcity. Cognitive radio offers a solution by enabling sharing of frequency bands to maximize spectrum utilization. This chapter explores spectrum allocation behavior, where secondary users allocate underutilized spectrum to maximize profit. Prior research suggests cooperative spectrum sharing benefits users, but real networks see selfish behavior to mitigate risks. The authors propose a dynamic duopoly game for competitive spectrum allocation. Secondary users employ bounded rationality to adapt strategies, aiming for Nash Equilibrium. One user considers competitors' past allocations to estimate future ones. The authors establish the existence and uniqueness of Nash Equilibrium, ensuring fairness. Local stability analysis of fixed points is conducted. Numerical simulations, including bifurcation diagrams, show that users integrating historical data and bounded rationality are more likely to reach Nash Equilibrium and maintain system stability.

Enhancing safety protocols in train stations requires proactive approaches to foresee and reduce any mishaps. In order to maximize safety in railroad environments, this study presents a novel machine learning framework that combines neural networks with random forest techniques. This hybrid model departs from traditional single-algorithm approaches by utilizing a variety of methods to enhance accident investigation and prediction. The chapter focuses on using this hybridized method, showcasing the functions of random forests and neural networks in predicting future events and assessing accident data from the past. The strategy of incorporating neural networks (multi-layer perceptron (MLP)) and random forests into the accident analysis framework is covered in the study. It also investigates if this model can be implemented in real-time and at scale within railway station safety systems. The results of the research reveal that the hybridized machine learning approach is a promising tool for improving safety standards in railroad stations.

Frequency of Itemsets plays a crucial role in analytics of retail industry, which delves into latent patterns in customer purchasing behavior. This paper presents an Apriori algorithm to extract associations among products in a given dataset, shedding light on frequently co-occurring items. By discerning these relationships, the business gains profound insights into customer preferences and tendencies, aiming not only to understand current purchasing behavior but also to identify potential cross-selling opportunities. As businesses rely on transactional data for insights, analysis reliability hinges on data quality. This study explores missing values, outliers, and data inconsistency, impacting market basket analysis accuracy. Leveraging the Apriori algorithm facilitates the revelation of robust product associations, enabling strategic optimization and heightened customer satisfaction. The gleaned insights inform targeted marketing, product placements, and inventory management, catalyzing more effective business optimization in the retail sector.

In terms of security, proactively identifying and mitigating network assaults continues to be essential to the protection of vital systems. This chapter presents a unique method for real-time network intrusion detection that makes use of the resilient gradient boosting technique known as the CatBoost algorithm. Conventional signature-based techniques are inadequate against advanced threats because they frequently fail to adjust to changing assault patterns. Because ML (machine learning), and specifically CatBoost, can handle high-dimensional data, nonlinear relationships, and categorical variables, it appears to be a promising answer. This study uses CatBoost to create an advanced model that can recognize complex patterns in network traffic data. Using a large dataset that includes both attack and regular cases, the system gains the ability to identify minute changes in network behavior that could be signs of impending danger. Through feature importance analysis, one can gain a deeper understanding of the critical elements that contribute to the detection of attacks and, consequently, improve network vulnerabilities. Accuracy, F1-score, recall, and precision are among the measures applied to evaluate the efficacy of the proposed CatBoost-based IDS (intrusion detection system). The CatBoost algorithm is superior to traditional methods in terms of successfully recognizing known and unknown attack types while avoiding false positives, as demonstrated by comparison analyses. In addition, it shows that the model is resilient to evasion tactics used by bad actors, which guarantees strong network infrastructure security. The results highlight the effectiveness of using CatBoost in network security applications and highlight the technology's potential as a preventative measure against online attacks that are constantly changing. More flexible and effective intrusion detection systems are made possible by this research, which advances machine learning-driven strategies for strengthening network defenses.

Software defined networks (SDN) and wireless cognitive radio networks (CRN) are examined within the context of dynamic spectrum management. The features include control and data plane separation, centralized control, adopting open-source standards, programmability, quality of service (QoS) management, and security. The transformative impact of network function virtualization (NFV) is explored with a perspective on its architecture and applications in SDN, internet of things (IoT), cloud computing, and blockchain. The security aspect of SDN with specific focus on mitigating denial-of-service (DoS) attacks and vulnerabilities associated with open flow protocol is also addressed. The cognitive-inspired security mechanisms adapt to evolving threats integrating machine learning (ML) and artificial intelligence (AI) based algorithms for dynamic threat detection and mitigation exemplified through case studies. Adoption of software-defined perimeter, zero trust, blockchain, and quantum-safe cryptography in future are discussed. Finally, SDN applications for IoT networks are discussed.

Machine learning applications are widely used in identifying fraudulent transactions. Still, the majority of the applications detect fraudulent activities only after their occurrence, rather than in real-time or near real- time. The significant rarity of fraudulent transactions compared to normal ones creates a highly imbalanced data, posing substantial challenges for effective fraud detection. Addressing this issue extends beyond traditional machine learning approaches and requires alternative strategies. This chapter discusses the importance of fraud detection methods and compares k-nearest neighbor (KNN), random forest approach, artificial neural network (ANN) and support vector machine (SVM) enhanced with quantum annealing solvers that are used to detect fraud in online banking transactions. This study focuses on choosing the optimal approach for various types of datasets, while considering the three key factors such as accuracy, speed, and cost.

Cognitive radios are used for dynamic spectrum management and interference identification. Automatic modulation classification is used for identification of modulation scheme of received signal. Modulated signals have a Cyclostationary property, through which they are identified and then further processed. Cyclostationary feature detection method detects the Cyclostationary features and then identifies the modulation type. The cyclic frequency domain profile that is extracted from spectral correlation function is used for the classification of modulation. Each modulation scheme has a different cyclic frequency domain profile.

Cognitive radio (CR) technology has gained attention in wireless communications for its ability to enhance spectrum utilization efficiency through the implementation of dynamic spectrum access. This technology enables secondary users (SUs), or unlicensed users, to access the radio spectrum of primary users (PUs) in a limited manner that doesn't interfere with the PUs. The implementation of spectrum handoff can enhance SU's throughput performance however it may cause high energy consumption. This, in turn, can reduce device lifetime and affect system performance, particularly for battery-operated CR devices. This chapter focuses on optimizing the available resources in the green CR system to maximize energy efficiency (EE) and throughput performance while maintaining system reliability. It also explores using energy beamforming techniques to enhance the throughput performance of multi-antenna CR systems. Furthermore, this chapter addresses the challenges associated with SU's spectrum sensing by introducing a new transmission protocol designed for smart sensing implementation.

The constantly growing volume of waste produced daily by people and businesses has made the management of waste in urban contexts a challenging issue. These days, it's common practice to combine deep learning and the internet of things model with efficient waste management tactics to offer flexible real-time data monitoring and categorization solutions. The proposed methodology involved IoT-enabled

Preface

Welcome to *Spectrum and Power Allocation in Cognitive Radio Systems*, a comprehensive reference book edited by Kannadhasan Suriyan, Dhaya R, R. Nagarajan, and Karthick Alagar.

Cognitive radio systems have emerged as a solution to the pressing need for more efficient utilization of limited spectrum resources in the face of burgeoning wireless services. The inefficiencies of traditional spectrum allocation strategies have led to the development of cognitive radio, which enables unlicensed users to access spectrum bands allocated to licensed users dynamically and without causing harmful interference to primary user transmissions.

The primary objective of this edited volume is to delve into the intricacies of cognitive radio systems, focusing on performance studies and the development of throughput- and energy-efficient resource allocation algorithms. Central to cognitive radio operation is spectrum sensing, which mitigates the inherent uncertainty in channel occupancy due to false alarms and miss-detections.

In contrast to conventional radio systems, cognitive radio represents a paradigm shift towards intelligent radio networks that optimally utilize underutilized spectrum. Through the identification of open channels and adaptive broadcasting, cognitive radio facilitates simultaneous usage of multiple frequencies, thereby enhancing spectrum efficiency.

This book serves as a valuable resource for understanding the sensing and interference processes of cognitive radio and elucidates its superior performance compared to traditional radio systems. It addresses the critical need for effective spectrum management in wireless networks and provides insights into passive and active spectrum awareness, spectrum management, and transmit power regulation.

As the demand for wireless devices and systems continues to grow, the scarcity of spectral resources poses a significant challenge. Cognitive radio technology offers a unique approach by conceptualizing the spectrum as a multidimensional space and enabling opportunistic access to underutilized bands. By transcending the limitations of fixed spectrum assignment strategies, cognitive radio maximizes spectrum efficiency while minimizing interference.

We hope that this book serves as a valuable resource for researchers, practitioners, and students interested in cognitive radio technology. It offers a comprehensive overview of the foundations, organization, and challenges of cognitive radio systems, paving the way for future advancements in wireless communication.

ORGANIZATION OF THE BOOK

Chapter 1: Advancements in Cognitive Radio Networks: Protocols, Architectures, Challenges, and Future Perspectives

Authored by Yamini B, Beslin Pajila P J, Anish T P, Gracelin Sheena B, Dinesh M G, Nalini M, and Siva Subramanian R from various esteemed institutions in India, this chapter provides a comprehensive exploration of Cognitive Radio Networks (CRNs) from foundational principles to state-of-the-art protocols and future prospects. It delves into the history, motivation, and scope of CRNs, covering dynamic spectrum access, waveform design, resource allocation, and networking protocols. Special emphasis is placed on cooperative communications and database-driven CRNs to address spectrum sharing economics and regulations. Furthermore, the chapter discusses next-gen technologies, energy-efficient designs, creative applications, and business models, concluding with insights into machine learning methods and future trends shaping CRN research and innovation.

Chapter 2: An Emergent Self Attention LSTM Module based Malicious User Detection in Cognitive Radio Networks

Sivasankari Jothiraj, Brindha M, and Persis Jeffrey V, from Velammal College of Engineering and Technology, India, introduce a novel approach for detecting malicious users in Cognitive Radio Networks (CRNs). By leveraging a combination of Self-Attention Mechanism and Long Short-Term Memory (LSTM) networks, the proposed model effectively captures complex temporal dependencies and spatial correlations in dynamic spectrum access patterns, enabling robust intrusion detection. Real-world CRN datasets are utilized to evaluate the model's performance under various attack scenarios, ensuring the integrity and security of CRNs against potential threats.

Chapter 3: Artificial Intelligence Techniques in Cognitive Radio Networks Applications

Authored by Kamalendu Pal from the University of London, United Kingdom, this chapter explores the integration of artificial intelligence (AI) techniques in cognitive radio applications to enhance cognitive radio network (CRN) cognition capabilities. The chapter reviews various AI techniques and their applications in CRNs, enabling systems to emulate human behaviors such as reasoning and learning. The discussions shed light on how AI-driven approaches contribute to tasks such as spectrum optimization, cognitive radio sensing, and resource management, paving the way for more efficient and intelligent CRN deployments.

Chapter 4: Automated Star Rating Generation for Analyzing the Performance of Public Department

Mangalapriya Sakthivel, Vennila S, Ilakkiya R, and Malini V from Vel Tech High Tech Dr. Rangarajan Dr. Sakunthala Engineering College, India, present a dynamic website integrating various functionalities to enhance sentiment analysis for analyzing the performance of public departments. Leveraging PHP and the Instagram API, the system gathers real-time data from Instagram posts and comments, enhancing the accuracy of sentiment analysis. By offering automated star ratings and comprehensive sentiment analysis, the proposed system provides valuable insights into public perception, facilitating informed decision-making and improved public service delivery.

Chapter 5: Commercial Fire Protection Automated Detection and Suppression

Authored by C. Annadurai, I. Nelson, Arun Venkatesh, and Karthik Sriram G from Sri Sivasubramaniya Nadar College of Engineering, India, this chapter emphasizes the importance of effective fire detection and suppression systems in safeguarding lives and assets. The proposed IoT-based fire management system integrates computer vision technology to revolutionize conventional firefighting methodologies. By prioritizing precise fire detection, accurate object identification, and optimized extinguishing strategies, the system minimizes response times and mitigates the catastrophic consequences of fire incidents, enhancing overall safety in commercial settings.

Chapter 6: Dynamic Spectrum Allocation in Cognitive Radio Networks A Fair Game with Asymmetric Information

Lhoussaine Daoudi and Mohamed Baslam from Sultan Moulay Sliman University, Morocco, propose a dynamic duopoly game model for competitive spectrum allocation in cognitive radio networks (CRNs). Through bounded rationality and historical data analysis, secondary users adapt strategies to achieve Nash Equilibrium, ensuring fairness and system stability. Numerical simulations demonstrate the effectiveness of integrating historical data and bounded rationality in reaching Nash Equilibrium, offering valuable insights into spectrum allocation strategies in CRNs.

Chapter 7: Hybrid Railway Safety Optimization with Neural Networks and Random Forests

Kaviyalakshmi M, Chithra D, Anusuya K, Divyabharathi V, Rohini S.M, and Dharani M.D from S.A Engineering College, India, present a novel machine learning framework for enhancing safety protocols in railway environments. By combining neural networks with random forest techniques, the hybrid model offers improved accident investigation and prediction capabilities. The study showcases the effectiveness of this approach in real-time accident analysis and underscores its potential for improving safety standards in railroad stations.

Chapter 8: Investigation and Prediction of Itemsets Frequency using Machine Learning Techniques

Sameera Banu M, Samyuktha S R, Gajalakshmi Duraikannu, and Arjun Paramarthalingam from University College of Engineering Villupuram, India, introduce an Apriori algorithm for analyzing itemsets frequency in retail analytics. By discerning robust product associations, the proposed approach facilitates targeted marketing, product placements, and inventory management, enhancing business optimization in the retail sector. The study addresses challenges such as missing values and outliers, ensuring the reliability of market basket analysis for informed decision-making.

Chapter 9: Machine Learning-based Network Attack Prediction

Kavitha M, Roscia Jeya Shiney J, Gopikannan P, and Mahalakshmi M from various institutions in India, propose a machine learning approach for real-time network intrusion detection using the CatBoost algorithm. By leveraging CatBoost's resilience to high-dimensional data and complex attack patterns, the model effectively identifies and mitigates network threats. The study highlights the significance of proactive threat detection in safeguarding vital systems against evolving cyber threats.

Chapter 10: NFV and Secure Cognitive SDN for Educational Backbone Network Deployment: Cognitive Enabled Routing in SDN

Muhammad Imran Majid, Faheem Khuhawar, Karrar Muhammad, Kamlesh Rathore, Aarif Ali, Muhammad Ali, Babu Mal, Umme Habiba, Shabana Muhammad, Faisal Shaikh, and Tariq Soomro from various institutions in Pakistan examine Software Defined Networks (SDN) and Wireless Cognitive Radio Networks (CRN) within the context of Dynamic Spectrum Management. The chapter explores the transformative impact of Network Function Virtualization (NFV) on SDN architecture and applications, emphasizing security measures to mitigate Denial-of-Service (DoS) attacks and vulnerabilities. Additionally, cognitive-inspired security mechanisms integrating Machine Learning (ML) and Artificial Intelligence (AI) algorithms are discussed, along with potential applications of SDN in IoT networks.

Chapter 11: Online Fraud Detection using Machine Learning Algorithms and Quantum Annealing Solvers

Akshaya B, Agnimita Das Bakshi, and Annie Micheal A from Sathyabama Institute of Science and Technology, India, propose a hybrid approach combining machine learning algorithms with quantum annealing solvers for online fraud detection in banking transactions. The chapter compares the efficacy of K-nearest neighbor (KNN), Random Forest, Artificial Neural Network (ANN), and Support Vector Machine (SVM) algorithms enhanced with quantum annealing solvers in detecting fraudulent activities. Through a comprehensive analysis of accuracy, speed, and cost considerations, the study provides insights into selecting optimal fraud detection strategies for diverse datasets.

Chapter 12: Predictive Analysis of Modulation Classification and Adaptive Modulator for Next Generation Networks

Rabia Bilal and Bilal Khan from institutions in Pakistan discuss the predictive analysis of modulation classification and adaptive modulation for next-generation networks. The chapter explores the application of Cognitive Radios for Dynamic Spectrum Management and Interference Identification, focusing on Automatic Modulation Classification to identify modulation schemes of received signals. By leveraging Cyclostationary Feature Detection Method and Spectral Correlation Function, the study enhances the classification accuracy of modulation types, contributing to efficient spectrum utilization in next-generation networks.

Chapter 13: QoS-Aware Smart Resource Allocation for Green Cognitive Radio Networks

Arifa Ahmed, Deepak Mishra, Ganesh Prasad, and Krishna Baishnab from various institutions in India present a chapter focusing on optimizing resources in green Cognitive Radio (CR) systems to maximize energy efficiency and throughput performance while ensuring system reliability. The study explores energy-efficient strategies such as energy beamforming techniques and addresses challenges associated with spectrum sensing through the introduction of a new transmission protocol designed for smart sensing implementation. By prioritizing energy efficiency and quality of service (QoS), the chapter offers valuable insights into sustainable resource allocation in CR networks.

Chapter 14: Revolutionizing Waste Management through the Integration of IoT and Deep Learning Technology

Authored by Sathana V, Sherwin Daniel, Sreekanth C B, Senthilvel K, and Santhoshraj D from K Ramakrishnan College of Engineering, India, this chapter proposes a methodology to revolutionize waste management by integrating IoT and deep learning technology. The chapter presents an IoT-enabled garbage management system that utilizes deep learning models for garbage detection and segregation. By leveraging TensorFlow on ResNet architecture, the system achieves accurate waste categorization, paving the way for efficient waste management practices and environmental sustainability.

Chapter 15: Spectrum Sensing and Sharing for Internet of Things

Poornima Pandian, Chithra Selvaraj, and Priyanga S from institutions in India examine the role of cognitive radio technology in enhancing spectrum utilization for Internet of Things (IoT) networks. The chapter discusses the limited availability of spectrum resources and the potential of cognitive radio to optimize spectrum usage through dynamic spectrum sensing and sharing. By enabling efficient frequency band usage and mitigating spectrum scarcity, cognitive radio technology facilitates the widespread adoption of IoT networks, thereby driving innovation and connectivity in the IoT ecosystem.

IN CONCLUSION

As editors of this edited reference book on *Spectrum and Power Allocation in Cognitive Radio Systems*, we are thrilled to present a comprehensive collection of chapters that delve deep into various aspects of cognitive radio networks (CRNs) and their applications. Through the collaborative efforts of esteemed researchers and scholars from around the globe, this book offers a rich tapestry of insights, methodologies, and advancements in the field.

From the foundational principles of CRNs to cutting-edge protocols, algorithms, and applications, each chapter contributes to a holistic understanding of cognitive radio technology and its transformative potential in wireless communications. The exploration of topics such as dynamic spectrum access, security mechanisms, machine learning integration, and resource optimization reflects the multidisciplinary nature of CRN research and its impact on diverse domains.

Furthermore, the diverse range of applications discussed in this book, including fraud detection, waste management, fire protection, and railway safety optimization, underscores the versatility and real-world relevance of cognitive radio technology. By bridging theoretical concepts with practical implementations, these chapters offer valuable insights for academics, researchers, practitioners, and industry professionals alike.

As we conclude this preface, we extend our heartfelt gratitude to all the contributing authors for their invaluable contributions and dedication to advancing the field of cognitive radio systems. We hope that this edited reference book serves as a valuable resource and inspires further research and innovation in the dynamic realm of cognitive radio technology.

Chapter 1
Advancements in Cognitive Radio Networks:
Protocols, Architectures, Challenges, and Future Perspectives

B. Yamini
SRM Institute of Science and Technology, India

P. J. Beslin Pajila
https://orcid.org/0000-0002-3905-2460
Vel Tech Rangarajan Dr. Sagunthala R&D Institute of Science and Technology, India

T. P Anish
R.M.K. College of Engineering and Technology, India

B. Gracelin Sheena
Sathyabama Institute of Science and Technology, India

M. G. Dinesh
EASA College of Engineering and Technology, India

M. Nalini
S.A. Engineering College, India

Siva Subramanian R.
https://orcid.org/0000-0002-7509-9223
R.M.K. College of Engineering and Technology, India

ABSTRACT

Cognitive radio networks (CRNs) revolutionise wireless communications by providing dynamic spectrum access and management to overcome static spectrum allocation. This overview explores CRNs from fundamentals to cutting-edge protocols and applications. It addresses cognitive radio theory, hardware, networking, and difficulties for academics and practitioners. Starting with CRNs' history and motivation, it describes research scope and qualities. The research examines dynamic spectrum access, waveform design, and resource allocation. Detailed networking protocol and standard analysis covers medium access control and interference management. Additionally, cooperative communications and database-driven CRNs are considered. To solve spectrum sharing economics and regulations, next-gen technologies, energy-efficient designs, creative applications, and business models are explored. Examining flexible CRN feature machine learning methods. The analysis concludes with future trends and guides cognitive radio network research and innovation.

DOI: 10.4018/979-8-3693-2893-4.ch001

1. INTRODUCTION

1.1 Background and Motivation

With an ever-growing need for spectrum resources to serve a wide range of applications, from mobile broadband to Internet of Things (IoT) devices, wireless communication has become an essential component of contemporary civilisation. However, since significant sections of the spectrum remain underutilised while other bands encounter congestion, old static spectrum distribution approaches often lead to inefficiency. The creation of Cognitive Radio Networks (CRNs), which attempt to dynamically alter spectrum utilisation to satisfy the various needs of users and applications, has been spurred by this inefficiency(Hong, X et al 2009). With CRNs, wireless communication is undergoing a paradigm change as radios are given the ability to perceive, adjust, and optimise their operational settings on their own in real time. CRNs may better use spectrum resources, increase overall network performance, and improve spectrum utilisation thanks to their cognitive capabilities(Saleem, Y et al 2014).

1.2 Cognitive Radio Networks: Definition and Features

The fundamental characteristic of a Cognitive Radio Network is its capacity to see, comprehend, and adjust to the radio frequency surroundings in an intelligent manner(Khattab et al 2012). In contrast to conventional static spectrum allocation, CRNs utilise dynamic spectrum access (DSA) algorithms, which minimise interference to main users while enabling radios to opportunistically access unoccupied spectrum bands. Spectrum sensing, spectrum mobility, spectrum sharing, and spectrum management are important features of CRNs. Using methods including energy detection, matched filtering, and cyclostationary feature recognition, spectrum sensing allows CRNs to find and identify unutilised spectrum bands, often known as spectrum gaps. Making choices about spectrum access and utilisation based on sensed data, user needs, and legal restrictions is known as spectrum management. CRNs may dynamically move between multiple frequency bands in order to minimise interference and maximise performance because of spectrum mobility. Finally, spectrum sharing techniques provide equitable and effective spectrum utilisation by facilitating cohabitation between main and secondary users.

1.3 Objectives and Scope

This survey's main goal is to provide a thorough overview of CRNs, including a range of topics such as theory, protocols, hardware, applications, difficulties, and future prospects. This study attempts to be a useful tool for scholars, and practitioners working in the subject of cognitive radio by synthesising current research and advancements. This survey's scope covers a broad variety of subjects, such as hardware architecture, networking protocols, spectrum sensing methods, developing applications, machine learning approaches, obstacles, and future trends. This survey intends to provide insights into the state-of-the-art, highlight possibilities and difficulties, and direct future research and innovation in the area of cognitive radio by thoroughly covering these subjects.

2. DYNAMIC SPECTRUM ACCESS AND MANAGEMENT

A crucial component of Cognitive Radio Networks (CRNs) is dynamic spectrum access and management, which permits effective spectrum resource use via clever adaptation and optimisation. The many methods and approaches used in CRNs for dynamic spectrum access and management are examined in this section.

Figure 1. Dynamic spectrum access and management

2.1 Methods of Spectrum Sensing

CRNs employ a technique called "spectrum sensing" to find chances for secondary access to the spectrum and to detect the presence of main users (Yucek et al 2009). A range of sensing methods are used, such as cooperative sensing, matched filtering, cyclostationary feature identification, and energy detection. The most used method is energy detection, in which radios utilise the energy levels inside a certain frequency range to identify spectrum gaps. In order to identify signals of interest, matched filtering compares incoming signals to waveforms that are known. Using the cyclostationary characteristics of signals, cyclostationary feature identification differentiates between main and secondary users. Cooperative sensing reduces the impacts of fading and shadowing by combining data from many radios, improving sensing reliability.

2.2 Spectrum Usage Measurements and Statistical Modelling

Effective spectrum management requires an understanding of spectrum consumption trends(Weiss et al 2019). Measurements of spectrum utilisation and occupancy provide important new perspectives on the regional and temporal variability of spectrum availability. To describe and forecast spectrum utilisation patterns, statistical modelling approaches including hidden Markov models, stochastic geometry, and Markov models are used. By capturing statistical relationships and changes in spectrum occupancy, these models make it easier to create effective strategies for spectrum access and allocation.

2.3 Interference Aggregation, Modulation, and Waveform Design

In order to maximise spectral efficiency and minimise interference in CRNs, waveform design and modulation methods are essential. Adapting transmission parameters according to channel circumstances and interference levels, cognitive radios use adaptive modulation algorithms including spread spectrum, adaptive modulation, and coding (AMC), and orthogonal frequency-division multiplexing (OFDM). Effective coexistence and spectrum sharing are made possible by interference aggregation methods like interference alignment and cancellation, which reduce interference from main and secondary users.

2.4 Sharing of the Dynamic Spectrum

In CRNs, effective cohabitation and cooperation between main and secondary users are made possible by dynamic spectrum sharing systems. Listen-before-talk (LBT), opportunistic spectrum access (OSA), and interference-aware MAC protocols are examples of spectrum access protocols that allow for the dynamic distribution of spectrum resources while maintaining equity and reducing disturbance to primary users. In order to locate and use accessible spectrum bands in an opportunistic manner, cognitive radios use database-driven techniques and spectrum sensing. They also dynamically modify transmission settings and protocols in response to regulatory limitations and channel circumstances.

2.5 Strategies for Allocating Resources

The goal of resource allocation techniques in CRNs is to maximise network performance and spectrum utilisation. Dynamic resource allocation strategies adaptively distribute spectrum resources according to user needs, traffic demands, and network circumstances. Examples of these systems include dynamic spectrum access (DSA), cognitive radio resource management (CRRM), and reinforcement learning-based algorithms. These tactics maximise QoS provision, reduce interference, and optimise spectral efficiency while improving total network capacity.

3. NETWORKING PROTOCOLS AND STANDARDS

In order to provide effective communication, spectrum management, and coordination amongst cognitive radios, networking protocols and standards are fundamental components of cognitive radio networks, or CRNs. This section explores the standards and protocols that control various CRN elements.

3.1 Cognitive Radio Network Architectures and Building Blocks

In order to facilitate dynamic spectrum access and management, CRNs display a variety of architectural styles and building elements. There are three types of architectural models: distributed, decentralised, and centralised. Each has benefits and drawbacks. Spectrum sensors, decision-making engines, databases for spectrum management, cognitive engines, and interfaces with legacy systems are all examples of building blocks. Together, these elements allow cognitive radios to perceive, pick up on, adjust to, and interact with their dynamic radio surroundings.

3.2 Cognitive Medium Access Control

Cognitive radios in CRNs are coordinated and have access to spectrum resources under the control of Cognitive Medium Access Control (MAC) protocols (Hu, S et al 2014). In order to dynamically modify transmission parameters, such as channel selection, transmission power, and contention window widths, depending on spectrum availability and interference levels, cognitive MAC protocols employ spectrum sensing capabilities. To provide effective spectrum sharing and cohabitation with main users, protocols such as contention-based MAC, opportunistic spectrum access, and dynamic spectrum access (DSA) are often used.

Figure 2. Networking protocols and standards

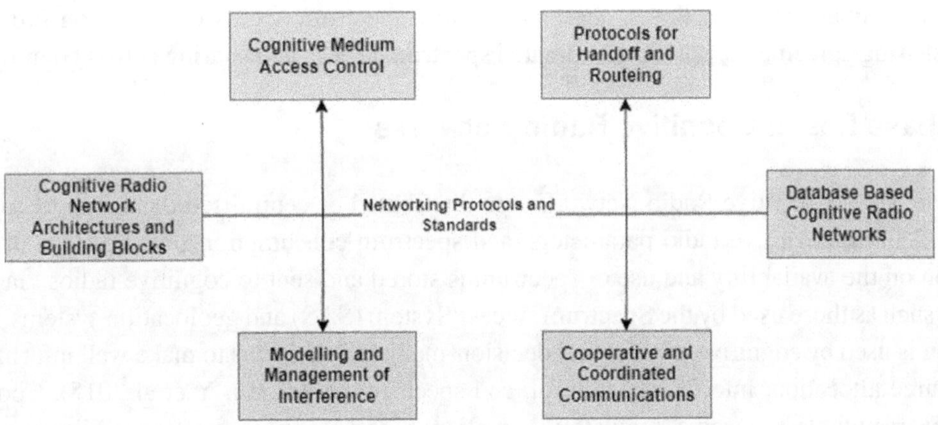

3.3 Modelling and Management of Interference

In order to maintain dependable communication and reduce interference in CRNs, interference control is essential. Stochastic geometry and interference temperature models are two examples of interference modelling approaches that quantify the effect of interference on network performance and provide information for interference-aware resource allocation strategies. By reducing the impact of interference from main and secondary users, interference management techniques such as interference reduction, interference cancellation, and cognitive interference alignment improve spectral efficiency and network capacity.

3.4 Protocols for Handoff and Routeing

In dynamic radio settings, smooth mobility and effective data routeing are made possible by handoff and routeing protocols in CRNs. In order to sustain connection and meet QoS criteria, cognitive radios must adaptively transition between available spectrum bands and access points. Network nodes can switch between spectrum channels and handoff protocols more smoothly. Examples of these protocols include mobility management protocols and cognitive handoff algorithms. In order to ensure effective data delivery in CRNs, routeing protocols, such as cognitive routeing algorithms and opportunistic routeing schemes, optimise data forwarding channels depending on spectrum availability, traffic circumstances, and network architecture.

3.5 Cooperative and Coordinated Communications

Cognitive radios may work together and exchange information to improve network performance and reliability via cooperative and coordinated communications. In order to increase spectral efficiency, reduce fading effects, and improve spectrum sensing accuracy, cooperative communication techniques such as cooperative beamforming, cooperative relaying, and cooperative spectrum sensing exploit geographical diversity and radio cooperation. In order to ensure equitable and effective spectrum utilisation, coordinated communication methods, such as dynamic spectrum access coordination protocols and spectrum sharing agreements, enable coordinated spectrum access and sharing across cognitive radios.

3.6 Database Based Cognitive Radio Networks

Database based Cognitive Radio Networks use distributed or centralised databases to control regulatory guidelines, cognitive radio parameters, and spectrum consumption statistics(Gao et al 2013). Information on the availability and use of spectrum is stored and sent to cognitive radios via spectrum databases, such as those used by the Spectrum Access System (SAS) and geolocation systems. Database information is used by cognitive engines and decision-making algorithms to make well informed choices on resource allocation, interference control, and spectrum access (Li, Y et al 2015). Coordination between diverse wireless systems, regulatory compliance, and effective spectrum utilisation are made possible by database-based CRNs.

4. ARCHITECTURE AND IMPLEMENTATIONS OF HARDWARE

Realising the cognitive capabilities of radios via dynamic spectrum access and effective communication is made possible in Cognitive Radio Networks (CRNs) in large part by hardware design and implementations. Aspects of hardware design and implementation of CRNs are examined in this section.

Figure 3. Architecture and implementations of hardware

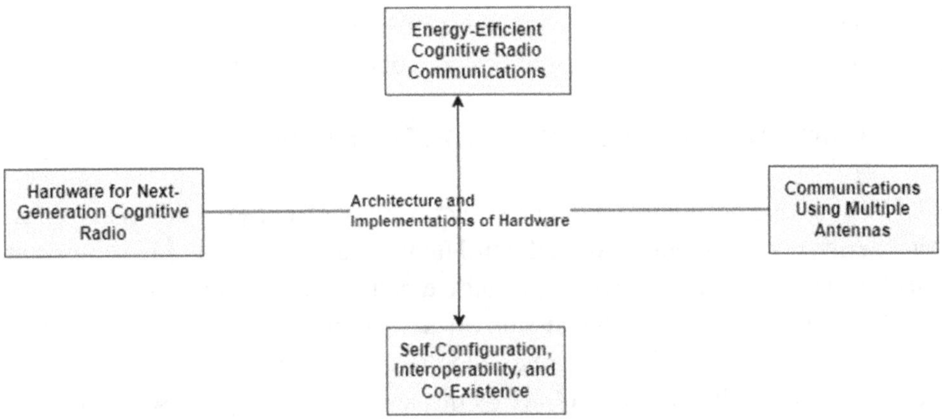

4.1 Hardware for Next-Generation Cognitive Radio

In order to enable higher cognitive functions, next-generation cognitive radio hardware incorporates developments in radio transceivers, computers, and sensors. Cognitive radios are built on top of software-defined radio (SDR) platforms, which allow hardware designs to be programmable and reconfigurable. Wideband tuneable radios, fast analog-to-digital converters (ADCs), and field-programmable gate arrays (FPGAs) for real-time signal processing are some of the key components of the hardware for the next generation of cognitive radios. These systems provide cognitive radios with the capacity to dynamically adapt to changing environmental circumstances and spectrum availability via agile spectrum sensing, adaptive modulation and coding, and dynamic spectrum access.

4.2 Communications Using Multiple Antennas

Enhancing spectral efficiency and connection dependability in CRNs requires multi-antenna-based communications, such as multiple-input multiple-output (MIMO) systems. Spatial diversity, multiplexing, and beamforming methods are made possible by many antennas at the transmitter and receiver in order to reduce fading effects, boost interference suppression, and raise data speeds. Phased array antennas and smart antennas are examples of advanced antenna arrays that provide adaptive beamforming and spatial selectivity, allowing cognitive radios to maximise signal transmission and reception in dynamic radio environments.

4.3 Energy-Efficient Cognitive Radio Communications

When designing cognitive radios, energy economy is very important since these devices are often battery-operated and have limited resources. The goal of energy-efficient cognitive radio communications is to reduce power use without sacrificing functionality or performance. Extending battery life and facilitating energy-efficient operation need low-power hardware components, such power management units, low-power ADCs, and energy-efficient transceivers. Sleep mode operation and dynamic voltage and frequency scaling (DVFS) are two examples of dynamic power management strategies that optimise power usage depending on activity levels and traffic load. Additionally, in order to reduce energy consumption without compromising performance, cognitive radio protocols and algorithms are made to adjust transmission parameters, schedule operations, and optimise resource allocation.

4.4 Self-Configuration, Interoperability, and Co-Existence

In order to facilitate a smooth integration and functioning with current wireless systems, cognitive radio hardware design places a premium on self-configuration, interoperability, and co-existence. Based on external factors and regulatory limitations, cognitive radios must automatically adjust their characteristics, including frequency bands, transmission power, and modulation techniques. Harmonisation initiatives, like IEEE 802.22 concerning cognitive radio-driven wireless regional area networks (WRANs), guarantee compatibility and interoperability across disparate wireless systems. Cognitive radios may live alongside main users and legacy systems while optimising spectrum use and reducing interference thanks to co-existence methods like spectrum sensing and interference mitigation strategies.

5. EMERGING APPLICATIONS AND BUSINESS MODELS

5.1 Cognitive Network Use Cases and Applications

Applications for Cognitive Radio Networks (CRNs) are many and span several sectors and industries. Among the noteworthy use cases are: Dynamic Spectrum Access (DSA): CRNs are perfect for wireless broadband access, Internet of Things deployments, and mission-critical communications because they allow for the effective use of spectrum resources in dynamic and diverse situations (Xing et al 2006). Public Safety and disaster Response: By dynamically distributing spectrum resources and instantly responding to changing circumstances, CRNs improve communication coverage and dependability in disaster situations. Smart Cities and Infrastructure: By offering dependable and robust communication infrastructure with dynamic spectrum access capabilities, CRNs support the implementation of smart grids, intelligent transportation systems, and environmental monitoring applications. Wireless Healthcare: CRNs satisfy the strict needs of healthcare systems by offering dependable, secure wireless connection with dynamic spectrum allocation to enable telemedicine, remote patient monitoring, and healthcare IoT applications.

Figure 4. Emerging applications and business models

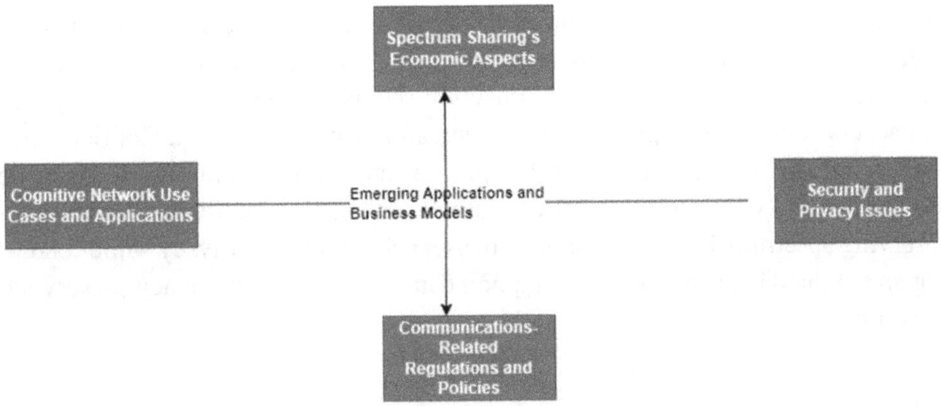

5.2 Spectrum Sharing's Economic Aspects

The legislative rules and commercial models pertaining to spectrum sharing in CRNs are significantly influenced by the economic factors involved. Increased spectrum availability, lessened spectrum congestion, and decreased infrastructure expenses are just a few of the financial advantages that come from sharing the restricted spectrum. Important economic factors consist of: Market-Based Approaches: Market-driven spectrum sharing methods, such leasing, trading, and auctions, allow for the effective distribution of spectrum resources and encourage investment in CRN installations. Cost-Benefit Analysis: Economic analysis assist stakeholders in determining the viability and possible advantages of spectrum sharing plans in CRNs. Examples of these analyses include cost-benefit analyses and spectrum valuation studies. Regulatory Policies: Stakeholder investment choices and the economic sustainability of CRN installations are influenced by the regulatory frameworks controlling use fees, licencing, and spectrum allocation.

5.3 Communications-Related Regulations and Policies

Regulatory regulations have a big influence on how CRNs are implemented and run. Important regulatory factors consist of: Spectrum Allocation and Licencing: To specify the rights and responsibilities related to spectrum use in CRNs, regulatory bodies assign spectrum bands and provide licences to main and secondary users. Interference Mitigation and cohabitation: To maintain cohabitation between primary and secondary users while minimising detrimental interference, regulatory laws need spectrum sharing protocols and interference mitigation technologies. Spectrum Management and Enforcement: To preserve order and equity in CRN installations, regulatory bodies regulate spectrum interference, monitor spectrum utilisation, and enforce adherence to spectrum use guidelines.

5.4 Security and Privacy Issues

To safeguard user data, maintain network integrity, and stop unwanted access, privacy and security are essential factors in CRNs. Important privacy and security factors consist of: Secure Spectrum Access: To guarantee secure spectrum access and stop unauthorised users from accessing spectrum resources, CRNs utilise access control, encryption, and authentication techniques. Data Confidentiality and Integrity: To safeguard sensitive information and guarantee data confidentiality and integrity both during transmission and storage, CRNs use encryption, authentication, and secure communication protocols. Privacy-Preserving Spectrum Sensing: To safeguard user anonymity and privacy while conducting spectrum sensing and exchanging spectrum consumption data, CRNs employ privacy-preserving spectrum sensing algorithms.

6. CHALLENGES AND ISSUES

In order for Cognitive Radio Networks (CRNs) to reach their full potential in wireless communication, a number of problems and obstacles must be resolved. This section lists a few of the major problems and suggests some possible fixes.

6.1 Cognitive Radio Design Difficulties

In order to design cognitive radios, a number of technological obstacles must be overcome, such as: Complexity: Complex algorithms and hardware are needed for cognitive radios to conduct spectrum sensing, decision-making, and dynamic spectrum access, which adds to the complexity of the design. Interference Mitigation: In order to guarantee dependable communication, cognitive radios must efficiently reduce interference from main and secondary users. This calls for sophisticated signal processing algorithms and interference management strategies. Energy Efficiency: To extend battery life and permit autonomous operation in resource-constrained contexts, energy-efficient hardware and algorithms are required. Power consumption is a crucial factor in cognitive radio design.

Figure 5. Challenges and issues

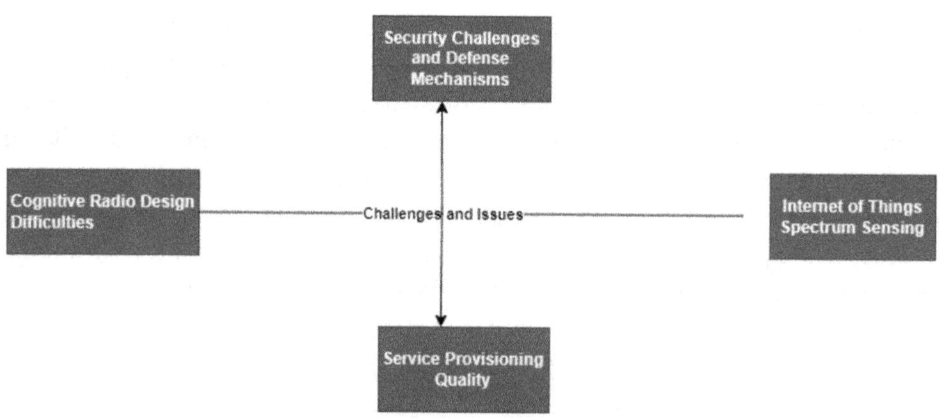

6.2 Security Challenges and Defense Mechanisms

Because spectrum sharing in CRNs is dynamic and open, security is a big problem. Among the principal security issues are: Spectrum Misuse: Unauthorised access to spectrum resources or communication disruption may occur when malicious users take advantage of flaws in spectrum sensing and access procedures. Privacy Protection: Activities related to spectrum sensing and sharing may divulge private information about users and their activities, hence posing privacy risks and necessitating the employment of privacy-preserving measures to safeguard user privacy. Authentication and Access Control: In order to prevent unauthorised access and guarantee network integrity, cognitive radios need strong authentication and access control techniques. They also need to authenticate and authorise users to use spectrum resources securely. Secure communication protocols, intrusion detection, authentication, encryption, and other defence measures may assist reduce security risks and shield CRNs from malevolent assaults.

6.3 Service Provisioning Quality

It is difficult to ensure quality of service (QoS) in CRNs since wireless channels are erratic and dynamic. Among the difficulties relating to QoS are: Dynamic Spectrum Access: It is difficult to ensure QoS criteria like throughput, latency, and dependability since spectrum availability and quality fluctuate across time and space. Resource Allocation: Dynamic and adaptive resource allocation methods are required to maximise network performance in order to fulfil the various QoS needs of various applications and users. Interference Management: To reduce interference and improve QoS, primary and secondary user interference may negatively impact CRN QoS. To counteract this, interference mitigation strategies and spectrum coordination mechanisms are needed. Sophisticated QoS provisioning techniques, including as traffic prioritisation, admission control, and QoS-aware routeing, may assist in efficiently meeting CRN QoS needs.

6.4 Internet of Things Spectrum Sensing

The vast number of low-power and resource-constrained Internet of Things (IoT) devices poses unique issues for spectrum sensing. Among the principal difficulties are: Scalability: In order to minimise computational and communication overhead and handle the enormous number of IoT devices, spectrum sensing techniques must scale. Energy Efficiency: In Internet of Things devices with limited battery capacity, energy is a valuable resource that is used during spectrum sensing activities. To extend battery life, energy-efficient sensing approaches are necessary. Reliability: Robust sensing algorithms and coordination mechanisms are required for spectrum sensing to be dependable in order to precisely identify spectrum opportunities and prevent collisions amongst IoT devices. Techniques for distributed, hierarchical, and collaborative sensing may help overcome these obstacles and make it possible for Internet of Things devices to share the spectrum effectively. These difficulties may be addressed and effective spectrum utilisation in millimeter-wave CRNs made possible by advanced sensor methods, beamforming, interference cancellation, and spectrum sharing protocols designed for mm-wave frequencies.

7. MACHINE LEARNING TECHNIQUES FOR COGNITIVE RADIO SYSTEMS

In order for radios in Cognitive Radio Networks (CRNs) to be able to adapt, learn, and optimise their behaviour in dynamic and unpredictable settings, machine learning (ML) approaches are essential. The use of different ML approaches in CRNs is examined in this section.

7.1 Distributed Adaptation and Optimization Methods

Cognitive radios can now automatically adjust their settings and function at their best in a decentralised way thanks to distributed adaptation and optimisation approaches that make use of machine learning capabilities. Among the important methods are: Reinforcement Learning (RL): Using input from the surroundings, RL algorithms allow cognitive radios to learn the best course of action via trial and error (Obite et al 2021). Radios may autonomously learn and adjust their transmission settings, channel selection, and spectrum access methods thanks to distributed reinforcement learning (RL) approaches including multi-agent RL and decentralised RL. This results in more effective spectrum utilisation and enhanced network performance (Wu, C et al 2010). Evolutionary Algorithms (EAs): Based on fitness criteria, EAs allow radios to develop and optimise their settings over time (Tuberquia et al 2018). Examples of EAs include particle swarm optimisation and genetic algorithms. Distributed EAs provide cooperative optimisation amongst cognitive radios, enabling them to jointly investigate and take advantage of spectrum possibilities and dynamically adjust to changing environmental circumstances (Pradhan et al 2014). Federated Learning (FL): By using FL approaches, cognitive radios may secure and maintain data privacy while learning from decentralised data sources (Chen et al 2021). Radios may adapt and optimise their behaviour based on dispersed information without centrally sharing sensitive data thanks to federated learning frameworks, which enable them to cooperatively train machine learning (ML) models using local data samples (Khan et al 2020).

7.2 Machine Learning Uses in Cognitive Radio

In CRNs, machine learning finds several uses, improving cognitive decision-making, resource allocation, interference control, and spectrum sensing. Among the important applications are: Spectrum Sensing and Occupancy Prediction: Radios can now reliably and efficiently detect spectrum opportunities and predict spectrum occupancy patterns based on historical data thanks to machine learning (ML) algorithms like support vector machines (SVMs), neural networks, and deep learning models. Resource Allocation and Optimisation: Cognitive radios can allocate spectrum resources, optimise transmission parameters, and dynamically manage interference to maximise network throughput, minimise latency, and guarantee that QoS requirements are met thanks to machine learning techniques like reinforcement learning, genetic algorithms, and supervised learning algorithms. Interference Mitigation and Coexistence: Cognitive radios can effectively mitigate interference from primary and secondary users, improving spectral efficiency and coexistence with legacy systems. These techniques include interference cancellation, cognitive interference alignment, and deep learning-based interference prediction models (Shitharth S et al 2023).

7.3 Spectrum Sensing and Management Based on Learning

Cognitive radios' spectrum awareness, flexibility, and decision-making skills are improved by learning-based spectrum sensing and management strategies that make use of machine learning algorithms. Among the important methods are: Deep Learning for Spectrum Sensing: Radios can learn intricate features and patterns from spectrum data thanks to deep learning models like recurrent neural networks (RNNs) and convolutional neural networks (CNNs). This increases the accuracy and resilience of spectrum sensing in busy and noisy environments.

Q-Learning for Flexible Spectrum Utilisation: Radios may discover the best spectrum access plans by experimenting and taking advantage of various activities depending on reward input from their surroundings according to Q-learning algorithms. In order to maximise spectrum utilisation while maintaining equity and coexistence with primary users, Q-learning-based spectrum access rules adjust to changing channel circumstances and interference levels. Online Learning for Adaptive Resource Management: Based on real-time network input, online learning algorithms allow cognitive radios to adaptively modify their resource allocation and management plans. Online learning frameworks enable the ongoing adjustment and refinement of cognitive radio settings, guaranteeing effective utilisation of the spectrum and flexibility in response to changing network circumstances. In conclusion, the cognitive capacities of radios in CRNs are greatly enhanced by machine learning approaches, allowing for autonomous adaptation, optimisation, and decision-making in dynamic and unpredictable settings. Cognitive radios can increase spectrum utilisation, improve network performance, and enable a wide range of wireless communication applications by utilising distributed adaptation and optimisation techniques, learning-based spectrum sensing and management techniques, and diverse applications of machine learning in CRNs.

8. FUTURE DIRECTIONS AND TRENDS

8.1 The Cognitive Networks of the Future

The goals of the next generation of cognitive networks will be to enhance cognitive capacities, optimise spectrum use, and provide smooth integration with newly developed technologies. Important patterns consist of: Cognitive Autonomy: Without human assistance, cognitive networks will become increasingly self-sufficient, using cutting-edge machine learning and artificial intelligence methods to make wise judgements and adjust to changing surroundings. Spectrum Sharing Innovations: Better cohabitation with main users and more effective spectrum utilisation are made possible by advancements in spectrum sharing methods, such as database-driven techniques and dynamic spectrum access (DSA) protocols. Edge Intelligence: In order to enable latency-sensitive applications and improve user experience, edge intelligence will be essential to cognitive networks. It will allow distributed decision-making and real-time adaption at the network edge (Kshirsagar PR et al 2020).

8.2 Evolving Standards and Business Models

The deployment and commercialisation of cognitive networks will be shaped by changing standards and business models, which will spur innovation and market expansion. Important advancements consist of: Standardisation Efforts: To enable broad adoption and guarantee interoperability across many systems and suppliers, standardisation groups will keep creating standards for cognitive radio technologies, interoperability, and spectrum sharing protocols. New Business Models: New business models that will allow stakeholders to more efficiently monetize spectrum resources and encourage investment in cognitive network installations include pay-per-use models, infrastructure sharing agreements, and spectrum sharing agreements. Regulatory changes: While guaranteeing adherence to spectrum laws and licencing requirements, regulatory changes would promote innovation and competition in the telecom sector and create a favourable atmosphere for the creation and implementation of cognitive networks.

8.3 Emerging Technologies and Applications

The development of cognitive networks will be driven by emerging technologies and applications, opening up new use cases and improving network capabilities. Important areas of growth consist of: 5G Integration: By integrating with 5G networks, cognitive networks will be able to take use of 5G capabilities and infrastructure, including massive MIMO, edge computing, and network slicing, to serve a variety of applications that have high throughput needs, low latency, and high reliability requirements. AI-Driven Optimisation: Cognitive networks will be able to better allocate resources, anticipate network behaviour, and adjust to shifting user needs as a result of advances in artificial intelligence (AI) and machine learning (ML). These developments will improve network performance and user experience (Artificial Intelligence et al 2023). IoT Integration: Cognitive networks will be able to support large-scale IoT deployments with a variety of connection needs, including low-power, wide-area, and dependable communication, thanks to integration with the Internet of Things. This will allow applications for smart cities, industrial IoT, and healthcare.

8.4 Research Opportunities and Open Challenges

Though they have come a long way, cognitive networks still have a number of obstacles to overcome and provide a wealth of prospects for future study. Important issues and areas for further study include: Spectrum Efficiency: To maximise spectral efficiency while maintaining fairness and coexistence with primary users, advances in spectrum sensing, access, and sharing approaches are needed. Improving spectrum efficiency remains a major problem. Security and Privacy: Research problems remain in addressing security and privacy issues related to cognitive networks, including safeguarding against hostile assaults, maintaining data confidentiality, and maintaining user privacy. Interoperability and Standards: Industry, academia, and regulatory agencies must work together to improve the interoperability of various cognitive network designs and standards, as well as to solve challenges pertaining to spectrum rules and licencing. Scalability and Robustness: It is still important to conduct research in this area to ensure that cognitive network solutions are both scalable and robust enough to support large-scale deployments and function reliably in a variety of environments. This requires improvements in distributed algorithms, network management, and fault tolerance mechanisms. In conclusion, there is a great deal of promise for the development of cognitive networks in the future thanks to developments in cognitive autonomy, novel approaches to spectrum sharing, changing industry standards and business models, new and developing technologies, and continuous research projects aimed at resolving unresolved issues and creating fresh chances for wireless communication innovation and expansion.

9. CONCLUSION

In this comprehensive review, we explored the diverse environment of Cognitive Radio Networks (CRNs), encompassing topics ranging from basic concepts to emergent trends and issues. The survey's key conclusions are as follows: Cognitive Radio Networks (CRNs) provide dynamic spectrum access and management capabilities, allowing for more efficient spectrum resource utilisation and network performance improvements. Spectrum sensing methods like as energy detection, matched filtering, and cooperative sensing are critical in detecting spectrum possibilities and allowing for dynamic spectrum access. Networking protocols and standards regulate communication, spectrum sharing, and coordination among cognitive radios, allowing for better spectrum utilisation and compatibility with older systems. Next-generation cognitive radio hardware, multi-antenna communications, and energy-efficient designs allow cognitive radios to achieve cognitive capabilities and adapt to changing surroundings. CRNs' emerging applications and business models cover a wide range of industries, including telecommunications, public safety, healthcare, and smart cities, fueling innovation and market development. Based on the outcomes of this study, numerous suggestions for further research in Cognitive Radio Networks are as follows: Advanced Spectrum Sensing methods: Create advanced spectrum sensing methods, such as machine learning-based methodologies and distributed sensing algorithms, to increase spectrum awareness and optimise spectrum utilisation. Strong Security and Privacy Mechanisms: Create strong security and privacy methods to defend against harmful assaults, maintain data confidentiality, and safeguard user privacy in dynamic and open spectrum contexts. Scalable and Adaptive Networking Protocols: Create scalable and adaptable networking protocols and standards to enable dynamic spectrum access, interoperability, and seamless integration with new technologies such as 5G and IoT. Energy-Efficient Hardware and Algorithms: Create energy-efficient hardware designs and algorithms in cognitive radios to reduce power consumption and increase battery life, allowing for autonomous operation and deployment in resource-constrained contexts. Interdisciplinary cooperation: Encourage interdisciplinary cooperation among academics, industry stakeholders, and regulatory agencies to overcome difficult issues and encourage innovation in CRNs, while adhering to regulatory standards and market expectations. To summarise, Cognitive Radio Networks have enormous potential to revolutionise wireless communication by allowing optimal spectrum utilisation, boosting network performance, and enabling a diverse variety of applications. Future research efforts should concentrate on tackling important problems, developing technology, and promoting cooperation in order to realise CRNs' full potential in the dynamic and ever-changing wireless communication ecosystem.

REFERENCES

Al-ani, A., Ul, S., Laghari, A., Manoharan, H., Selvarajan, S., & M. Uddin, M. (2023). *Improved Transportation Model with Internet of Things Using Artificial Intelligence Algorithm*. Tech Science Press. .10.32604/cmc.2023.038534

Chen, Z., Xu, Y. Q., Wang, H., & Guo, D. (2021). Federated learning-based cooperative spectrum sensing in cognitive radio. *IEEE Communications Letters*, 26(2), 330–334. 10.1109/LCOMM.2021.3114742

Gao, Z., Zhu, H., Liu, Y., Li, M., & Cao, Z. (2013, April). Location privacy in database-driven cognitive radio networks: Attacks and countermeasures. In 2013 [IEEE.]. *Proceedings - IEEE INFOCOM*, 2751–2759.

Hong, X., Chen, Z., Wang, C. X., Vorobyov, S. A., & Thompson, J. S. (2009). Cognitive radio networks. *IEEE Vehicular Technology Magazine*, 4(4), 76–84. 10.1109/MVT.2009.934672

Hu, S., Yao, Y. D., & Yang, Z. (2014). Cognitive medium access control protocols for secondary users sharing a common channel with time division multiple access primary users. *Wireless Communications and Mobile Computing*, 14(2), 284–296. 10.1002/wcm.2185

Khan, L. U., Alsenwi, M., Yaqoob, I., Imran, M., Han, Z., & Hong, C. S. (2020). Resource optimized federated learning-enabled cognitive internet of things for smart industries. *IEEE Access : Practical Innovations, Open Solutions*, 8, 168854–168864. 10.1109/ACCESS.2020.3023940

Khattab, A., Perkins, D., & Bayoumi, M. (2012). *Cognitive radio networks: from theory to practice*. Springer Science & Business Media.

Kshirsagar, P. R., Manoharan, H., & Al-Turjman, F. (2020). Design and Testing of Automated Smoke Monitoring Sensors in Vehicles. *IEEE Sensors Journal*, XX, 2020. 10.1109/JSEN.2020.3044604

Li, Y., Zhou, L., Zhu, H., & Sun, L. (2015). Privacy-preserving location proof for securing large-scale database-driven cognitive radio networks. *IEEE Internet of Things Journal*, 3(4), 563–571. 10.1109/JIOT.2015.2481926

Obite, F., Usman, A. D., & Okafor, E. (2021). An overview of deep reinforcement learning for spectrum sensing in cognitive radio networks. *Digital Signal Processing*, 113, 103014. 10.1016/j.dsp.2021.103014

Pradhan, P. M., & Panda, G. (2014). Comparative performance analysis of evolutionary algorithm based parameter optimization in cognitive radio engine: A survey. *Ad Hoc Networks*, 17, 129–146. 10.1016/j.adhoc.2014.01.010

Saleem, Y., & Rehmani, M. H. (2014). Primary radio user activity models for cognitive radio networks: A survey. *Journal of Network and Computer Applications*, 43, 1–16. 10.1016/j.jnca.2014.04.001

Shitharth, S., Yonbawi, S., Manoharan, H., Alahmari, S., Yafoz, A., & Mujlid, H. (2023). Physical Stint Virtual Representation of Biomedical Signals With Wireless Sensors Using Swarm Intelligence Optimization Algorithm. *IEEE Sensors Journal*, 23(4), 3870–3877. 10.1109/JSEN.2022.3233407

Tuberquia, M., & Hernandez, C. (2018). New approaches in cognitive radios using evolutionary algorithms. *Iranian Journal of Electrical and Computer Engineering*, 8(3), 1636. 10.11591/ijece.v8i3.pp1636-1646

Weiss, M. B., Werbach, K., Sicker, D. C., & Bastidas, C. E. C. (2019). On the application of blockchains to spectrum management. *IEEE Transactions on Cognitive Communications and Networking*, 5(2), 193–205. 10.1109/TCCN.2019.2914052

Wu, C., Chowdhury, K. R., Di Felice, M., & Meleis, W. (2010, May). Spectrum management of cognitive radio using multi-agent reinforcement learning. In *AAMAS* (pp. 1705-1712).

Xing, Y., Chandramouli, R., & Mangold, S. (2006). Dynamic spectrum access in open spectrum wireless networks. *IEEE Journal on Selected Areas in Communications*, 24(3), 626–637. 10.1109/JSAC.2005.862415

Yucek, T., & Arslan, H. (2009). A survey of spectrum sensing algorithms for cognitive radio applications. *IEEE Communications Surveys and Tutorials*, 11(1), 116–130. 10.1109/SURV.2009.090109

Chapter 2
An Emergent Self-Attention LSTM Module-Based Malicious User Detection in Cognitive Radio Networks

Sivasankari Jothiraj
https://orcid.org/0000-0002-8054-1020
Ultra College of Engineering and Technology, Madurai

M. Brindha
Velammal College of Engineering and Technology, India

V Persis Jeffrey
Velammal College of Engineering and Technology, India

ABSTRACT

Cognitive radio networks (CRNs) represent a dynamic and intelligent paradigm for efficient spectrum utilization by allowing unlicensed users, known as secondary users, to opportunistically access underutilized spectrum bands. However, the open and shared nature of the spectrum in CRNs exposes them to various security threats and attacks. In this context, the need for robust intrusion detection mechanisms becomes paramount to safeguard the integrity of the network. This work proposes a novel approach for detecting attackers in cognitive radio networks using a combination of self-attention mechanism and long short-term memory (LSTM) networks. The self-attention LSTM model is designed to capture complex temporal dependencies and spatial correlations in the dynamic spectrum access patterns, making it well-suited for the inherent variability and uncertainty in CRNs. The proposed approach is evaluated using real-world CRN datasets, considering various attack scenarios such as spectrum sensing data falsification and malicious spectrum access.

DOI: 10.4018/979-8-3693-2893-4.ch002

1. INTRODUCTION

In recent times, radio spectrum has become very scarce in availability due to a huge volume of dependency on its bandwidth. This has resulted in the non-availability of spectrum for certain essential services. Moreover user utility will subsequently increase the delay and waiting time for service provisions. Increased delay will result in a deterioration of the QoS value. Intelligent means of utilizing the scarcely available radio frequency spectrum was inspired in this research work to bring about the creation of cognitive radio networks (CRNs) (Akyildiz *et al.* 2009). Figure 2 provides a conceptual architecture of the CR network.

Figure 1. Conceptual architecture of cognitive radio network

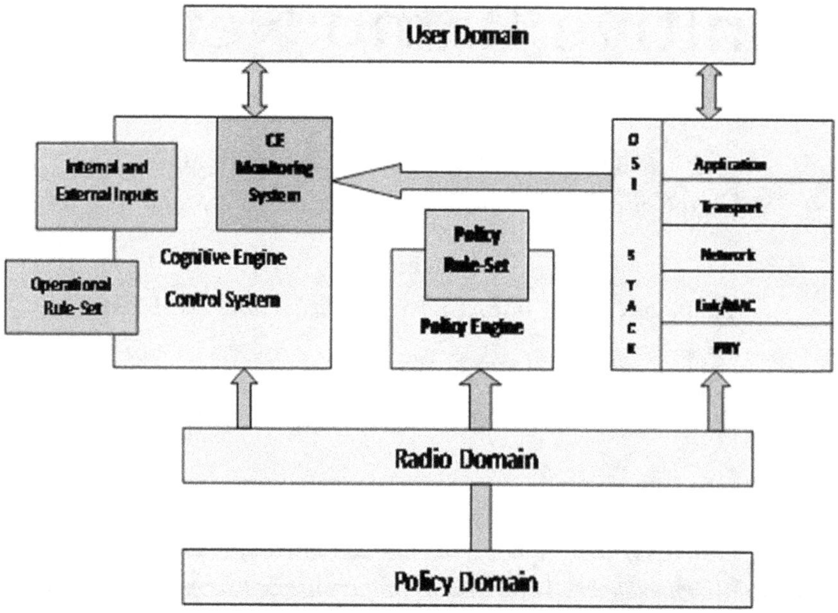

(www.wirelessinnovation.org)

Cognitive radio networks could precisely be defined as an intelligent mechanism of sensing and allocation of channels in the available spectrum for wireless communication applications. Cognitive radio networks are broadly categorized into two types: fully cognitive, and spectrums sensing cognitive networks. The node senses the available channel for any vacancy and allocates a channel to the requesting user, based on the feedback. As shown in Figure 1 it could be seen that the cognitive engine is part of the OSI network model along with its accessories, with the sensing and policy engine forming the core of the CR network. The primary duty of monitoring the bandwidth channel for any vacancy lies in the sensing at the radio domain interface. It also keeps track of the user's incoming requests that may be authorized or unlicensed. In both cases, based on the sensed spectrum availability, the cognitive engine decides upon the allocation of the channel to the desired user, with respect to the configuration parameters that identify it to be licensed or unlicensed.

In the CR network, several susceptible attacks have occurred in the spectrum availability and access Khanna et al. (2022). These attacks collapse the spectrum sensing data in the CR network by blocking or jamming the process. It is not easy to protect licensed primary users from an attacker node. The attacks can be done on all five layers of the CR network. These attacks reduce the spectrum stability, sharing and accessing limits in the CR network.

The CR attacks are disaggregated by the target's five layers of TCP/IP, starting with a physical layer attack and ending with application layer attacks Salahdine et al. (2020) Furthermore, cross-layer attacks on any layer can reach another layer.

Spectrum sensing can be accessed in CR via an unallocated spectrum band and an open-air medium as a physical layer channel. This interaction can open a gadget to an assortment of safety issues, which can be taken advantage of by an aggressor during the range detecting process. The primary user emulation (PUE') attack, jamming attack (JaA), primary user location (PUL') attack, eavesdropping (EvD), objective function (OF') attack, and learning attack (LeA) are the most frequently used attacks on CR channels and prevent access to the actual physical layer of the CR network.

The malicious nodes are hidden and modify their behaviour to mimic that of a primary user. This node is capable of transmitting a few special signals in the licensed band, much like a legal primary user. This can cause the CR user to incorrectly detect the primary user's availability. This attack convinced the CR user that the primary user was still active on a channel.

The allocation of radio spectrum for all wireless communication devices has turned out to be a daunting approach due to the increasing demand for wireless applications. Additionally, the investigation conducted in this thesis is related to a scenario of 5G IoT network implementation. In which huge volumes of real-time data are expected to arrive and travel through the network, the communication devices may need an increased allocation of bandwidth. Especially, the devices involved in handling, storing, and processing of multimedia content demand a higher bandwidth. In a scarcely available spectrum, the implementation of cognitive radio networks and their concepts to the 5G IoT scenario is going to be quite challenging and testing in nature. A typical illustration of a cognitive radio network, overlapping with the primary transmitter and a base station scenario is depicted in Figure 2.

Figure 2. Illustration of a cognitive radio network scenario

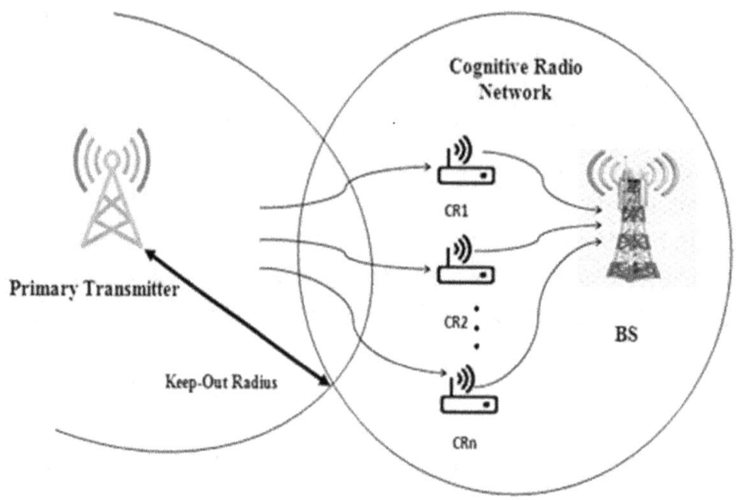

(Akyildiz et al. 2009)

As depicted in Figure 2, it could be understood that the cognitive radio networks (CRN) offer a useful and intelligent method of bandwidth utilization from the scarcely available radio spectrum. In the above visualized model, cognitive radio networks are found to converge within the primary transmitter coverage area. The functionality of the cognitive radio network in the above scenario is two-fold which can be a cooperative method or a spectrum overlay process. During this process, both the PUs and the SUs utilize the same bandwidth with suitable techniques for providing services to consumers to minimize the interference of SUs on the licensed users. The SUs can perform the opportunistic access of bandwidth by monitoring and identifying white holes which remain unutilized by the PUs. However, in both methods, demerits exist in the form of interference due to cooperative utilization of radio spectrum, demanding appropriate filtering techniques without degrading the quality of services. In addition, latency remains to be a critical issue due to a coexisting nature, even though PUs is given the highest priority in the CR. In the opportunistic method the SUs must be removed and moved to a new band once the PUs becomes active after a period of idle behavior. It increases both latency and uncertainty about the availability of a new white hole in the bandwidth spectrum at the required time.

Figure 3. Illustration of spectrum hole scenario in CRN

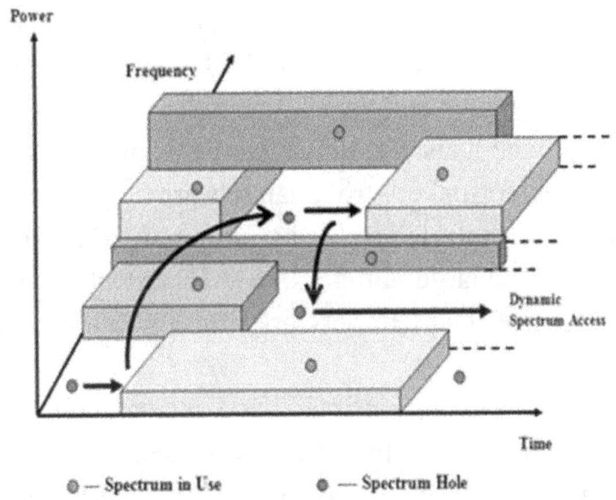

(Akyildiz et al. 2009)

In both the above cases, a common method of bandwidth allocation by the cognitive radio network is working through an efficient and continuous spectrum sensing technique. Monitoring the available radio band is done for identifying vacancies that arise then and there when the PUs switches over from active state to idle state and vice-versa. The concept of spectrum holes is depicted in Figure 3 shown above.

Figure 3 clearly illustrates the presence of spectrum holes popularly known as white holes in a cognitive radio network sensing scheme. It could further be elaborated that in CRN sensing, available bandwidth is allocated to the primary users who represent the licensed class of consumers. However, the PUs may not be active for the entire duration of time and may switch over to the OFF state. Once the required service provision by way of information processing, call transfer, voice transmission, and processing is over PU may change its state. In such cases, the bandwidth utilized by them is left vacant resulting in the creation of white holes or spectrum holes. This is being continuously monitored or sensed by the sensing algorithm, implemented in the CRN. Sensed spectrum is to be allocated to the queued up secondary user on a priority basis until the period of time when the PU switches back to the active state. This condition is once again detected by the sensing algorithm which has to make arrangements instantaneously to stop the access of bandwidth by SUs and relocate them to a newly available white hole, if any. This entire process has to be carried out quickly to avoid delay or latency. The PU enjoys the highest priority in terms of access to the spectrum and under no circumstances; it can be denied access to the bandwidth. On the other hand, the SUs who are removed and relocated attract an interruption of services. The removal and relocation have to be carried out at a very minimal time so as not to affect the degradation of the quality of service, provided to the consumers.

The CR task is adaptive in nature, as it is determined by certain network parameters. Several radio parameters are involved, including the centre frequency, the power, the bandwidth, the modulation type, the channel access protocol, the coding rate, the frame size, and the encryption type. The CR is determined by these parameters, and the optimal choice is determined by the application's requirements. It is used to resolve one or more OF's by maximizing throughput and minimizing power consumption, among other things. The attackers attempted to exploit the techniques' flaws in order to estimate and manipulate these parameters. While assessing the OF' esteem corresponding to the necessities of client applications on the organization, it is feasible to show up at unfortunate arrangements

Because of the dispersed correspondence, the CR network is basic for associating hubs in a multi-bounce CR network in a few covering ways. The DLL-based Macintosh control frameworks are traded to safeguard the range in a complex and multi-state association. At the point when numerous CR hubs are expected to impart simultaneously, the range can be designated to a particular coordinated information channel. At long last, the normal control channels are clogged. The pernicious hub is utilized to give a disseminated instead of concentrated status by immersing the control channel with counterfeit Macintosh control outlines. Due to the large number of MAC control frames, this procedure may result in a decrease in network performance. The CR Network's throughput decreases to near zero as a result of the CCSD' attack.

Figure 4 gives a clear picture of a typical primary and secondary user in a coexisting radio environment with the licensed band, surrounded by a host of unlicensed users around it. As discussed in the preceding sections, the efficiency of the CRN depends on an efficient spectrum sensing technique. It must be very precise to identify the spectrum holes in order to allow the SUs to have opportunistic spectrum access to the radio band. And thereafter, the SUs have to be moved when the PU becomes active in its state of operation.

As visualized in Figure 5, it could be taken into consideration as a network consisting of K sub bands with the frequency of ω_v, where v = 1, 2, 3....K. K denotes the set of available bands from the radio spectrum of PUs.

PUs - Primary User SUS - Secondary Users

Pr.BS - Primary Base Station Sr.BS - Secondary Base

Pr.TxR - Primary Transmitter Station FC - Fusion Center

Figure 4. Illustration of a co-existing CRN model

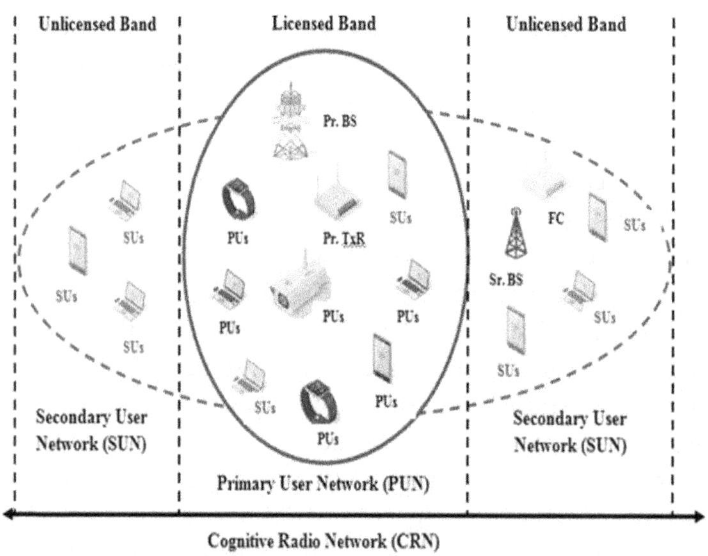

(Hong et al. 2014)

A wideband communication model is taken as the investigation model in this research work as the application is centered on a 5G communication network. If P is the number of sub-bands at time instance T_c, the condition of the spectrum allocation after utilization of P users with the sub band is $P < K$.

The inspiration for this work comes from the basic requirement for hearty interruption discovery components in CRNs to defend their uprightness and execution. The common range climate of CRNs requests insightful arrangements equipped for knowing typical organization conduct from peculiar exercises presented by assailants. The proposed approach coordinates the qualities of Self-Consideration LSTM to catch complex worldly conditions and spatial connections inborn in unique range access designs.

2. LITERATURE SURVEY

Almuqren at al. (2024) shows the evidence of significance of deep learning algorithm in the spot of detection of various security attacks associated with the cognitive radio networks.

Khanna et al. (2024) and Turkyilmaz et al. (2023) exhibit the utilization of machine learning algorithm for security attack detection in cognitive radio networks. Saifuddin et al. (2017) proposed a method for developing and detecting the Primary User Emulation (PUE) attack in cognitive radio networks. The main advantage of the CR network is its efficient use of opportunistic and rare spectrum resources. Attacks on knowledgeable radio networks based on primary user emulation pose a serious threat to their deployment.

Fulai Liu et al. (2016) proposed a novel attacker identification algorithm with the possibility of a dynamic attack that could significantly reduce the likelihood of a false alarm and significantly increase the likelihood of overall detection. Meanwhile, the weight coefficient is investigated for energy detection (ED) to increase the robustness of the system .The reproduction results show that the proposed calculation performs better compared to the past one.

Armi et al. (2018) researched harmful user attacks and their effects on perceiving results in Cognitive Radio Networks (CRN). They consider several harmful users who mimic the basic user (PU) signal. Destructive clients are haphazardly conveyed over a default distance around the optional client (SU).

Hu et al. (2012) Investigate a type of attack on a cognitive radio (CR) network called a highly active band (MAB) attack, in which an attacker or malicious CR node is most common within a multi-band CR network. Detects or determines the active band. Denial of service (DoS) targets it with an attack.

Nguyen Thanh et al. (2015) examined the Primary User Emulation (PUE) attack, which is a significant security issue in Cognitive Radio (CR) networks. To limit selfish PUE attackers, channel monitoring processes must be implemented to determine the identity of the active user and the identity of the self-interested PUE attacker.

Akbari et al. (2018) describea Knowledge Radio Network (CRN) with N Secondary Users (SUs), M Real Primary Users (PUs) and Single Attackers, where the primary threat is the Primary User Emulation (PUE) attack. It is expected to be that K active PUs coexist in the communication range of the CRN, where the parameter $K \leq$ is unknown to the PUE detection system.

Yuan etc. (2012) Propose a faith-based defence strategy against PUE attacks in CR networks, eliminating the deployment of additional sensor networks and expensive hardware used in previous literature. In the proposed approach, each secondary user calculates local function and compatibility function, counts messages, exchanges messages with neighbouring users, and calculates beliefs until convergence occurs.

Magdalene et al. (2017) has focused its efforts at CRN on a specific security threat called the Byzantine invasionSpectrum Sensing Data Falsification (SSDF) is the name of this attack. By sending false sensing reports, malicious users attempt to take control of the FC in SSDF attacks. Decentralized networks have increased their attack flexibility, which makes this important.Raj Kumari et al. (2018) proposed a non-consensual Distributed Spectrum Sensing (NCSS) scheme in which a node determines the situation with PU by examining the crude energy level of the PU signal from its h-jump neighbours.A message passing scheme is used and compared to the current consensus algorithm.

Sharma et al. (2013) suggest a better scheme for rotation and selection of cluster heads. Environmental scenarios related to mobile communications, such as fading and shadowing, are considered. Comparing the Comparing the proposed strategy to previously published methods demonstrates its superior performance. for cooperative cognitive radio networks. A new technique has been proposed to improve the selection and rotation of cluster heads to improve the spectrum sensing performance of secondary users.

3. PROPOSED SYSTEM

Cognitive Radio Networks (CRNs) represent a dynamic paradigm in wireless communication, allowing unlicensed users to opportunistically access underutilized spectrum bands. However, the open and shared nature of the spectrum in CRNs makes them susceptible to security threats. To address this challenge, the study proposes an advanced intrusion detection system leveraging a combination of Self-Attention Mechanism and Long Short-Term Memory (LSTM) networks.

Figure 5. Functional architecture of proposed system

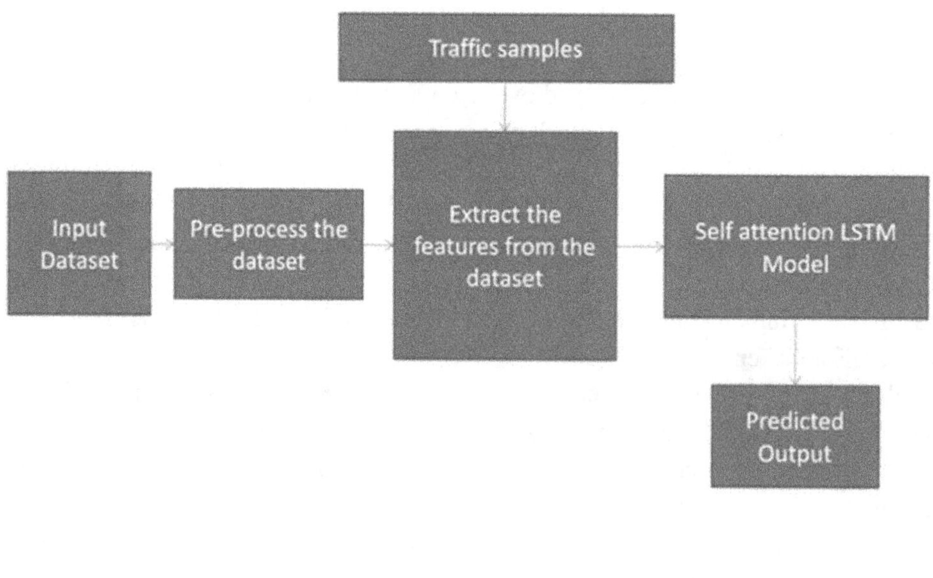

a. Data Preprocessing Module

The system initiates with data preprocessing, involving cleaning, normalization, and feature extraction from CRN datasets. Temporal sequences are prepared for input into the subsequent modules.

b. Self-Attention Mechanism Module

The Self-Attention Mechanism is employed to enhance the model's adaptability. It permits the model to zero in on significant segments of the range, powerfully gauging the significance of various elements during the growing experience.

c. Long Short-Term Memory (LSTM) Network Module

Designed to capture long-term dependencies in temporal dynamics, the LSTM module is crucial for remembering and learning patterns over extended time intervals.

d. Attack Scenario Generation Module

Simulating various attack scenarios, this module generates synthetic attack data such as spectrum sensing data falsification and malicious spectrum access. The combination of attack and normal data facilitates comprehensive model training and evaluation.

e. Training Module

The combined dataset undergoes training to optimize model parameters and weights. The Self-Attention LSTM model learns to differentiate normal and attack patterns, ensuring a robust intrusion detection capability.

f. Evaluation Module

Real-world CRN datasets, encompassing diverse attack scenarios, are employed to evaluate the model's performance. Metrics such as accuracy, precision, recall, and false-positive rates are analyzed to assess the model's effectiveness.

g. Real-Time Processing Module

Optimizing the model for real-time processing is essential for deployment in dynamic CRN environments. Techniques like hardware acceleration or parallel processing are considered to ensure scalability.

h. Results Visualization Module

This module generates visualizations such as confusion matrices, ROC curves, and precision-recall curves, providing insights into the model's strengths and limitations.

4. RESULTS AND DISCUSSION

The proposed Self-Attention LSTM model demonstrates effectiveness in accurately detecting and classifying attackers while maintaining a low false-positive rate. Results on real-world CRN datasets validate its robust performance under diverse attack scenarios.

Figure 6. Dataset generations

	A	B	C	D	E	F	G	H	I	J	K	L	M	N	O	P	Q	R	S	T	U
1	dt	switch	src	dst	pktcount	bytecount	dur	dur_nsec	tot_dur	flows	packetins	pktperflow	byteperflow	pktrate	Pairflow	port_no	tx_bytes	rx_bytes	tx_kbps	rx_kbps	tot_kbps
2	11425	1	10.0.0.1	10.0.0.8	45304	48294064	100	7.16E+08	1.01E+11	3	1943	13535	14428310	451	0	3	1.44E+08	3917	0	0	0
3	11605	1	10.0.0.1	10.0.0.8	126395	1.35E+08	280	7.34E+08	2.81E+11	2	1943	13531	14424046	451	0	4	3842	3520	0	0	0
4	11425	1	10.0.0.2	10.0.0.8	90333	96294978	200	7.44E+08	2.01E+11	3	1943	13534	14427244	451	0	1	3795	1242	0	0	0
5	11425	1	10.0.0.2	10.0.0.8	90333	96294978	200	7.44E+08	2.01E+11	3	1943	13534	14427244	451	0	2	3688	1492	0	0	0
6	11425	1	10.0.0.2	10.0.0.8	90333	96294978	200	7.44E+08	2.01E+11	3	1943	13534	14427244	451	0	3	3413	3665	0	0	0
7	11425	1	10.0.0.2	10.0.0.8	90333	96294978	200	7.44E+08	2.01E+11	3	1943	13534	14427244	451	0	1	3795	1402	0	0	0
8	11425	1	10.0.0.1	10.0.0.8	45304	48294064	100	7.16E+08	1.01E+11	3	1943	13535	14428310	451	0	4	3665	3413	0	0	0
9	11425	1	10.0.0.1	10.0.0.8	45304	48294064	100	7.16E+08	1.01E+11	3	1943	13535	14428310	451	0	1	3775	1492	0	0	0

Figure 6 depicts the data set generation can also be referred as sensing matrix which will be fed to the pre-processing stage.

Figure 7. Training process

```
*Python 3.7.6 Shell*
File  Edit  Shell  Debug  Options  Window  Help
Epoch 7/10
1/1 [==============================] - ETA: 0s - loss: 0.6426 - accuracy: 0.94
4
1/1 [==============================] - 0s 17ms/step - loss: 0.6426 - accurac
 0.9444 - val_loss: 0.6722 - val_accuracy: 0.8000
Epoch 8/10
1/1 [==============================] - ETA: 0s - loss: 0.6318 - accuracy: 1.00
0
1/1 [==============================] - 0s 26ms/step - loss: 0.6318 - accurac
 1.0000 - val_loss: 0.6657 - val_accuracy: 0.8000
Epoch 9/10
1/1 [==============================] - ETA: 0s - loss: 0.6211 - accuracy: 1.00
0
1/1 [==============================] - 0s 24ms/step - loss: 0.6211 - accurac
 1.0000 - val_loss: 0.6594 - val_accuracy: 0.8000
Epoch 10/10
1/1 [==============================] - ETA: 0s - loss: 0.6105 - accuracy: 1.00
0
1/1 [==============================] - 0s 16ms/step - loss: 0.6105 - accurac
 1.0000 - val_loss: 0.6532 - val_accuracy: 0.8000
```

Figure 7 shows the training process to train the model for the large volume of sets processed in the previous stages. It marks the first step in the detection process.

Figure 8. Confusion matrix

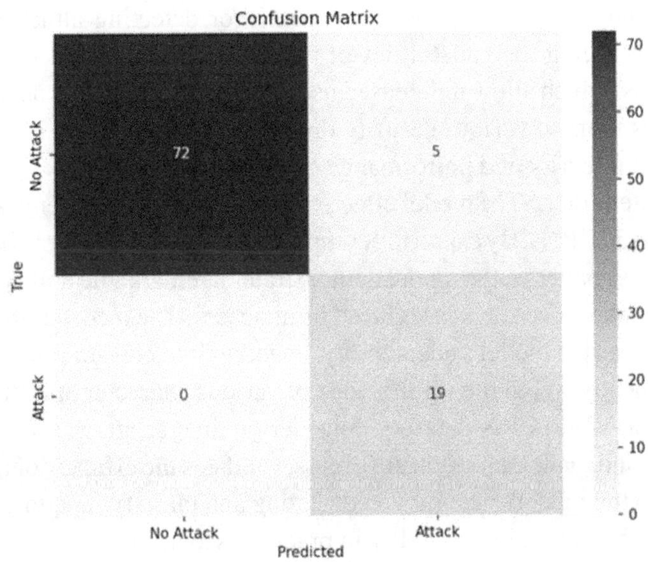

Two classes were trained with labels Attack and No Attack. Confusion matrix was evaluated in accordance with the above said labels and it is depicted in the figure 1.8. Almost four parameters was estimated namely Accuracy, Precision, Recall and F1-Score. The comparison chart of those four parameters was compared with machine learning algorithms such as SVM, Adaboost and deep learning algorithm as CNN. Of all the techniques the proposed algorithm forecasts its significance and seems to be superior to the existing algorithms. The comparison chart is shown in the Figure 9.

Figure 9. Comparison chart

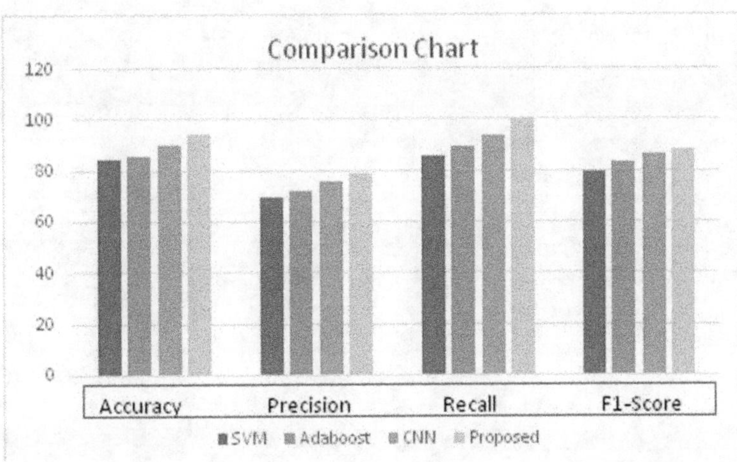

5. CONCLUSION

In conclusion, this study introduces a novel approach for detecting attackers in Cognitive Radio Networks (CRNs) by leveraging the capabilities of a Self-Attention LSTM model. The dynamic and intelligent nature of CRNs, which allow unlicensed users to opportunistically access underutilized spectrum bands, also exposes them to various security threats. Robust intrusion detection mechanisms are essential to safeguard the integrity and performance of these networks.

The proposed Self-Attention LSTM model effectively addresses the challenges posed by the inherent variability and uncertainty in CRNs. By capturing complex temporal dependencies and spatial correlations in dynamic spectrum access patterns, the model demonstrates a remarkable ability to discern normal network behavior from anomalous activities introduced by attackers. The incorporation of the self-attention mechanism further enhances the model's adaptability, enabling it to concentrate on relevant segments of the spectrum and dynamically weigh the significance of various characteristics throughout the learning process.Evaluation on real-world CRN datasets, considering diverse attack scenarios such as spectrum sensing data falsification and malicious spectrum access, validates the efficacy of the proposed approach. thus the results showcase the model's accuracy in detecting and classifying attackers while maintaining a low false-positive rate, indicating its reliability in practical scenarios.

Importantly, the scalability and real-time processing capabilities of the Self-Attention LSTM model make it well-suited for deployment in dynamic CRN environments. This suggests a practical applicability of the proposed approach in ensuring the continued reliable and secure operation of cognitive radio technologies. As the landscape of security threats evolves, this study contributes to the enhancement of security measures in CRNs, addressing the imperative need for advanced intrusion detection mechanisms in the face of emerging challenges.

REFERENCES

Abeer, F. (2018). Hierarchical Cluster-Based Cooperative Spectrum Sensing in Cognitive Radio Networks Using Adaptive Threshold. *IEEE 61st International Midwest Symposium on Circuits and Systems (MWSCAS)*. IEEE.

Akbari, K., & Abouei, J. (2018). Signal Classification for Detecting Primary User Emulation Attack in Centralized Cognitive Radio Networks. Electrical Engineering (ICEE), Iranian Conference On. 10.1109/ICEE.2018.8472515

Akyildiz, I. F., Lee, W. Y., & Chowdhury, K. R. (2009). CRAHNs: Cognitive radio ad hoc networks. *Ad Hoc Networks*, 7(5), 810–836. 10.1016/j.adhoc.2009.01.001

Ali, A., Yaqoob, I., Ahmed, E., Imran, M., Kwak, K. S., Ahmad, A., Hussain, S. A., & Ali, Z. (2018). Channel Clustering and QoS Level Identification Scheme for Multi-Channel Cognitive Radio Networks. *IEEE Communications Magazine*, 56(4), 164–171. 10.1109/MCOM.2018.1700752

Almuqren, L., Maray, M., Alotaibi, F. A., Alzahrani, A., Mahmud, A., & Rizwanullah, M. (2024). Optimal Deep Learning Empowered Malicious User Detection for Spectrum Sensing in Cognitive Radio Networks. *IEEE Access : Practical Innovations, Open Solutions*, 12, 35300–35308. 10.1109/ACCESS.2024.3367993

Armi, N., Gharibi, W., Khan, W. Z., Zangoti, H., Rizvi, S., & Wael, C. (2017). Error detection of malicious user attack in cognitive radio networks. *2017 International Conference on Radar, Antenna, Microwave, Electronics, and Telecommunications (ICRAMET)*. IEEE. 10.1109/ICRAMET.2017.8253152

Caso, G., Soleimani, H., De Nardis, L., Tosti, A., & Di Benedetto, M. G. (2014). SENSIC: Mobility-aware cluster-based cooperative spectrum sensing for cognitive radio networks. *2014 IEEE International Conference on Ultra-WideBand (ICUWB)*. IEEE. 10.1109/ICUWB.2014.6958959

Hong, X., Wang, J., Wang, C.-X., & Shi, J. (2014). Cognitive radio in 5G: A perspective on energy-spectral efficiency trade-off. *IEEE Communications Magazine*, 52(7), 46–53. 10.1109/MCOM.2014.6852082

Hu, N., Yao, Y.-D., & Mitola, J. (2012). Most Active Band (MAB) Attack and Countermeasures in a Cognitive Radio Network. *IEEE Transactions on Wireless Communications*, 11(3), 898–902. 10.1109/TWC.2012.011812.110927

I., E.E. (2023). Clement JC. GRU-SVM Based Threat Detection in Cognitive Radio Network. *Sensors (Basel)*, 23(3), 1326. 10.3390/s2303132636772366

Khanna, A., Rani, P., Sheikh, T. H., Gupta, D., Kansal, V., & Rodrigues, J. J. (2022). Blockchain-based security enhancement and spectrum sensing in cognitive radio network. *Wireless Personal Communications*, 127(3), 1899–1921. 10.1007/s11277-021-08729-0

Kohlstruck, C., Engel, M., & Gotzhein, R. (2018). *The Selective Clustering Energy Detector for Cognitive Radio Networks – Conceptual Design and Experimental Assessment. 2018 14th International Wireless Communications & Mobile Computing Conference*. IWCMC.

Liu, F., Zhou, Y., Sun, Z., Du, R., & Sheng, J. (2016). Dynamic attack probability based Spectrum Sensing against Byzantine attack in Cognitive Radio. *2016 2nd IEEE International Conference on Computer and Communications (ICCC)*. IEEE.

Magdalene, A. H. S., & Thulasimani, L. (2017). Fuzzy Clustering Means (FCM) for Mitigating Spectrum Sensing Data Falsification (SSDF) Attack in Cognitive Radio Networks. *2017 IEEE International Conference on Computational Intelligence and Computing Research (ICCIC)*. IEEE. 10.1109/ICCIC.2017.8524462

Nguyen Thanh, N., Ciblat, P., Pham, A. T., & Nguyen, V.-T. (2015). Surveillance Strategies Against Primary User Emulation Attack in Cognitive Radio Networks. *IEEE Transactions on Wireless Communications*, 14(9), 4981–4993. 10.1109/TWC.2015.2430865

Rajkumari, R., & Marchang, N. (2018). Secure Non-Consensus Based Spectrum Sensing in Non-Centralized Cognitive Radio Networks. *IEEE Sensors Journal*, 18(9), 3883–3890. 10.1109/JSEN.2018.2816595

Ramli, A., & Grace, D. (2010). RF signal Strength based clustering protocols for a self-organizing cognitive radio network. *2010 7th International Symposium on Wireless Communication Systems*.

Saifuddin, K. M., Reza, K. F., Fahim, M. A. N. I., & Alam, S. S. (2017). Detection of primary user emulation attack in cognitive radio environment. *2017 6th International Conference on Informatics, Electronics and Vision & 2017 7th International Symposium in Computational Medical and Health Technology (ICIEV-ISCMHT)*. IEEE.

Salahdine, F., & Kaabouch, N. (2020). Security threats, detection, and countermeasures for physical layer in cognitive radio networks: A survey. *Physical Communication*, 39, 101001. 10.1016/j.phycom.2020.101001

Saleem, Y., Yau, K.-L. A., Mohamad, H., Ramli, N., Rehmani, M. H., & Ni, Q. (2017). Clustering and Reinforcement-Learning-Based Routing for Cognitive Radio Networks. *IEEE Wireless Communications*, 24(4), 146–151. 10.1109/MWC.2017.1600117

Sharma, P., & Abrol, V. (2013). Optimized cluster head selection & rotation for cooperative spectrum sensing in Cognitive Radio Networks. *2013 Tenth International Conference on Wireless and Optical Communications Networks (WOCN)*. IEEE. 10.1109/WOCN.2013.6616260

Turkyilmaz, Y., Senturk, A., & Bayrakdar, M. E. (2023). Employing machine learning based malicious signal detection for cognitive radio networks. *Concurrency and Computation*, 35(2), e7457. 10.1002/cpe.7457

Yuan, Z., Niyato, D., Li, H., Song, J. B., & Han, Z. (2012). Defeating Primary User Emulation Attacks Using Belief Propagation in Cognitive Radio Networks. *IEEE Journal on Selected Areas in Communications*, 30(10), 1850–1860. 10.1109/JSAC.2012.121102

Zhang, J., Fan, W., Yao, F.-Q., Zhao, H., & Li, Y. (2010). Cluster-based distributed topology management in Cognitive Radio Ad Hoc networks. *2010 International Conference on Computer Application and System Modeling (ICCASM 2010)*. IEEE.

Chapter 3
Artificial Intelligence Techniques in Cognitive Radio Networks Applications

Kamalendu Pal
https://orcid.org/0000-0001-7158-6481
University of London, UK

ABSTRACT

Cognitive radios (CRs) are intelligent wireless sensor devices that sense, learn, and adapt to real-time operating environments. One of the essential characteristics of a CR-based communication network (also known as cognitive radio networks (CRNs)) is to optimize radio spectrum usage. CRNs have two types of users: Primary users (PUs) and secondary users (SUs). PUs are licensed users who can utilize the available spectrum at any instance of their licensed time. SUs are unlicensed users who can utilize the spectrum only when PUs are not utilizing the allocated spectrum band. This chapter reviews various artificial intelligence (AI) techniques in cognitive radio applications to enhance CRN cognition capability. AI enables systems to solve problems by emulating human behaviours (e.g., reasoning, and learning), and these behaviours are studied in applications related to the tasks of cognitive radio

1. INTRODUCTION

With the advent of a growing number of communication devices and the demand for wireless communication (Pal, 2022a) (Pal, 2022b), radio spectrum scarcity happens and mandates the establishment of methods for developing and designing technical ways to access the available radio spectrum efficiently (Pal, 2023). The radio spectrum is a shared resource assigned to various licensed holders according to the fixed spectrum assignment policy. It is also well known that a considerable portion of the spectrum is not utilized in different times and places. Cognitive radio (CR) and the concept of its network were proposed to solve this problem by opportunistically utilizing the spectrum during the absence of their legal owners' use. Academics and practitioners are putting forward their ideas and views on how cognitive radio networks (CRNs) will play a significant role in under-utilizing spectrum resources to meet the continuous demand for wireless applications (Mitola, 1999) (Lo, 2009) (Pal, 2023). Radio spectrum

DOI: 10.4018/979-8-3693-2893-4.ch003

utilization can be maximized, and different communication *network coding schemes* (NC) are deployed to better transmit data packets over the CRN networks.

The distribution of data packets from a source node to multiple destination nodes over a communication network, known as *multicasting*, is an area that has received immense attention in recent decades. In multicasting, the fundamental problem dealt with by researchers is the effective scheduling of data streams. A novel solution proposed for this type of problem is known as *network coding* (NC), in which information is encoded at the node level and then spread over the data communication network, thereby improving the scheduling of data streams (Pal, 2023). In addition, coding schemes can enhance the throughput and radio spectrum utilization efficiency of wireless data communication (Pal, 2023). Network coding has various benefits, for example, increasing the throughput by communicating more information with fewer packet transmissions. Besides, NC can send more information for less bandwidth by reducing the number of transmissions, and there is no need for packet retransmission, thereby saving the available bandwidth. NC also reduces packet loss by applying coding at the source and intermediate nodes instead of only at a source node. In this way, protecting against packet loss, NC helps avoid link failures by sharing network resources among different flows. NC helps improve network performance by providing optimal routing (i.e., data travel path) and enhancing security by coding the packets (Pal, 2023). Network coding practice has been used in different types of networks and their applications. For example, NC has been used for wireless sensor networks, vehicular ad-hoc networks, multicast networks, peer-to-peer networks, file distribution, service discovery, and distributed content storage networks (Pal, 2023).

CRNs enable cognitive users (or secondary users) to sense the environment to identify spectrum holes, analyze the parameters, and make dynamic resource allocation management decisions. These capabilities are realized by integrating artificial intelligence (AI) techniques into the heart of CR applications. AI enables cognitive users to solve problems by emulating human biological processes such as learning, reasoning, decision-making, self-adaption, self-organization, and self-stability. In this way, CRNs have evolved to solve the fixed spectrum assignment problem by dynamically assigning spectrum to secondary users. By dynamic utilization of spectrum bands by SUs, the efficiency of CRN increases, and the spectrum scarcity problem gets reduced. In this radio spectrum sharing mechanism, when a PU requests to access its licensed range, the SUs preferentially using the same spectrum should move over to other unused scopes to safeguard the transmission of the PU and continue their data delivery (Lo, 2009). However, several data transmission-related throughput degradations may occur in this transition process (Lane, 2023).

Various surveys have been proposed in the literature for the applications of AI techniques in CRNs. In (Abbas et al., 2015), a survey on different learning techniques such as fuzzy logic (FL), genetic algorithms (GAs), neural networks (NNs), game theory (GT), reinforcement learning (RL), support vector machine (SVM), case-based reasoning (CBR), decision tree (DT), entropy, Bayesian, Markov model (MM), multiagent systems (MAS), and artificial bee colony (ABC) algorithm was presented. They discussed their strengths and limitations according to spectrum sensing and decision-making. In (He et al., 2010), the state of the art in using AI in CR to promise reasonable selections for deploying a practical CR and the relative advantages of various proposed techniques in differing applications were surveyed. The survey's techniques include artificial neural networks (ANNs), metaheuristic algorithms (MEAs), hidden Markov models (HMMs), rule-based reasoning (RBR), ontology-based systems (OBSs), and case-based systems (CBSs). In (Bkassiny et al., 2013), a survey on different learning techniques such as RL, GT, NNs, SVM, and MM were presented. They also discussed their strengths, weaknesses, and challenges. In (Gavrilovska et al., 2013), the authors considered GT, RL, and reasoning approaches such as Bayesian networks (BNs), FL, and CBR. In (Salau et al., 2015), a survey of the AI techniques, GA, ANN, HMMs, and MEAs were presented, and their strengths, weaknesses, and open research issues were examined.

In this chapter, the use of AI techniques is studied in applications related to the main challenges of CRNs, including spectrum sensing, spectrum sharing, spectrum mobility, and spectrum management regarding dynamic spectrum access, resource allocation, parameter adaptation, and optimization problems. The aim is to provide a single source as a survey paper to help researchers better understand the recent implementations of AI approaches to different cognitive radio designs. AI techniques mentioned in this paper include ANNs, Markov chains (MCs), FL, GT, SVM, RBSs, OBSs, CBSs, MASs, simulated annealing, and tabu search. As well as bio-inspired algorithms, including GA, differential evolution (DE), particle swarm optimization (PSO), bacterial foraging optimization (BFO), ant colony optimization (ACO), cat swarm optimization (CSO), ABC, different soft-computing influenced algorithms (Thomas, 2007) (Niyato & Hossain, 2008), and knowledge-based methods (e.g., Rule-Based Reasoning, Case-Based Reasoning) are considered.

While all of these applications are recognized as in the domain of cognitive radio, there is much disagreement on precisely what is and what is not a cognitive radio, with everyone having their own interpretation and definition. Nonetheless, the following attributes are commonly expected from cognitive radio and form a baseline set of assumptions for the remainder of this chapter.

- Spectrum Sensing: In the spectrum sensing phase, the CR node detects the unused spectrum.
- Spectrum Analyze: In the spectrum analysis phase, the CR node analyzes the available spectrum.
- Spectrum Use Decision: Spectrum use decision phase; CR node selects the best available channel. It includes understanding the environment and capability of the radio (awareness), making informed decisions on actions (reasoning), and learning the impact of these actions on the performance of the radio, as well the performance of the network in which the radio is embedded (learning).
- Adapt: Finally, adapt the selected channel.

Figure 1. Working principle of capabilities of CRN

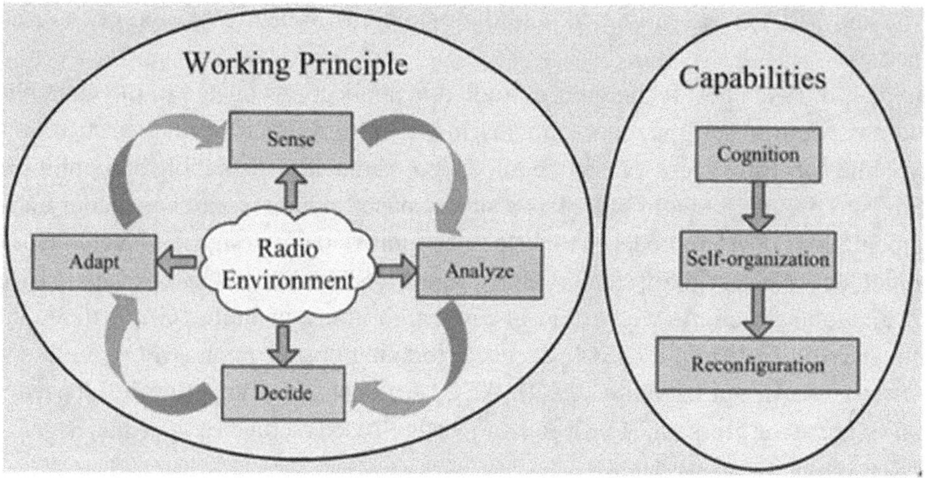

Figure 1 presents these phases in a diagrammatic representation. These four attributes can be embodied in a common element known as a cognitive engine (CE). This paper defines a CE as an "intelligent" agent that looks at the cognition tasks in a CR, where intelligence denotes behaviour consistent with a specified goal. The CE can be implemented as an independent entity interacting with the radio transceiver [e.g., reconfigurable radio transceiver implemented with software-defined radio (SDR)] or as a collection of interacting entities, with each entity fulfilling a specific role. Given the input from its environment or user (observation), the CE analyses and classifies the situation, finds the appropriate response to the stimulus (cognition), and conducts the decision (reconfiguration). For example, this response can adapt radio parameters such as the channel coding scheme, the modulation scheme, and the operation frequency, given user requirements, current environment conditions, and previous experiences at the CE.

Implementing the various aspects of a CE is an active area of research in which considerable attention is given to each of these three attributes. This chapter focuses on the practical issues associated with realizing cognition in a radio, which we view as an application of artificial intelligence (AI) to the radio domain, while only considering the attributes of observation and reconfiguration as needed for clarity. Specifically, this chapter surveys the state of the art in using AI in CR to ascertain available choices for implementing a practical CR and the relative merits of various proposed techniques in differing applications. This chapter takes the following structure. Section 2 reviews AI techniques proposed for use in a CE and presents examples of their applications. Section 3 discusses practical issues with CR, such as convergence time, implementation complexity, stability, and how to manage multiple coexisting AIs, within a single CE. Section 4 reviews two CEs developed using different AI techniques. Finally, Section 5 concludes with some observations on the different AI techniques.

2. OVERVIEW OF RADIO SPECTRUM MANAGEMENT

The radio spectrum is a subcategory of the electromagnetic radiation that carries radio waves. The visible light, infrared, ultraviolet, x-rays, and gamma rays are also portions of the electromagnetic spectrum. In particular, radio spectrum refers to frequencies ranging from 3 kHz to 300 GHz, as presented in Table 1. Besides, these frequencies support a wide range of radio communication systems, including narrow and broadband mobile telecommunications, broadcasting, aeronautical and marine communications, remote control and monitoring, and communications for public bodies such as Défense and emergency services.

Table 1. Classification of the radio frequency spectrum

Band Name	Frequency	Wavelength
Very Low Frequency (VLF)	3 – 30 kHz	33 – 10 km
Low Frequency (LF)	30 – 300 kHz	10 – 1 km
Medium Frequency (MF)	300 – 3000 kHz	1000 – 100 m
High Frequency (HF)	3 - 30 MHz	100 – 10 m
Very High Frequency (VHF)	30 – 300 MHz	10 – 1 m
Ultra-High Frequency (UHF)	300 – 3000 MHz	1000 – 100 mm
Super High Frequency (SHF)	3 – 30 GHz	100 – 10 mm
Extremely High Frequency (EHF)	30 – 300 GHz	10 – 1 mm

The radio spectrum's key characteristics are the propagation features and the amount of information carried by signals. Generally, signals transmitted using higher frequencies reach shorter distances but have a higher bandwidth. These physical characteristics of the radio spectrum limit the range of applications for which a particular band is suitable. Some spectrums, such as the UHF, bear to be convenient for a wide variety of services and are, therefore, in great demand, as shown in Table 2. In addition, every frequency band of the radio spectrum can be used for different licensed and unlicensed services. In order to ensure that competing applications for radio spectrum do not interfere with each other, it is essential to coordinate the use of radio spectrum nationally and internationally.

In addition, every frequency band of the radio spectrum can be used for different licensed and unlicensed services. In order to ensure that competing applications for radio spectrum do not interfere with each other, it is essential to coordinate the use of radio spectrum nationally and internationally.

Table 2. Radio spectrum and some of its applications

Band Name	Communication Systems
Very Low Frequency (VLF) and Low Frequency (LF)	Naval broadcast and emergency services.
Medium Frequency (MF)	AM radio, Aeronautic services.
High Frequency (HF)	Short wave broadcasting, Amateur radio, Maritime, Remote control
Very High Frequency (VHF)	Analog TV, Remote control, Aeronautic, FM radio, Digital audio broadcasting
Ultra-High Frequency (UHF)	GSM, WCDMA, LTE, WiMAX, UWB, WiFi, Bluetooth, Wireless sensors, Analog TV, Digital TV, Meteorological satellite, DCP, Aeronautic, Satellite digital radio, Remote control

continued on following page

Table 2. Continued

Band Name	Communication Systems
Super High Frequency (SHF)	UWB, WiFi, WiMAX, DCP, Radar satellite broadcast
Extremely High Frequency (EHF)	WiMAX, Radar satellite broadcast

LEGENTS:
AM: Amplitude modulation, TV: Television, FM: Frequency modulation, GSM: Global system for mobile communications, WCDMA: Wideband code LTE: Long-term evolution, WiMAX: Worldwide interoperability for microwave access, UWB: Ultra-wideband, WiFi: Wireless fidelity, DCP: Digital cordless phones.

2.1 Radio Spectrum Regulation and Usage

Spectrum management regulates radio frequencies (RFs) to ensure efficient and effective use of radio spectrum. Conventionally, the radio spectrum is managed through national, regional, and international bodies, usually called regulators. For example, in the USA, the radio spectrum is used by the National Telecommunications and Information Administration (NTIA) for federal government users (e.g., military use) and the Federal Communications Commission (FCC) for commercial use. The Office of Communication (OFCOM) in the UK manages the civilian use of the radio spectrum. At the regional level, for instance, in Europe, the European Conference of Post and Telecommunications (CEPT) provides detailed guidance on radio frequency allocation, licensing, and technical criteria to national regulatory authorities within the European Union. The International Telecommunications Union drives radio spectrum regulation across countries and regions through the World Radiocommunication Conferences (WRC). WRC provides a framework and guidance for global spectral use for wireless applications that span multiple countries.

Specific spectrum measurement studies have evaluated the effectiveness of exclusive and static spectrum allocation. In this respect, it has been revealed that the radio spectrum is significantly underutilized during some periods in a particular geographic area. The unused spectrum bands in the temporal and spatial domains are called spectrum holes or white spaces. These idle or underutilized RF bands offer a great opportunity for wireless communications. Allowing unused or not fully used parts of the spectrum owned by licensed users to become available to unlicensed users can alleviate the spectrum scarcity problem and increase spectrum utilization. The following sections present some opportunities for AI-based techniques for radio spectrum management in CRNs.

3. ARTIFICIAL INTELLIGENCE TECHNIQUES

AI enables systems to solve problems by emulating human biological processes such as learning, reasoning, decision-making, self-adaptation, self-stability, and self-organization. The intelligent equipment will learn from its environment to exploit its experience. However, programming equipment capable of adapting to all situations and evolving according to new constraints is challenging. The ambient radio environment allows cognitive radios to adapt and learn their wireless transmission. Thus, AI must be implemented and adopted at the heart of cognitive radio technology. Various AI techniques are used and implemented in CRNs, including artificial neural networks (ANNs), Markov chains (MCs), fuzzy logic (FL), game theory (GT), support vector machine (SVM), rule-based systems (RBSs), case-based systems (CBSs), ontology-based systems (OBSs) muti agent systems (MASs), and evolutionary algorithms such as genetic algorithm (GA), Differential evolution (DE), particle swarm-based optimization (PSO), Artificial bee colony (ABC), simulated annealing (SA), and tabu search (TS).

3.1 Artificial Neural Networks

Artificial Neural Networks (ANNs) are networks of connected elementary processors operating in parallel distributed processing. Each elementary processor (artificial neuron) is characterized by inputs xi(i = 0…n), weights wij(j = 0…n), activation function F(x,w), the internal state of activation a = F(x,w), transition function f(a), and the output s = f (a). Each artificial neuron computes a single output based on the received information. They are inspired by biological brain behavior.

Three types of ANNs are used in CRNs: multilayer perceptron networks in linear (MLPN) and non-linear (MNPN) forms. These networks consist of many layers of computational frameworks, interconnected in a feed-forward manner. Each neuron represented in one layer has directed interconnections to the neurons of the subsequent layer. In MLPN, each layer is a linear combination of the previous layer's outputs. In MNPN, the units of these networks usually use a sigmoid function to initiate activation. The third one is the radial basis function network (RBFN), which has two layers and is a particular class of multilayer feed-forward networks. Each unit in the hidden layers employs a radial basis function, such as a Gaussian kernel, as the activation function. The radial function is centered at the point specified by the weight vector associated with the unit (Jain et al., 1996).

3.2 Markov Chains

Markov Models (MMs) are a mathematical formalism that generalizes short-path approaches in a stochastic environment. These models include state concepts that summarize the situation of the agent at each moment, the action that influences the dynamics of the state, and the reward associated with each state transition. They represent the temporal dependencies in a sequence of variables S0…St…, When each variable only depends on the previous one in the sequence, we say we have a Markov chain: P(St |S0, …, St-1) = P(St |St-1).

3.3 Game Theory

Game theory (GT) is a mathematical decision technique that specifies the behaviours of multiple rational entities (players) in strategic situations to maximize their rewards. A game mathematically modelled by three variables G = (N, S, U). N is the set of agents, S is the set of strategies, and U is the set of utility functions. Each player acts according to their strategy to maximize their utility. *Nash equilibrium* of a game is described as a point at which the utility function of each player does not increase if the player deviates from that point (Bkassiny & Jayaweera, 2013). Four categories of game theoretic approaches (Wang et al., 2010) can be used to model players' behaviour in a wireless environment: cooperative games, noncooperative games, action games, and stochastic games.

- In a cooperative game, the players are jointly grouped as coalitions and improve their utility function.
- In a noncooperative game, each player aims to choose the optimal strategy to maximize his reward. The famous concept of this category is the Nash equilibrium.

- In auction games, the players are the buyers who must select the appropriate bidding strategy in order to maximize their utility. An auctioneer conducts them.
- A stochastic game (Wang et al., 2010) extends the Markov decision process by considering the interactive competition among agents. A stochastic game has a set of states and a collection of actions, one for each player.

3.4 Fuzzy Logic

Fuzzy Logic (FL) is suitable for handling inaccuracy, uncertainty, and ambiguity in intelligent systems. In a FL system, the values are represented by fuzzy rather than classical sets. The advantage of this representation is to simulate human interpretations. The central components of an FL system are a fuzzifier, a fuzzy inference system, and a defuzzifier.

Fuzzification is the step that consists of the fuzzy quantification of the actual values. The fuzzy inference system implements an inference engine of the "if…then" type, presenting the relationship between the inputs and outputs. In the defuzzification, the output of the fuzzy system is mapped into non-fuzzy values as the system's output.

3.5 Support Vector Machine

Support vector machine (SVM) is a binary categorization or classification method based on a linear classifier called hyperplane. This classifier is assumed to separate the data into two classes such that the distance between the closest points of each class to the hyperplane is maximized [14]. This method uses an efficient training dataset to learn the model's parameters. It is based on kernel functions that allow the optimal separation of data.

3.6 Rule Based Systems

The rule-based system (RBS) is a classical method of knowledge representation. The rules are extracted from a specific area and used to make decisions. An RBS consists of a rule base (RB) that contains the list of rules and the inference engine (IE) to take actions based on the RB (Pal, 1998).

3.7 Case-Based Systems

The case-based system (CBS) is an analogical reasoning. The solution to new problems is selected from previous similar cases. The knowledge base is called the case base, where cases represent past experiences (Pal, 1998). Upon new solutions obtained from case adaptation, the case database is updated with the new cases. Usually, CBS involves a 4-stage cycle (Pal, 1998) (Pal & Campbell, 1997) (Pal, & Palmer, 2000) (Pal, 1998): retrieve, reuse, revise, and retain.

3.8 Ontology-Based Systems

An ontology-based system (OBS) is a structured set of entities and concepts representing the meaning of a field of information and relationships among these concepts. Every field creates ontologies to minimize complexity and organize information into data and knowledge. In OBS, the ontology is used to reason about the domain's attributes of interest (He et al, 2010). Ontology consists of:

- Classes: a set of objects of the area,
- Instances: individuals belonging to classes of the area,
- Attributes: properties of objects,
- Relations: links between entities,
- Ontology language: to facilitate machine processing, such as XML, Resource Description Framework (RDF), and Web Ontology Language (OWL) (Pal, 2017).

3.9 Genetic Algorithm

GA is considered the earliest form of MEAs based on natural selection and survival of the fittest. In the GA, the population is generated by randomly selecting a group of individuals called chromosomes. A chromosome is made up of a fixed number of genes. An individual chromosome is then evaluated using a fitness function. Individuals with higher fitness are highly likely to be selected for crossover operation. Two chromosomes are then selected to reproduce one or more new chromosomes. These later suffer mutation operation. Depending on the user's requirements, this process continues for several generations or until an appropriate solution is obtained.

3.10 Differential Evolution

Differential evolution (DE) is a stochastic optimization metaheuristic inspired by GAs and evolutionary strategies combined with a geometric research technique. The DE method generates the initial population by a uniform random draw on each variable's values. The user specifies the variables' lower and upper bounds according to the problem's nature. After initialization, the algorithm performs a series of transformations on the individuals in an evolution process. The DE standard uses three techniques (mutation, crossing, and selection) as GAs. The algorithm successively applies these three operations to each generation to produce a test vector.

3.11 Particle Swarm-Based Optimization Method

The particle swarm-based optimization (PSO) method is an evolutionary computation inspired by the choreographed behaviour and dynamic motion of swarms of birds or fishes experimenting with the idea of "collective intelligence" in biological systems (Kennedy & Eberhart, 1995). Unlike the GA, the basic PSO algorithm has no crossover and mutation operators. In PSO, a particle population is created and randomly distributed over the search space. The position of each particle is evaluated using the fitness function. Each particle stores the optimal solution it has achieved so far as personal best. The optimal solution, achieved so far by the population, is stored as global best. The position of each particle is updated towards its personal and global best. The fitness of each particle's new position is evaluated using the fitness function. If the fitness of a particle's current position is better than that of its personal best position, then the personal best position is updated. The global best position is also updated, which provides the final global solution to the optimization problem.

3.12 Artificial Bee Colony

Artificial bee colony (ABC) is a self-organization behavioural model presented by bee colonies (Baykasoglu et al., 2007). The feeding behaviour of bees inspired this model. The bees leave their colony to look for promising food sources. After finding a good food source, a bee returns to the hive to inform other bees via an agitation dance, representing a communication tool. Through this agitation dance, the bee transmits three crucial information: the distance, the direction, and the quality of the source food to other bees. The bee uses this especially agitation dance to convince other bees to be the followers and return to the food source. Therefore, the more the bees are attracted to the source food, the more this later is a promising source.

3.13 Simulated Annealing

Simulated annealing (SA) is a generalized Monte Carlo optimization technique in which a temperature parameter is introduced (Kirkpatrick et al., 1983). The historical analogy is inspired by the annealing of metals in metallurgy: a metal cooled too quickly has many microscopic defects. However, if it is cooled slowly, the atoms rearrange, the defects disappear, and the metal has a very ordered structure. The first phenomenon is equivalent to a local optimum in a combinatorial optimization problem, and the second is equivalent to a global optimum.

The performance of this approach depends, among others, on the cooling rule used (i.e., the decrease of the temperature parameter T). The fast-cooling leads to a local optimum that can be of poor quality. The slow cooling requires high computation time, and the adjustment of the various parameters (initial temperature, number of iterations per temperature step, and temperature decrease) can be long and difficult.

3.14 Tabu Search

Tabu search (TS) was introduced by Glover (Glover, 1986) and showed its performance on many optimization problems. The principle of the algorithm is to examine, at each iteration, the neighbourhood of the current solution and to select the best one. Applying this principle, the method allows returning to solutions that seem less exciting but may have a better neighbourhood. However, the risk is to cycle between two solutions. In order to avoid this phenomenon, the approach prohibits visiting a recently visited solution. For this, a tabu list of the attributes of the last visited solutions is maintained. Each new solution is removed from the most formerly visited solutions list. The search for the following current solution is done in the neighbourhood of the current solution without considering the solutions of the tabu list.

3.15 Multiagent Systems

A multiagent system (MAS) is a distributed system consisting of a set of reactive or cognitive entities interacting with each other (Pal & Karakostas, 2014) and located in a familiar environment. MAS is characterized by: (i) each agent has his skills, (ii) no global control of the multiagent system, (iii) decentralized data, (iv) adding new agents to MAS do not affect the others, which explains the scalability and modularity, (v) learning and adaptation, and (vi) interaction capabilities.

4. APPLICATION OF ARTIFICIAL INTELLIGENCE TECHNIQUES IN CRNs

In this section, the state of the art of research addressing AI techniques to CRNs is presented. They are grouped based on the significant tasks of CRN, as shown in Table 3.

Table 3. Research addressing artificial intelligence techniques

AI Technique	Spectrum Sensing	Spectrum Sharing	Spectrum Handoff	Spectrum Management			
				RA	OP	PA	DSA
ANN	P			P			P
MM	P		P	P			P
FL	P		P	P			P
GT		P		P			P
MHA				P	P	P	P
SVM	P			P			
RBS							P
OBS							P
CBS					P	P	
MAS		P					P

LEGEND:
P = Characteristic Present; RA = Resource Allocation; OP = Optimization Problem; PA = Parameter Adaptation; DSA = Decision Support Application

4.1 Spectrum Sensing Task

Spectrum sensing is a binary hypothesis testing to determine the presence or absence of a primary radio signal. Spectrum sensing aims to detect the spectrum hole, which means a band of frequencies that PU is not utilizing at a particular time and geographic location. The three known spectrum sensing methods in CRNs are matched filter detection (MFD), energy detection (ED), and cyclostationary detection (CSD). MFD requires prior knowledge about the PU signal, which will not always be possible. However, no prior information is needed for the CSD method, and it can extract information about the primary signal waveform. However, this method is complex to implement. ED is the most common spectrum sensing technique because this method does not need any prior knowledge about the unknown signal. It is less complex and takes minimum sensing time; however, at the same time, it is susceptible to uncertainty in noise power, and it cannot differentiate between PU and secondary user (SU) signals.

Considering the limitations of the previous sensing schemes, spectrum sensing for cognitive radio is investigated from a binary classic aspect and combined with various proposed AI tools. The aim of using AI techniques is to optimize the spectrum sensing time, the spectrum sensing error caused by false alarm rate, and miss detection probability. It also improves detection accuracy and enhances performance stability.

A neural network model is designed for spectrum sensing (Pattanyak et al., 2013). The proposed ANN model predicts the channel's status for a fixed distance in the TV band. Depending on the channel capacity predicted, the channel status is decided. The parameters used for training are the distance, SNR, channel capacity, and spectral efficiency over a TV band. In (Vyas et al., 2017), the authors suggested a novel hybrid spectrum sensing scheme. The proposed scheme combines classical energy detection, Likelihood Ratio Test statistic (LRS-G2), and ANN. The researchers reviewed the performance of the proposed scheme on several real-world primary signals of various radio technologies, and it has been found that for all those radio technologies, the proposed scheme outperforms the classical and improved ED.

The disadvantages of the cyclostationary feature detection technique and the Matched filter technique are overcome by the ANN model developed in (Varatharajana et al., 2012). The authors proposed a double threshold method for spectrum sensing. They have combined the ANN with the cyclostationary and the ED techniques. ANN is selected to train the samples of signal data and detect the existence of the primary user, specifically the backpropagation algorithm. Therefore, the ANN has been adopted in spectrum sensing for CRNs, and it is usually combined with a cyclostationary scheme to provide reliable signal classification and efficient PU signal prediction because of their ability to be trained at any time.

The Markov model models random processes changing from one state to another over time. In Markov models, the states are visible to the observer. However, some states are hidden or not explicitly visible in the HMM. HMMs have been used for spectrum sensing in CR (Kim et al., 2007) (Ghosh et al., 2009). A research group (Kim et al., 2007) proposed using the HMM to process signal cyclostationary upcoming modern attributes for primary signal detection in CR. In a research article (Ghosh et al., 2009), the authors validated the currently available Markov chain model for wireless channel use with real-time measured data in the paging band and formulated the spectrum-sensing problem using an HMM. The HMM-based prediction method has been widely used in CR networks. In a research article (Renault et al., 2017), the authors investigated the use of Non-Stationary Hidden Markov (NSHMM) and Hidden Bivariate Markov (BHMM) Models through simulations and real-time applications to predict RF channel occupancy in cognitive radio systems.

The results exhibited good potential for enabling successful secondary use of a large fraction of the available spectrum. In (Roy & Muralidhar, 2015), HMM-based channel state prediction was described. The predicted channel states are output in conjunction with corresponding probabilities that are helpful to subsequent decisions. The results show that the proposed approach for channel state prediction is practical and can be utilized together with traditional spectrum sensing techniques for spectrum sensing. Also, it can be utilized to provide predictive information to upper-level modules of cognitive radio.

FL is another attractive technique, particularly in cases where target problems are challenging to model with traditional mathematical methods but are simultaneously easier for human people to understand. In cooperative spectrum sensing, the decisions on the presence of the PU are based on observation results from several cognitive radio nodes. The fusion controller assumes the final decision-making process.

A fuzzy-based fusion rule is proposed in (Jacob et al., 2015) (Elgadi et al., 2017) (Yang et al., 2007). In (Jacob et al., 2015), the considered parameters are energy and SNR received from the neighbouring nodes. The Fuzzy rule has shown a better performance under the probability of detection and probability of false alarm compared to the "AND" and "OR" rules. In (Elgadi et al., 2017), the authors proposed a novel distributed cooperative spectrum sensing approach by deploying an FL for local decision-fusion; the environmental properties and SU mobility are utilized in the decision-making process. The proposed method is assessed on an industrial-measured power dataset. The results have shown high robustness against instantaneous changes in SU velocity levels, reasonable probability of detection at extremely low SNRs at different mobility levels, and performance that surpasses state-of-the-art research under different SNR and velocity levels.

In (Yang et al., 2007), the technique considers the sensing results' reliability at different cognitive radio nodes. The credibility of the sensing node is determined in a training stage with fuzzy evaluation. The FL is an efficient tool for the CRN environment where there is limited or no information about certain environmental factors. It is used in cooperative spectrum sensing to provide additional flexibility to existing combining methods. SVMs have also been applied to CRNs. Support vector machine theory mentions that a line easily separates two data groups. However, it is challenging to classify linearly in practice. Thus, a kernel function is proposed to map the input low-dimensional vector into a high-dimensional feature space for linear classification (Bkassiny & Jayaweera, 2013).

In (Zhang & Zhai, 2011), spectrum sensing and real-time detection are used in SVM. The sample data was classified as primary users or not for training and testing purposes on the proposed SVM classification model. In (Hu et al., 2019), the SVM classification technique is considered for cooperative spectrum sensing. The SVM scheme is performed in three steps: (i) gathering training samples, (ii) obtaining support vectors from the training samples, and (iii) a test sample was put into the SVM to classify the PU signals.

In (Awe et al., 2013), authors apply SVM classification techniques to eigenvalue-based spectrum sensing for cognitive radios. The applications of SVM to CRNs have been mostly limited to signal classification problems (Bkassiny & Jayaweera, 2013). SVMs produce efficient classifiers with high prediction accuracy and less overfitting, even if training examples contain errors. SVMs typically outperform ANNs for limited sample data but require prior knowledge of the observed process distribution and labelled data.

5. FUTURE RESEARCH DIRECTIONS

This chapter has presented that the relevance of AI techniques varies by application and implementation. The decision in choosing one or some AI techniques over other techniques for cognitive radio design can depend on, on the one hand, the application requirement, the available prior knowledge, the accuracy, the robustness, and the computation complexity; on the other hand, on the hardware capabilities such as processor speed, and memory use. In the future, this research will investigate the advantages and disadvantages of AI techniques in CRN applications.

6. CONCLUSION

AI techniques lie at the heart of CR, and understanding the trade-offs in the selection and design of AI processes is critical to a successful CR design. This paper has reviewed several AI techniques proposed to provide cognition capability in a cognitive engine. This chapter has seen that AI techniques have been applied to numerous CR applications, but many implementations remain rudimentary due to the field's interdisciplinary nature and perhaps because products are just beginning to appear.

The appropriateness of AI techniques varied by application and implementation. The decision to choose one or some AI techniques over other techniques in CRN design must be made based on the application requirement, considering the trade-offs among response time, processing complexity, training sample availability, and robustness. In addition, the learning capability of the AI technique needs to be considered and exploited in designing any industrial application software, as learning is critical to the performance of autonomously deployed CRNs.

REFERENCES

Abbas, N., Nasser, Y., & Ahmad, R. (2015). Recent advances on artificial intelligence and learning techniques in cognitive radio networks. *EURASIP Journal on Wireless Communications and Networking*, 2015(1), 2015. 10.1186/s13638-015-0381-7

AIT. (2023). *About.* Artificial Intelligence Topics. https://aitopics.org/search

Akyildiz, I. F., Lee, W. Y., & Chowdhury, K. R. (2009). CRAHNs: Cognitive radio ad hoc networks. *Ad Hoc Networks*, 7(5), 810–836. 10.1016/j.adhoc.2009.01.001

Awe, O. P., Zhu, Z., & Lambotharan, S. (2013). Eigenvalue and support vector machine techniques for spectrum sensing in cognitive radio networks. *Proceeding of the Conference on Technologies and Applications of Artificial Intelligence*. IEEE. 10.1109/TAAI.2013.52

Baby, S. M., & James, M. (2016). A Comparative Study on Various Spectrum Sharing Techniques. *Procedia Technology*, 25, 613–620. 10.1016/j.protcy.2016.08.152

Baldo, N., & Zorzi, M. (2008). *Learning and adaptation in cognitive radios using neural networks.* In 5th IEEE CCNC, Las Vegas, NV. 10.1109/ccnc08.2007.229

Baldo, N., & Zorzi, M. (2008). Learning and adaptation in cognitive radios using neural networks. *Proceeding of Conference Record – IAS Annual Meeting (IEEE Industry Applications Society).* IEEE.

Balieiro, A., Yoshioka, P., Dias, K., Cavalcanti, K., & Cordeiro, C. (2013). Adaptive spectrum sensing for cognitive radio based on multiobjective genetic optimisation. *Electronics Letters*, 49(17), 1099–1101. 10.1049/el.2013.1173

Baykasoğlu, A., Ozbakır, L., & Tapkan, P. (2007). Artificial bee colony algorithm and its application to generalized assignment problems. *Swarm intelligence: focus on ant and particle swarm optimization*, 129-156.

Bkassiny, M. & Jayaweera, S. K. (2012). A Survey on machine-learning techniques in cognitive radios. *IEEE Communications Surveys & Tutorials, 15*, 1136-1159.

Bkassiny, M., Li, Y., & Jayaweera, S. K. (2013). A survey on machine learning techniques in cognitive radios. *IEEE Communications Surveys and Tutorials*, 15(3), 1136–1159. 10.1109/SURV.2012.100412.00017

Blum, C., & Roli, A. (2003, September). Metaheuristics in combinatorial optimization: Overview and conceptual comparison. *ACM Computing Surveys*, 35(3), 268–308. 10.1145/937503.937505

Cattoni, A. F., Ottonello, M., Raffetto, M., & Regazzoni, C. S. (2007). *Neural networks mode classification based on frequency distribution features*. In 2nd Int. Conf. CrownCom, Orlando, FL.

Chu, S. C., & Tsai, P. W. (2007). Computational intelligence based on the behaviour of cats. *International Journal of Innovative Computing, Information, & Control*, 3, 163–173.

Elgadi, R., Hilal, A. R., & Basir, O. (2017). A fuzzy logic approach for cooperative spectrum sensing in cognitive radio networks. *Proceeding of IEEE 30th Canadian Conference on Electrical and Computer Engineering (CCECE)*. IEEE. 10.1109/CCECE.2017.7946841

Federal Communication Commission. (2008). *Adopts Rules for Unlicensed Use of Television White Spaces*. FCC. http://hraunfoss.fcc.gov/edocs_public/attachmatch/DOC-286566A1.pdf

Fehske, A., Gaeddert, J., & Reed, J. H. (2005). A new approach to signal classification using spectral correlation and neural networks, in *Proc. 1st IEEE Int. Symp. New Frontiers DySPAN*, (pp. 144–150). IEEE. 10.1109/DYSPAN.2005.1542629

Gavrilovska, L., Atanasovski, V., Macaluso, I., & DaSilva, L. (2013). L. (2013). Learning and reasoning in cognitive radio networks. *IEEE Communications Surveys and Tutorials*, 15(4), 1761–1777. 10.1109/SURV.2013.030713.00113

Ghosh, C., Cordeiro, C., Agrawal, D. P., & Rao, M. B. (2009). Markov chain existence and hidden Markov models in spectrum sensing. *Proceeding of IEEE International Conference on Pervasive Computing and Communications,* (pp. 1-6). IEEE. 10.1109/PERCOM.2009.4912868

Glover, F. (1986). Future paths for integer programming and links to artificial intelligence. *Computers & Operations Research*, 13(5), 533–549. 10.1016/0305-0548(86)90048-1

Gonzalez, C. R. A., Dietrich, C. B., Kragh, F. E., Sayed, S., Volos, H. I., Gaeddert, J. D., Robert, P. M., & Reed, J. H. (2009, October). Open-source SCA-based core framework and rapid development tools enable software-defined radio education and research. *IEEE Communications Magazine*, 47(10), 48–55. 10.1109/MCOM.2009.5273808

Greenis, P. (2009). Smart communications for smart grids, in *Proc. Nanotech Conf. Expo*, Houston, TX, May 3–7, 2009.

Hasegawa, M., & Nguyen, T. Ha, Miyamoto, G., Murata, Y. & Kato, S. (2007). Distributed optimization based on neurodynamic for cognitive wireless clouds. *ProceedingIEEE 18th Int. Symp. PIMRC*. IEEE.

Haykin, S. (1999). *Neural Networks: A Comprehensive Foundation*. Prentice-Hall.

Haykin, S. (2005, February). Cognitive radio: Brain-empowered wireless communications. *IEEE Journal on Selected Areas in Communications*, 23(2), 201–220. 10.1109/JSAC.2004.839380

He, A., Gaeddert, J., Tirado, L. M., & Reed, J. H. (2010). A survey of artificial intelligence for cognitive radios. *IEEE Transactions on Vehicular Technology*, 59(4), 1578–1593. 10.1109/TVT.2010.2043968

Hu, S., Yao, Y. D., & Yang, Z. (2014). MAC protocol identification using support vector machines for cognitive radio networks. *IEEE Wireless Communications*, 21(1), 52–60. 10.1109/MWC.2014.6757897

IEEE10582 (2023). *Standard for Definitions and Concepts for Dynamic Spectrum Access: Terminology Relating to Emerging Wireless Networks, System Functionality, and Spectrum Management.* IEEE. https://standards.ieee.org/ieee/1900.1/10582/

IEEE. (2023). *Working Group on Wireless Regional Area Networks*. IEEE. https://www.ieee802.org/22/

Jacob, J., Jose, B. R., & Mathew, J. (2015). A fuzzy approach to decision fusion in cognitive radio. *Procedia Computer Science*, 46, 425–431. 10.1016/j.procs.2015.02.040

Jain, A. K., Mao, J. C., & Mohiuddin, K. M. (1996, March). artificial neural networks: A tutorial. *Computer*, 29(3), 31–44. 10.1109/2.485891

Kennedy, J., & Eberhart, R. (1995). Particle swarm optimization. *Proceeding of the IEEE International Conference on Neural Networks*, (pp. 1942-1948). IEEE. 10.1109/ICNN.1995.488968

Kim, K., Akbar, I. A., Bae, K. K., Urn, J. S., Spooner, C. M., & Reed, J. H. (2007). Cyclostationary approaches to signal detection and classification in cognitive radio. *Proceeding of the 2nd IEEE international Symposium on New Frontiers in Dynamic Spectrum Access Networks*, (pp. 212-215). IEEE. 10.1109/DYSPAN.2007.35

Kiran, M., Coakley, S., Walkinshaw, N., McMinn, P., & Holcombe, M. (2008). Validation and discovery from computational biology models. *Bio Systems*, 93(1-2), 141–150. 10.1016/j.biosystems.2008.03.01018487010

Kirkpatrick, S., Gelatt, D. C., & Vechhi, M. P. (1983). Optimization by simulated annealing. *Science*, 220(4598), 671–680. 10.1126/science.220.4598.67117813860

Lane, W. (2023). *Topic 9: Cognitive Radio Potential for Public Safety*. FCC. http://www.fcc.gov/

Li, M., Hei, Y., & Qiu, Z. (2017). Optimization of multiband cooperative spectrum sensing with modified artificial bee colony algorithm. *Applied Soft Computing*, 57, 751–759. 10.1016/j.asoc.2017.03.027

Lo, R. (2009). *Moving towards cognitive NetOps*. In JSTeF, San Diego, CA.

Lu, C., Fitzek, F. H. P., & Eggers, P. C. F. (2007, September). A cooperative scheme enabling spatial reuse in wireless networks. *Wireless Communication Magazine*, 15(5), 457–471. 10.1007/978-1-4020-5979-7_23

McHenry, M. (2005). *NSF Spectrum Occupancy Measurements Project Summary*. Shared Spectrum.

Mitola, J. (1999). Cognitive radio for flexible mobile multimedia communication, In *Proceedings of the IEEE International Workshop on Mobile Multimedia Communication (MoMuC)*. IEEE.

Mitola, J., & Maguire, G. Q. (1999). Cognitive radio: making software radios more personal. *IEEE Personal Communications*. IEEE.

Neel, J. (2007). Synthetic symmetry in cognitive radio networks. *Proc. Forum Tech. Conf. Product Expo. SDR*, Denver.

Neel, J., & Amanna, A. (2009). Introduction to cognitive radio technology for transportation systems. *Proc. SDR Forum Workshop Smart Communication Transport Systems.*, Houston, TX.

Ning, Z., Yu, Y., Song, Q., Peng, Y., & Zhang, B. (2015). Interference aware spectrum sensing mechanisms in cognitive radio networks. *Computers & Electrical Engineering*, 42, 193–206. 10.1016/j.compeleceng.2014.10.011

Niyato, D., & Hossain, E. (2008, December). Spectrum trading in cognitive radio networks: A market-equilibrium-based approach. *IEEE Wireless Communications*, 15(6), 71–80. 10.1109/MWC.2008.4749750

Orcay, O., & Ustundag, B. (2008). *Pattern recognition in cognitive communication*. In 23rd ISCIS, Istanbul, Turkey.

OSSIE. (2023). *Wireless @ Virginia Tech*. Open Source SCA Implementation—Embedded. http://wireless.vt.edu/

Pal, K. (1999). *An approach to legal reasoning based on a hybrid decision-support system*. The University College London, Gower Street, London.

Pal, K. (2018). Ontology-Based Web Service Architecture for Retail Supply Chain Management. In *Proceedings of the 9th International Conference on Ambient Systems, Networks and Technologies*, Porto, Portugal. 10.1016/j.procs.2018.04.101

Pal, K. (2022a). Cryptography and Blockchain Solutions for Security Protection of Internet of Things Applications. In Sahoo, B., & Yadav, S. (Eds.), *Information Security Practices of the Internet of Things, 5G, and Next-Generation Wireless Networks*. IGI Global Publication. 10.4018/978-1-6684-3921-0.ch008

Pal, K. (2022b). Security Issues and Solutions for Resource-Constrained IoT Applications Using Lightweight Cryptography. In Verma, S., Vyas, V., & Kaushik, K. (Eds.), *Cybersecurity Issues, Challenges, and Solutions in the Business World*. IGI Global Publication. 10.4018/978-1-6684-5827-3.ch010

Pal, K. (2023). Influence of AI on Cognitive Radio Applications. K. Kumar (Ed.), *Applications of Artificial Intelligence in Wireless Communication Systems*. IGI Global Publication.

Pal, K. & Campbell, J. A. (1997). An Application of Rule-Based and Case-Based Reasoning within a Single Legal Knowledge Based System. *Data Base for Advances in Information Systems, 28*(4), 48-63.

Pal, K., & Karakostas, B. (2014). *Multi Agent-Based Service Framework for Supply Chain Management*. In 5th International Conference on Ambient Systems, Networks and Technologies (ANT-2014), Belgium.

Pal, K., & Palmer, O. (2000). A decision-support system for business acquisitions. *Decision Support Systems*, 27(4), 411–429. 10.1016/S0167-9236(99)00083-4

Pattanayak, S., Venkateswaran, P., & Nandi, R. (2013). Artificial intelligence-based model for channel status prediction: A new spectrum sensing technique for cognitive radio. *International Journal of Communications, Network and Systems Sciences*, 6(3), 139–148. 10.4236/ijcns.2013.63017

Pattnaik, S. S., Bakwad, K. M., Devi, S., Panigrahi, B. K., & Das, S. (2011). Parallel bacterial foraging optimization. *Handbook of Swarm Intelligence, 8*, 487-502.

Pradhan, P. M., & Panda, G. (2013). Cooperative spectrum sensing in cognitive radio network using multiobjective evolutionary algorithms and fuzzy decision making. *Ad Hoc Networks*, 11(3), 1022–1036. 10.1016/j.adhoc.2012.11.007

Qadir, J. (2013). Artificial intelligence based cognitive radio networks routing for cognitive radio networks. *IEEE Communications Surveys and Tutorials*, 2013.

Qineti, Q. (2007). Cognitive Radio Technology a Study for Ofcom—Volume 1, Feb. 12, 2007. [Online]. Available: https://www.ofcom.org.uk/research/technology/research/emer_tech/cograd/

Reed, J. H. (2005). *Development of a Cognitive Engine and Analysis of WRAN Cognitive Radio Algorithms—Phase I*. Wireless @ Virginia Tech, Virginia Polytech. Inst. State Univ.

Renault, L., Rodrigues, L., & Pinto, E. L. (2017). HMM models and estimation algorithms for real-time predictive spectrum sensing and cognitive usage. *Proceeding of xxxv simposio brasileiro de telecommunicators Procesamiento de sinais*. Sociedade Brasileria de Telecomunicaoes.

Rondeau, T. W., Le, B., Rieser, C. J., & Bostian, C. W. (2004). *Cognitive radios with genetic algorithms: Intelligent control of software defined radios*. Forum Tech. Conf. SDR, Phoenix, AZ.

Roy, P. P., & Muralidhar, D. M. (2015). Hidden Markov Model based channel state prediction in cognitive radio networks. *International Journal of Engineering Research & Technology (Ahmedabad)*, 4(2), 391–394.

Salau, H. B., Aibinu, A. M., Onwuka, E. N., Onumanyi, A. J., & Bima, M. E. M. E. (2015). *Applications of artificial Intelligence techniques in cognitive radios: a survey*. Research Gate. https://www.researchgate.net/publication/307935017

SDR Forums. (2023). *Software Defined Radio Forum, Cognitive Radio Work Group*. SDR Forums. http://www.sdrforum.org/pages/aboutTheForum/currentProjects.asp

Tabaković, Z. (2011). *A survey of cognitive radio systems*. Semantic Scholar. https://pdfs.semanticscholar.org/19c8/471e97301e408c3e3fe08eb8f5027ef9eed6.pdf

Thomas, R. (2007). *Cognitive networks*. [Ph.D. dissertation, Virginia Polytech. Inst.]

Ustundag, B., & Orcay, O. (2008). *Pattern based encoding for cognitive communication*. 3rd International Conference CrownCom, Singapore.

Varatharajana, B., Praveen, E., & Vinoth, E. (2012). Neural network aided enhanced spectrum sensing in cognitive radio. *Procedia Engineering*, 38, 82–88. 10.1016/j.proeng.2012.06.012

Vyas, M. R., Patel, D. K., & Benitez, M. L. (2017). Artificial neural network based hybrid spectrum sensing scheme for cognitive radio. *Proceeding of IEEE 28th Annual International Symposium on Personal, Indoor, and Mobile Radio Communications (PIMRC)*, (pp. 8-13). IEEE.

Wang, B., Wu, Y., & Liu, K. J. (2010). Game theory for cognitive radio networks an overview. *Computer Networks*, 54(14), 2537–2561. 10.1016/j.comnet.2010.04.004

Yang, A., Cai, Y., & Xu, Y. (2007). A fuzzy collaborative spectrum sensing scheme in cognitive radio. *Proceeding of International Symposium on Intelligent Signal Processing and Communication Systems (ISPACS)*. Research Gate.

Zhang, D., & Zhai, X. (2011). SVM-based spectrum sensing in cognitive radio. *Proceeding of the 7th International Conference on Wireless Communications, Networking and Mobile Computing*. Research Gate.

Zhang, Z., & Xie, X. (2007). Intelligent cognitive radio: Research on learning and evaluation of CR based on neural network, In proceedings *ITI 5th ICICT*, Dhaka, Bangladesh, December 16–18, 2007, 33–37.

Zhu, X. L., Liu, Y. A., Weng, W. W., & Yuan, D. M. (2008). *Channel sensing algorithm based on neural networks for cognitive wireless mesh networks*. 4th Int. Conf. WiCOM, Dalian, China.

Chapter 4
Automated Star Rating Generation for Analyzing the Performance of Public Department

Mangalapriya Sakthivel
https://orcid.org/0009-0006-6552-1204
Vel Tech High Tech Dr. Sakunthala Dr. Rangarajan Engineering College, India

S. Vennila
Vel Tech High Tech Dr. Sakunthala Dr. Rangarajan Engineering College, India

S. R. Ilakkiya
Vel Tech High Tech Dr. Sakunthala Dr. Rangarajan Engineering College, India

S. V. Malini
Vel Tech High Tech Dr. Sakunthala Dr. Rangarajan Engineering College, India

ABSTRACT

In today's digital era, the role of social media cannot be overstated as it serves as a crucial platform for exchanging data and allows individuals to express their thoughts and opinions. One of the key procedures in this context is sentiment analysis, which involves analyzing the sentiments and polarity of people's thoughts to determine whether a post or comment has received positive, negative, or neutral feedback. This procedure plays a significant role in understanding the overall sentiment surrounding a topic or post. To improve accessibility and streamline operations, this team has developed a dynamic website that integrates various functionalities to enhance the sentiment analysis process. The system utilizes PHP and leverages the Instagram API, enabling them to gather real-time data from Instagram posts and comments. By doing so, they aim to enhance the accuracy of sentiment analysis, providing users with a comprehensive understanding of the sentiments expressed. The proposed system offers several benefits.

DOI: 10.4018/979-8-3693-2893-4.ch004

1. INTRODUCTION

In the contemporary era of digital connectivity, the Internet has emerged as a ubiquitous platform for information dissemination, fostering a remarkable surge in individuals actively voicing their personal ideas and publicly sharing information. This unprecedented proliferation of user- generated content has become an invaluable goldmine for analysts and researchers, presenting an abundant reserve of data for informed decision-making strategies. Playing a pivotal role in facilitating seamless data exchange, dynamic websites empower individuals to freely articulate their thoughts and actively participate in vibrant discussions with a diverse range of peers. Recognizing the immense potential inherent in this vast repository of information, our team has meticulously crafted a dynamic website that empowers individuals to provide invaluable feedback on diverse social welfare departments, Within this interactive platform, users are afforded the opportunity to submit their feedback alongside essential location information, ensuring contextual relevance in the evaluation process. Leveraging cutting-edge sentiment analysis techniques, we meticulously scrutinize and evaluate the received feedback to generate precise and accurate star ratings for each submission. By harnessing the power of tweets, we extract rating values embedded within users' concise posts, maximizing the utilization of available data sources. Delving deeper into the intricacies of sentiment analysis, our system employs a sophisticated sentence structure analysis framework. This comprehensive approach enables us to unravel the nuanced layers of word meaning. discern synonymous expressions, and effectively gauge polarity fluctuations even in the presence of negation words. Furthermore, our system takes into account the influence of adverbs, nouns, and adjectives on word polarity, thereby fine- tuning the analysis process to ensure utmost accuracy.

Extensive research substantiates the efficacy of sentiment analysis in detecting and extracting sentiments, moods, and attitudes expressed by individuals and cohesive groups. Within our system, we harness the power of Instagram Public Rating Analysis (PRA), an influential technique encompassed within the text and opinion mining domain Muralidharan (2011) – J.Wilson (2005). This particular approach revolves around meticulously analyzing the ratings conveyed within tweets, culminating in the creation of a robust XAMPP model .

This model serves as a pivotal tool for assessing the accuracy and reliability of the sentiment analysis performed within our system. Central to our approach is the meticulous scrutiny of users' tweets and hashtags, affording us a holistic understanding of diverse scenarios encompassing each criterion under evaluation. In pursuit of enhanced accuracy and comprehensive insights from users' posts, our system seamlessly integrates PHP, the Instagram API, and Text Blob, a powerful text processing library. These invaluable tools empower us to derive rich insights and make well-informed decisions, capitalizing on the sentiments expressed within the feedback. Through the holistic fusion of these cutting-edge technologies, we aim to deliver a dependable and efficient sentiment analysis system, enabling organizations to glean invaluable insights from user feedback with unparalleled accuracy and meticulous attention to detail. Our main objective is to conduct a target-oriented analysis in order toidentify opinions or attitudes towards various topics or entities. In addition, we also aim to recognize and understand the emotions expressed or evoked by text, without any specific attachment to a particular target. This allows us to gain a comprehensive understanding of sentiment and emotional responses. One of our key goals is to determine the level of investment from your specific audience in embracing their thoughts regarding specific entities Shiv dhar (2023)- Shiv Naresh (2023). This involves analyzing the text data to gain insights into how people feel about your incident. By utilizing sentiment analysis techniques, we can quickly process the input data and generate sentiment scores for each relevant review topic, aspect, or entity.

Based on the analysis and the acquired sentimental scores from the reviews, we will generate a star rating that reflects the overall sentiment and perception of your audience. This rating provides a concise representation of the sentiment associated with the specific entities or topics under consideration. To facilitate decision making, we will summarize and consolidate the obtained results in a user- friendly file format. This format will be designed to he uncomplicated and easy to interpret, enabling you to make informed decisions based on the sentiment and emotional responses expressed in the text data.

Instagram PRA, which falls under the category of text and opinion mining, plays a crucial role in analyzing tweet ratings. By leveraging the power of a XAMPP model, this approach aims to train the model using the gathered data and subsequently evaluate its accuracy. The ultimate goal is to create a robust and reliable model that can be utilized in future applications, based on the outcomes obtained. The sentiment analysis process begins by carefully examining the text present in various corpora or tweets. To extract the relevant information, the method employs a text extraction mechanismthat retrieves the text from input documents or tweet sources. This step ensures that the subsequent analysis is performed on the appropriate textual content. Once the text is extracted, the method proceeds with preprocessing techniques to eliminate noisy words or terms that may introduce inconsistencies or irrelevant information. This preprocessing step is crucial in refining the data and enhancing the overall quality of the analysis. By removing noise, the method obtains a cleaner feature set, enabling more accurate sentiment analysis. In order to understand the underlying sentiment or emotion expressed in the text, the method employs advanced techniques such as stemming and tagging. Stemming helps identify the root words from the noise-free feature set, allowing for a more comprehensive analysis of sentiment Kavya (2019) – Akshi kumar (2021). Tagging further enhances this process by assigning appropriate labels or markers to different elements of the text, aiding in the identification of sentiment-related patterns, Building upon the noise-free feature set, the method proceeds to estimate similarity measures for various classes of sentiment or emotion. By assessing the similarity values, the method can effectively perform sentiment detection, accurately categorizing the text into different sentiment categories. This step plays a critical role in uncovering the sentiment expressed in the text and providing valuable insights for further analysis.

2. LITERATURE SURVEY

[1]Sentiment analysis, also known as opinion mining or emotional intelligence (El), involves extracting valuable information from unstructured and disorganized text found on various social platforms and online sources like Twitter, WhatsApp, feedback, blogs, and comments, Opinion mining encompasses the creation of rule-based automated systems that analyze data using predefined rules. Additionally. machine learning principles are employed to develop automated systems for conducting sentiment analysis. In certain cases, a hybrid model combining both rule-based and machine learning algorithms is utilized to achieve more accurate results in sentiment analysis.[2] The development of a natural language programming (NLP) tool is a challenging task due to the intricate nature of natural language (NL), NLP systems can be fragile and susceptible to failure when encountering unfamiliar vocabulary or new sentence structures that are not supported by the system. Without informative feedback, the time required to complete programming tasks may increase, leading to user discouragement. Recognizing the importance of clear and actionable feedback, we propose an innovative hybrid system for generating automatic feedback. This system combines a formal rule-based approach with a data-based multi-label classification (MLC) method to predict the user's intent and provide examples of successful methods

to accomplish the intended task or similar tasks. We assess the effectiveness of the feedback system through two phases of end-user studies, demonstrating its usefulness in assisting users in troubleshooting their inputs and successfully completing their tasks.[3] In order to enhance student learning, educational assessments must incorporate feedback. With the advent of Artificial Intelligence (AI), educational evaluation has evolved, much like education itself has transformed through technological advancements. Despite the increasing popularity of online tests in the past decade, only a limited amount of research has explored the utilization of Al for generating feedback. In addition to non-verbal feedback provided through a learning analytics (LA) dashboard or a digital score report, written feedback can be generated using NLP-based algorithms from a model. Furthermore, this research presents suggestions for leveraging AI in the generation of feedback. By contributing to our understanding of the feedback creation process, this study aims to facilitate the future development of digital feedback.[4] One of the issues with surveys is that they often fail to capture important emotional responses, leading to a lack of critical feedback. Our research has revealed that customers frequently rate companies highly in surveys, even when they encounter significant issues with their products or services. This crucial negative feedback tends to go unnoticed. As a result, these surveys can inadvertently cause companies to lose customers without understanding the reasons behind it. However, there is a wealth of valuable data available if one knows where and how to analyze it. In general, the content of these commentsprovides a more reliable indicator of customer behavior. Unfortunately, these comments are often overlooked, or if used, they are typically analyzed after the scores have been computed.[5] For students to learn, feedback is essential. The frequency, format, and timeliness of feedback resulting from educational evaluations have all evolved gradually along with technological breakthroughs that have made it possible to embrace digital learning environments with assessment capabilities. Teachers can create a variety of feedback. supporting student learning thanks to cutting-edge technologies powered by artificial intelligence (AI). Although digital technologies have been adopted quickly in education, prior research on educational feedback has mostly. concentrated on the theoretical foundations of feedback practices, which has limited its coverage of Al-based technologies. This study intends to identify and organize possible areas for the use of AI for feedback purposes, to educate academics and practitioners about the current and future of Al applications in feedback practices, and to provide a framework for future research.

[6] An efficient strategy to assist student learning is the giving of educational feedback.Educational academics have turned to the creation of artificial intelligence (AI) systems in order to provide feedback to many pupils. Although Al-driven feedback has had some degree of success, there are still a number of issues that come up when using Al in this situation. Concerns about responsibility, explainability of AI algorithms, and privacy present difficulties and foster mistrust between educational. stakeholders and Al initiatives. To close the gap between the demands of educational stakeholders and the efficient application of AI in giving reliable feedback, these issues must be solved.

3. ALGORITHM

A problem-solving approach that relies on a set of predefined rules or logical statements is referred to as a rule-based algorithm, also called a rule-based system or expert system. These rules, typically created by experts in a specific field, are utilized to make decisions and solve problems based on the available facts. The basic functioning of a rule-based algorithm can be described as follows:

1.Information Base: The algorithm employs an initial knowledge base that contains a collection of rules. Each rule consists of a condition (also known as an antecedent) and an associated action (also known as a consequent). The action represents the desired course of action when the condition is met, while the condition outlines the prerequisites that must be fulfilled for the rule to be applied.

2.Rule Assessment: Starting from the top of the knowledge. base, the algorithm evaluates each rule to determine if its condition holds true based on the most recent data or information available. The conditions can involve logical operators such as AND, OR, and NOT, and they can range from simple to complex.

3.Rule Implementation: If a rule's condition is satisfied, the algorithm executes the corresponding action or a combination of actions associated with that rule. These actions can include making a decision, providing a recommendation, or initiating a specific procedure.

4.Iteration: The algorithm continues the process of evaluating and applying rules in a loop until a specific condition is met. This condition could be achieving the desired outcome or reaching a point where no more rules are available for application.

Rule-based algorithms have been utilized in various fields such as artificial intelligence, expert systems, decision support systems, and business logic systems. They are particularly beneficial in scenarios where there are well-defined standards or a need to capture and automate human expertise. However, it is important to note that rule-based algorithms have certain limitations. They heavily rely on the predefined rules and may struggle to handle situations that do not align with those rules. The effectiveness of a rule-based system is also influenced by the accuracy and comprehensiveness of the rules as well as the quality of the input data.

To determine the sentiment polarity of user sentiments, rule- based sentiment analysis algorithms utilize a predefined set of rules to automatically categorize incoming data. Natural Language Processing (NLP) techniques are employed to implement these rules, including stemming, part-of-speech tagging, parsing, lexicons, and tokenization. Here is an example illustrating how a rule-based sentiment analysis model could be applied:

The model assigns different emotions to lists of words, with positive words such as “great,” “love,” and “amazing” categorized as positive, while negative terms like “horrible,” “worst,” and “bad” are considered negative. The analysis of an input sentence involves examining the words in the sentence and keeping track of the number of polarized words Based on the number of words that fall into the most extreme categories such as “positive.” “negative,” or “neutral”-this count is then divided into buckets of various kinds.

4. RULE-BASED ALGORITHM IN ACTION

Depending on how sophisticated you want your sentiment analysis model to be, rule-based or automatic algorithms can be used. A rule-based algorithm can be used in sentiment analysis to ascertain the sentiment or emotional tone expressed in a given text. An overview of the general operation of a rule- based algorithm in sentiment analysis is provided below:

1. Preprocessing: Stop words, special characters, and other noise are removed from the text data during preprocessing. Lemmatization or stemming may also be used in this phase to break down words to their most basic components.
2. Lexicon Creation: Words or phrases that are connected to particular sentiment polarities-positive, negative, or neutral-are compiled into a sentiment lexicon. Each word or phrase in the lexicon is given a sentiment score that represents the strength of that sentiment.
3. Rule-based Matching: The algorithm scans the text that has been previously processed and compares the terms in the text to the words in the lexicon. If a match is discovered, sentimentanalysis will take into account the word or phrase's sentimentscore.
4. Rule-based Scoring: To ascertain the general sentiment of the text, the system employs a number of rules or heuristics. As an illustration, it might take into account the quantity and strength of positive and negative phrases, the presence of negations or intensifiers, and the context in which the words are used. These guidelines aid in determining the text's overall sentiment score.
5. Sentiment Classification: The algorithm divides the text into positive, negative, and neutral sentiment groups based on the overall sentiment score. Depending on the particular implementation or requirements, the threshold for classifying the attitude may change.
6. Post-processing: More post-processing actions can be taken to polish the sentiment analysis findings. This could entail applying domain-specific rules, taking into account the sentiment of the surrounding context, or examining sentiment transitions within the text.

Algorithms for rule-based sentiment analysis are frequently utilized where transparency and interpretability are required. They enable domain knowledge to be incorporated into the sentiment analysis process and allow human specialists to set the rules. They might not, however, be as good at capturing subtle linguistic variations or context-specific attitudes as machine learning techniques like supervised learning or deep learning models.

We can use explainable ML. approaches on the developed predictive models to discover the performance-influencing elements based on each prediction. Finally, utilizing the predicted student performance and recognized factors for students, automatic and intelligent feedback is produced. The primary novelty is that we combine a prediction technique. with an explainable ML approach, which, in contrast to other studies, enables fine-grained insights that assist the delivery of precise data-driven actionable feedback to students and explains the "why" of the predictions. In other words, we offer a method that provides students with more useful knowledge than can be obtained by simply notifying them of the forecast. Based on the suggested explainable ML-based strategy, we create a data-driven feedback and intelligent action recommendation (DFIAR) dashboard in this work.

Rule-based systems allow for the explicit representation of information in the form of rules. These rules are easily understood and interpreted by humans and machines alike. making the system visible and auditable. Deterministic decision-making is used in rule-based systems. Each rule represents a distinct condition and action, allowing for a detailed examination of the incoming data. Transparency can be useful in crucial situations where decision justification is critical. Rules are simply adjusted or extended to reflect changes in system needs or domain knowledge. New rules can be added or current rules can be changed without having an impact on the overall system architecture. Rule-based systems may be rapidly built and implemented. By specifying a set of rules, developers can design working systems without the requirement for lengthy algorithm development or training processes.

5. PROCESS OF WORK FLOW

6. METHODOLOGY

Data extraction: Data extraction is an important stage in automatic star rating production since it entails gathering useful information from comments and places that can be utilized to provide a suitable star rating. Instagram API is used to download the tweets, and 9684 training and 8987 testing tweets are downloaded. Pre-processing is performed on the acquired data to remove notations and extraneous information. This can include eliminating hashtags, mentions, URLs, and special characters, and then extracting features from the data to train a machine-learning model. This can include things like the frequency with which certain words or phrases are used, the employment of emoticons, and the sentiment of other people mentioned in the post. The algorithm analyses the text and provides a sentiment label based on patterns, and it may download data in any file format. After downloading the data, you can use it for backups or other purposes.

Text parsing: Text parsing is the process of analyzing a text's grammatical structure in order to grasp its meaning. It can be used to extract essential elements or properties from comments and locations that are indicative of sentiment and customer experiences for a specific department. Text parsing is the process of breaking down a sentence or paragraph into its constituent pieces, such as words, phrases, clauses, and sentences, and analyzing their relationships to one another. This can be accomplished through the use of several techniques such as part-of-speech (POS) tagging, named-ntity recognition (NER), and dependency parsing. Part-of- speech tagging is the process of assigning a part of speech to each word in a sentence, such as a noun, verb, adjective, and so on... This information can be utilized to find crucial terms and phrases in the feedback that are indicative of sentiment and experiences. Named entity recognition is the process of identifying and classifying named entities in text, such as people, locations, organizations, and things. This information can be used to identify specific items or services that people are referring to in their feedback and to understand the context in which they are being discussed. Dependency parsing entails examining the relationships between words in a sentence. This data can be utilized to decipher the meaning of the comments and to discover relevant elements or properties that are indicative of sentiment. demands of educational stakeholders and the efficient application of Al in giving reliable feedback, these issues must be solved.

- **Text pre-processing:** Text pre-processing is cleaning and translating raw text input into a format that may be easily analyzed by NLP algorithms. Text pre-processing often consists of multiple processes, including:

- **Tokenization:** Tokenization is the process of breaking down text into individual words or tokens, which are then utilized as input for further analysis. Case normalization is the process of converting all text to lowercase or uppercase in order to standard the data and reduce the number of unique tokens.
- **Stop word removal:** Removing frequent words such as "the", "and", and "is" that add little significance to the text.
- **Stemming or Lemmatization:** Reducing words to their base form (stemming) or converting them to their dictionary form (lemmatization) to reduce the number of unique tokens and increase analysis accuracy.
- **Spell checking and correction:** Detecting and repairing frequent spelling problems to assure analysis accuracy.
- **Remove special characters and punctuation:** Remove non-alphanumeric characters such as hashtags, emojis, and punctuation marks.
- **Removing numbers:** Removing numerical data, such as dates or phone numbers, that may not contribute value to the analysis.

Depending on the application and the nature of the text data being analyzed, the specific pre-processing techniques employed may differ. Text preprocessing is essential for accurate and efficient text data analysis and can increase the performance of machine learning models and other NLP techniques.

Feature Representation: Several strategies for feature representation in Al-based automatic star rating generation are available, including:

- **Bag of words:** This method involves expressing comments as a collection of words, with each word regarded as a feature and given a weight based on its frequency in the comments. This method is simple and effective, but it ignores the word relationships in the comment. Word embeddings: In this technique, words in the comments are represented as dense vectors in a high-dimensional space. In this space, each word is represented by a point, and the distance between points depicts the semantic ties between the words. This method is excellent for capturing word relationships in comments as well as finding noteworthy keywords and phrases.
- **Sentiment lexicons:** This technique involves identifying the sentiment of feedback by using pre-defined sentiment lexicons or dictionaries. Each word in the feedback is assigned a sentiment score based on its position in the lexicon, and the sentiment ratings are then utilized to assign a star rating. By employing the proper feature representation approaches, it is possible to obtain more accuracy in star rating production and gain better insights into customer sentiment.

  Sentiment analysis: Sentiment analysis, often known as opinion mining, is a computational approach for identifying and extracting subjective information from text data, such as individual opinions, sentiments, attitudes, and emotions. It entails examining the written content to see if it delivers a favorable, negative, or neutral mood. To automatically classify and quantify sentiment in text data, sentiment analysis approaches rely on various natural language processing (NLP) and machine learning algorithms. Traditional methodologies such as rule-based systems and lexicon-based analysis can be used, as well as more modem techniques such as machine learning algorithms and deep learning models. Sentiment analysis is a technique often employed in automated star rating production to analyze the sentiment or emotion displayed in customer evaluations or comments. It entails extracting the text's polarity, which can be represented as positive, negative, or neutral sentiment, and translating it into a numerical number that can be utilized as input for machine learning algorithms. Sentiment analysis can be used in automated star rating generation to assess the overall sentiment of the reviews and feedback, which can then be utilized to generate an appropriate star rating. For example, if the sentiment analysis suggests that the reviews are overwhelmingly positive, the program may assign a higher star rating. Similarly, if the sentiment analysis suggests that the reviews are overwhelmingly unfavorable, the program may assign a lower star rating.

7. CONCLUSION

We proposed a dynamic website for easy access to all departments on one webpage. It will be beneficial to see a fast review of the different departments. We employ text blobs in this method to determine the polarity of the text (positive, negative, or neutral). Instagram PRA is a type of text and opinion mining. It focuses on analyzing tweet ratings and feeding the data to a XAMPP model to train it and then test its correctness so that we may use this model in the future based. on the results. The sentiment analysis method starts by reading the text in any corpus or tweets. The technique takes the text from the input document and extracts it. In addition, theapproach employs preprocessing to reduce noisy words or concepts. The approach finds root words by using the noise removal feature.

REFERENCE

Bhagyashri Wagh, J.V. (2007). Wankhade Sentiment Analysis on Twitter Data usingNaive Bayes. *International Journal of Advanced Research in Computer and Communication Engineering*.

Cogburn, D. L., & Espinoza-Vasquez, F. K. (2011, February). From networked nominee to networked nation: Examining the impact of web 2.0 and social media on political participation. and civic engagement in the 2008 Obama campaign. *Journal of Political Marketing*, 10(1-2), 189–213. 10.1080/15377857.2011.540224

Dhar, S. (2018). *Methods for Sentiment Analysis Computer Engineering*. VIVA Institute of Technology, University of Mumbai, India.

Glassford, A. (2015) *Multiclass Emotion Analysis of Social Media Posts*. Stanford University.

Kumar, A. (2015). *Emotion Analysis of Twitter using Opinion Mining Dept. of Computer Engineering*. Delhi Technology University.

Liu, B. (2012, April). Sentiment analysis and opinion mining: Synthesis Lectures on Human Lang. *Tech.*, 5(1), 1–167.

Mathur, R. (2024). *Pathak- Analyzing Sentiment of Twitter Data using Machine Learning Algorithm*. Academmia.

Muralidharan, L. Rasmussen, D., & Patterson. (2011). *Hope for Haiti: an analysis of facebook and twitter usage during the earthquake relief efforts*. Public Relations Rev.

Ravinder, A., Aakarsha, C., Kohlia, S., & Pratyush Ahujaa, S. G. (2019). *The Impact of Features Extraction on the Sentiment Analysis Noida, India*. Semantic Scholar.

Suppala, K., & Rao, N. (2019). Sentiment Analysis Using Naïve Bayes Classifier. *International Journal of Innovative Technology and Exploring Engineering (UITEE)*.

Wilson, T., Wiebe, J., & Hoffmann, P. (2005). *Recognizing contextual polarity in phrase-level sentiment analysis*. Proc. Conf. Human Lang. Tech. and Empirical Methods in Natural Lang. Process., Vancouver, B.C., Canada. 10.3115/1220575.1220619

Yang, C. (2014). *Context-aware learning for sentence-level sentiment analysis with posterior regularization*. Proc. 52nd Annu. Meeting on Assoc. for Computational Linguistics, Baltimore, MD.

Chapter 5
Commercial Fire Protection Automated Detection and Suppression

C. Annadurai
Sri Sivasubramaniya Nadar College of Engineering, India

I. Nelson
Sri Sivasubramaniya Nadar College of Engineering, India

Arun Venkatesh
Sri Sivasubramaniya Nadar College of Engineering, India

Karthik Sriram G.
Sri Sivasubramaniya Nadar College of Engineering, India

ABSTRACT

The advancement of effective fire detection and suppression systems holds utmost importance in protecting lives and assets in real-world scenarios. By leveraging technologies such as smoke detectors, heat sensors, and sophisticated algorithms, these systems swiftly identify potential fire outbreaks. Real-time monitoring facilitates prompt responses, thereby reducing the likelihood of extensive damage. As global efforts intensify towards bolstering safety protocols, the ongoing evolution of fire detection and suppression systems emerges as a crucial factor in mitigating the catastrophic consequences of fire incidents. The proposed project entails an IoT-based fire management system that integrates computer vision technology to revolutionize conventional firefighting methodologies. This comprehensive system prioritizes precise fire detection, accurate object identification, and optimized extinguishing strategies based on temperature analysis.

DOI: 10.4018/979-8-3693-2893-4.ch005

1. INTRODUCTION

Conventional fire detection and suppression systems have historically served as the cornerstone of fire safety protocols, relying on traditional sensors and manual intervention. However, these systems often suffer from issues such as delayed response times and limited coverage in complex environments. Human-dependent activation and the challenge of detecting early fire indicators present significant hurdles. The integration of Internet of Things (IoT) technology into fire safety systems represents a significant leap forward, offering tailored applications to enhance detection and suppression processes. By leveraging IoT-enabled devices such as intelligent sensors and interconnected detectors, fire safety systems can now access real-time data and facilitate seamless communication. This transformative technology enables automated response mechanisms, drastically reducing response times to critical levels. Additionally, remote monitoring and management capabilities empower proactive risk identification, ensuring swift action in the event of a fire. In essence, the infusion of IoT applications into fire detection and suppression systems not only addresses shortcomings of traditional methods but also ushers in a new era of intelligent, responsive, and automated fire safety solutions.

2. RELATED WORK

A. Gaur, A. Singh, A. Kumar, et al. (2020) conducted a comprehensive literature review on video flame and smoke-based fire detection algorithms. Their work, published in Fire Technology, offers insights into the latest advancements in fire detection technology, contributing to enhanced fire safety measures.

C.-L. Lai and J.-C. Yang (2008) developed an advanced real-time fire detection system for video surveillance, presented at ISCAS 2008. Their work aimed to enhance fire detection in surveillance videos, employing computer vision and image processing methods.

B. U. Toreyin, Y. Dedeoğlu, U. Gündükbaÿ, and A. E. Çetin (2006) introduced a computer vision-based method for real-time fire and flame detection. Their study, published in Pattern Recognition Letters, contributes to advancements in pattern recognition technology for immediate fire hazard identification.

H. Wu et al. (2019) introduced an intelligent fire detection approach using cameras and computer vision methods, published in Process Safety and Environmental Protection. Their study focused on developing a system capable of visually analyzing images captured by cameras to detect fire-related patterns, contributing to improved fire detection capabilities.

H. Yar, T. Hussain, Z. A. Khan, et al. (2021) introduced a vision sensor-based real-time fire detection system designed for resource-constrained IoT environments. Their study, published in Computational Intelligence and Neuroscience, contributes to the development of efficient fire detection solutions tailored for IoT applications.

He et al. (2002) investigated smoke venting and fire safety in an industrial warehouse. Their study, published in Fire Saf. J., analyzed various aspects including smoke venting strategies and their effectiveness in improving fire safety measures within the warehouse environment. The research highlighted key considerations for enhancing fire safety practices in industrial settings.

Ho and Kuo (2009) presented a real-time video-based fire smoke detection system, focusing on detecting fire and smoke using video input in real-time. Published in the proceedings of AIM 2009, the paper discusses system details, performance, capabilities, and limitations for fire and smoke detection applications.

J. Pincott, P. W. Tien, S. Wei, et al. (2022) proposed indoor fire detection strategies utilizing computer vision-based methods. Their work, published in the Journal of Building Engineering, contributes to advancements in fire safety technology for indoor environments.

K. Muhammad, S. Khan, M. Elhoseny, et al. (2019) proposed an efficient fire detection method tailored for uncertain surveillance environments. Published in IEEE Transactions on Industrial Informatics, their research contributes to the development of robust fire detection systems, enhancing safety in diverse surveillance scenarios.

L.-Y. Lv et al. (2022) investigated smart fire-warning materials and sensors, outlining design principles, performances, and applications. Their research, published in Materials Science and Engineering: R, Reports, contributes to the development of innovative fire safety technologies with potential applications across various industries.

M. Maier et al. (2016) explored the concept of the Tactile Internet, discussing its vision, recent progress, and open challenges in IEEE Communications Magazine. The authors provided insights into the potential of the Tactile Internet and highlighted areas for further development and research.

M. Sajjad et al. (2017) developed a Raspberry Pi assisted face recognition framework for law-enforcement services in smart cities. Published in Future Generation Computer Systems, their work contributes to the integration of Raspberry Pi technology for enhanced law enforcement and public safety in urban environments.

P. Foggia et al. (2015) conducted a study on real-time fire detection for video-surveillance applications. Their research proposed a method using a combination of experts based on colour, shape, and motion, contributing to advancements in video technology for fire safety.

P. Sridhar et al. (2020) proposed an efficient rule-based algorithm for fire detection in real-time videos in the Journal of Computational Theoretical Nanoscience. Their algorithm aims to detect fires accurately and swiftly in video streams, potentially aiding in early fire detection and prevention.

P. V. K. Borges and E. Izquierdo (2010) investigated a probabilistic approach for vision-based fire detection in videos. Their work, published in IEEE Transactions on Circuits and Systems for Video Technology, presents a method contributing to the field of video technology for enhanced fire detection capabilities.

S. Geetha et al. (2021) conducted a survey on machine vision-based fire detection techniques in Fire Technology. The authors reviewed various methods and technologies used in fire detection using machine vision, providing insights into the state-of-the-art approaches, and identifying potential areas for improvement.

S. Vinod, S. Bhupendra Singh, et al. (2013) conducted a study on face detection using Haar cascade classifier with simple and complex background images, implemented through OpenCV. Published in the International Journal of Advanced Technology in Engineering and Science, their research contributes to advancements in face detection technology.

T.-H. Chen, P.-H. Wu, and Y.-C. Chiou (2004) proposed an early fire-detection method based on image processing. Their research, presented at the International Conference on Image Processing, contributes to advancements in fire safety technology by utilizing innovative image processing techniques for timely fire detection.

Ukil, A., Braendle, H., and Krippner, P. (2011) conducted a comprehensive review of distributed temperature sensing technology and its applications. Published in the IEEE Sensors Journal, their work provides valuable insights into the advancements and diverse applications of distributed temperature sensing systems.

W. Phillip Ian et al. (2006) presented a study on facial feature detection using Haar classifiers in the Journal of Computing Sciences in Colleges. Their research focused on utilizing Haar classifiers for facial feature detection, contributing to the field of computer vision and pattern recognition.

3. PROPOSED WORK

The proposed system integrates a camera for fire detection, while Raspberry Pi orchestrates the extinguishing process using interconnected hardware components. Employing User Datagram Protocol (UDP), the integrated fire detection and extinguishing mechanisms ensure a swift and accurate response to fire hazards. Methodologically, careful consideration has been given to real-time situations, such as power failure, to ensure system resilience. In the event of power failure, the system seamlessly transitions to backup power sources, ensuring continuous operation and uninterrupted fire detection and suppression capabilities.

The DC Water Pump along with a Servo Motor is used for fire suppression.s To tackle smoke dispersion, DC fans are deployed to vent out smoke, ensuring improved visibility and breathable air. Additionally, LED indicators strategically guide occupants along safe pathways to exits, enhancing evacuation procedures. This method combines advanced data acquisition, processing, and deployment of targeted extinguishing mechanisms, all while addressing the critical need for smoke clearance and safe evacuation routes.

This integrated approach combines Computer Vision, network communication, and intelligent hardware to revolutionize fire response. Early detection through Computer Vision, real-time collaboration via interconnected networks, and autonomous risk assessment by intelligent hardware enhance fire detection, suppression, and mitigation efforts. Ultimately, this approach maximizes safety for occupants and responders while minimizing property damage.

Figure 1. System block diagram

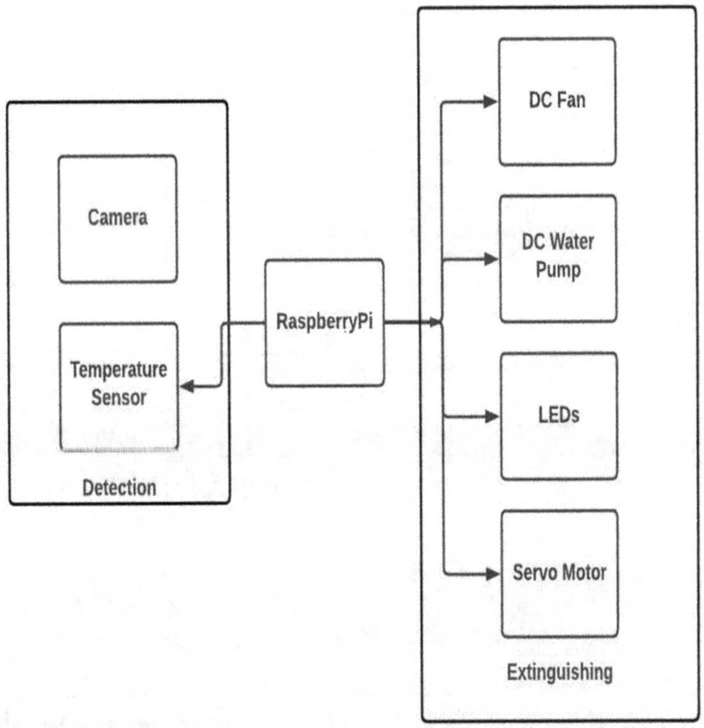

Figure 2. Flowchart of proposed method

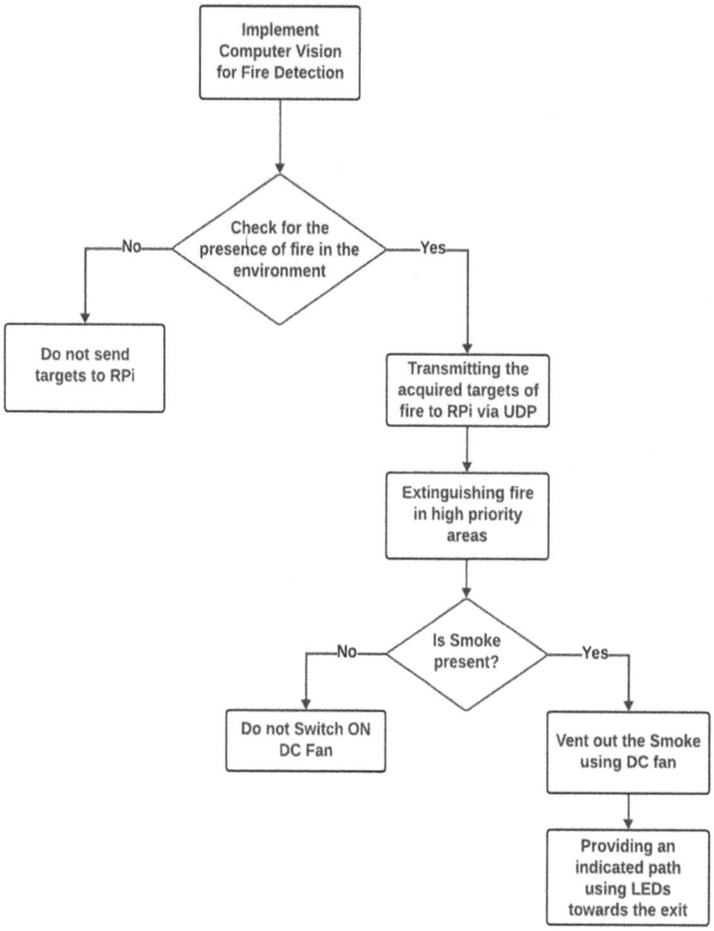

3.1 Hardware Specifications

The following aims to provide a thorough understanding of the specifications of the sensors, actuators, and central processing unit, such as the Raspberry Pi, which form the backbone of the smart fire detection and suppression system.

3.1.1 Raspberry Pi

- Processor: Broadcom BCM2711, Quad-core Cortex-A72 @ 1.5GHz.
- RAM: 2GB, 4GB, or 8GB LPDDR4.
- Storage: MicroSD card slot.
- Connectivity: USB 3.0, USB 2.0, Gigabit Ethernet, Dual-band Wi-Fi, Bluetooth 5.0, 2 × HDMI.
- Video/Audio: H.265 4Kp60 decode, H.264 1080p60 decode, OpenGL ES 3.0, 3.5mm audio jack.

- GPIO: 40-pin header.
- Power: 5V/3A via USB Type-C or GPIO.
- Dimensions: 85.6mm x 56.5mm x 17mm.
- OS Support: Raspberry Pi OS (Linux), others.

Figure 3. Raspberry Pi

3.1.2 DS18B20 Sensor

- Temperature Range: -55°C to +125°C
- Accuracy: ±0.5°C (-10°C to +85°C)
- Interface: Single-wire digital (1-Wire protocol)
- Supply Voltage: 3.0V to 5.5V
- Current Consumption: Active: 1mA, Standby: <1μA
- Package Types: TO-92, SOP-8, Waterproof Probe
- Compatibility: Works with Raspberry Pi, microcontrollers
- Operating Temperature: -55°C to +125°C

Figure 4. DS18B20 sensor

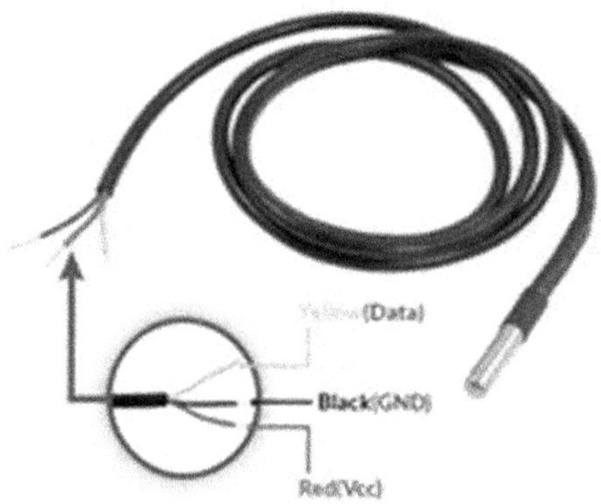

3.1.3 DC Fan

- Voltage: Commonly 5V, 12V, or 24V.
- Size: Varies (e.g., 80mm, 120mm).
- Speed (RPM): Variable.
- Airflow (CFM): Variable.
- Static Pressure: Variable.
- Noise Level (dBA): Variable.
- Bearing Type: Varies (e.g., sleeve, ball, fluid dynamic).
- Connector Type: 2, 3, or 4-pin.
- Dimensions: Varies (e.g., 25mm, 38mm).
- Power Consumption: Variable.

Figure 5. DC fan

3.1.4 DC Water Pump

- Voltage: Commonly 5V, 12V, or 24V.
- Flow Rate: Varies (measured in L/h or GPH).
- Head Height: Maximum vertical lift.
- Power Consumption: Measured in watts.
- Connection Type: Input and output hose connectors.
- Noise Level (dB): Operational noise.
- Pump Type: Centrifugal or diaphragm.
- Dimensions: Size and form factor.
- Materials: Construction materials.
- Temperature Range: Operational temperature limits.

Figure 6. DC water pump

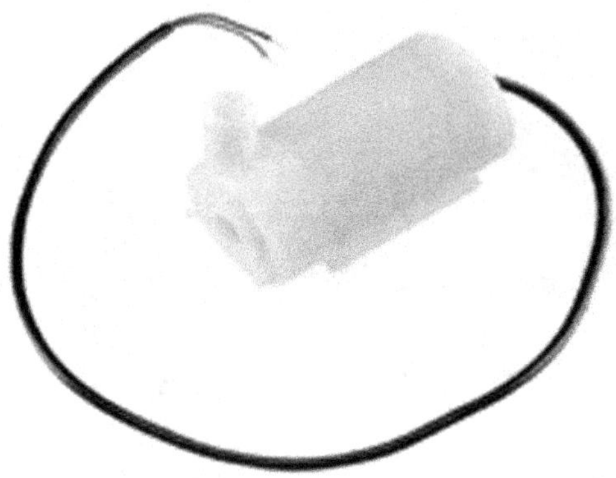

3.1.5 Relay

- Input Power: 5V/3A via USB Type-C or GPIO.
- GPIO: 40-pin header.
- Dimensions: 85.6mm x 56.5mm x 17mm.
- Control: Low trigger signal.
- Compatibility: Raspberry Pi OS (Linux), others.

Figure 7. Relay

3.1.6 Servo Motor

- Voltage: 4.8V - 6V.
- Torque: 1.8 kg·cm at 4.8V, 2.2 kg·cm at 6V.
- Speed: 0.12 sec/60° at 4.8V, 0.10 sec/60° at 6V.
- Rotation Angle: 180 degrees.
- Weight: 9 grams.
- Dimensions: 22.2mm x 11.8mm x 31mm.
- Operating Temperature: -30°C to +60°C.
- Dead Band Width: 10 microseconds.
- Operating Pulse Voltage: 3V - 7V.
- Plug Type: JR (Fits standard servo connectors).

Figure 8. Relay

4. RESULTS AND DISCUSSION

For effective fire detection, Cascade Trainer GUI is utilized, which simplifies the training process for the fire recognition system. Initially, a primary directory is established to streamline workflow management. Within this directory, the creation of two distinct subfolders is crucial: one designated for fire-displaying images (termed the positive folder) and another for non-fire images (referred to as the negative folder). Diligent organization and storage of relevant images in these folders significantly contribute to the system's learning capabilities, aiding in the differentiation between fire and non-fire scenarios. Subsequently, determining the number of training stages is imperative, indicating the depth of the system's learning iterations. The model undergoes training using the categorized images, enabling it to discern the distinctive features of fire and non-fire instances. This training phase culminates in the creation of an XML file, encapsulating the acquired knowledge and empowering the system to identify fire instances in real-time situations.

The User Datagram Protocol (UDP) serves as a communication protocol utilized for transmitting data across networks. Successfully integrating the fire detection system with the Raspberry Pi via UDP facilitates seamless communication between the Raspberry Pi and the fire detection system. This integration enables efficient real-time data exchange and coordination between the two systems.

Utilizing the Haar cascade technique, the coordinates of fire locations are extracted from the video feed. Subsequently, these coordinates are transmitted to the extinguishing system, managed by the Raspberry Pi, utilizing the User Datagram Protocol (UDP). Upon receiving the coordinates, the servo motor mechanism adjusts itself to align with the specified direction, effectively extinguishing the flame. Additionally, the DS18B20 sensor is tasked with monitoring the temperature at the obtained coordinates, ensuring comprehensive monitoring of fire incidents.

The DC water pump, integrated with a relay, was successfully implemented within the system. Activation of the DC water pump was automated, occurring as required to suppress fires effectively.

Figure 9. Fire detection using Haar Cascade

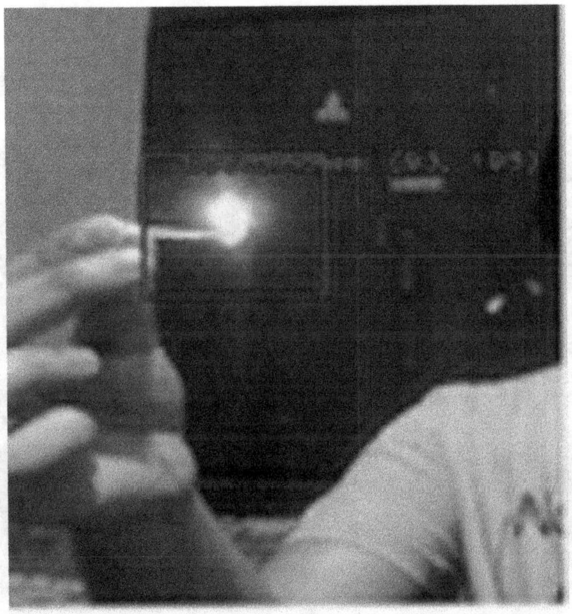

Figure 10. Coordinates of fire location

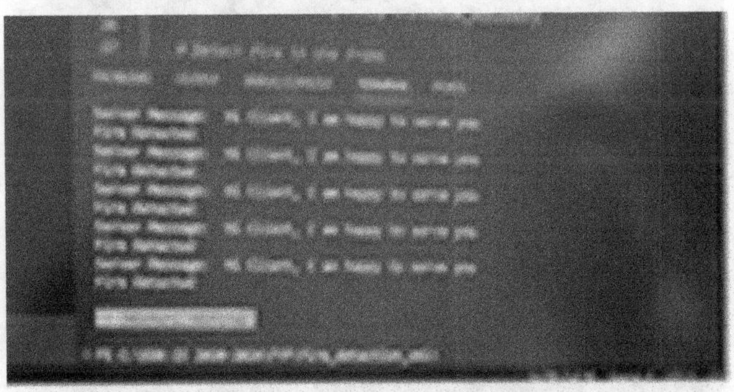

Figure 11. Successful connection with user datagram protocol

Figure 12. Fire coordinates received at Raspberry Pi

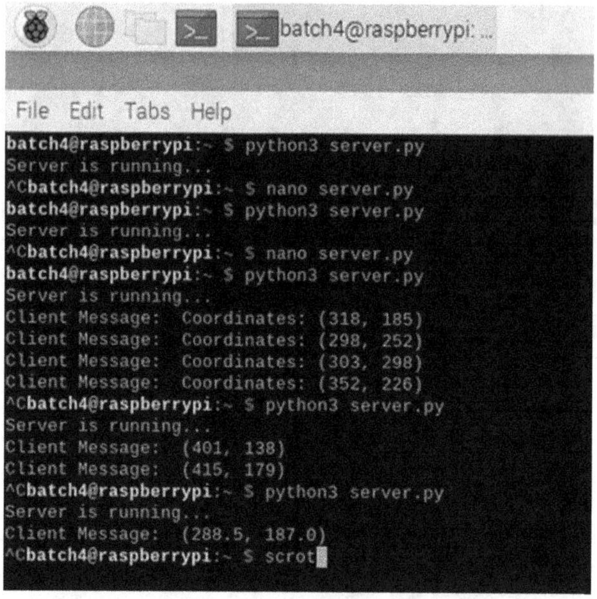

Figure 13. Temperature monitoring

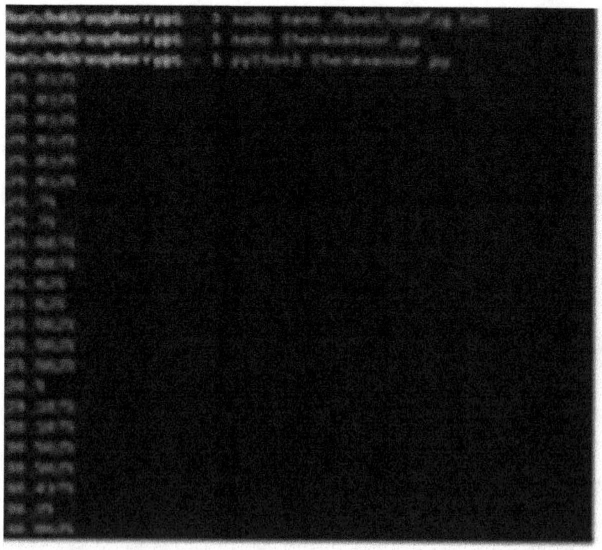

Figure 14. Circuit connection

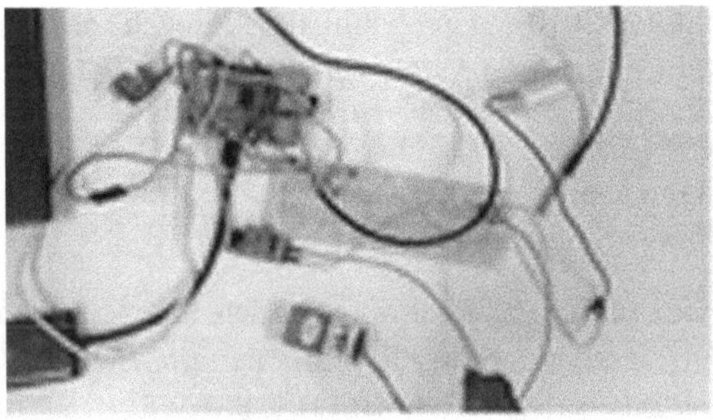

5. CONCLUSION

Concluding, the project demonstrates a notable progression in fire safety by integrating computer vision technology employing Haar cascade for fire detection. The effective linkage established between the fire detection and extinguishing systems via User Datagram Protocol (UDP) ensures uninterrupted communication. Leveraging flame coordinates to govern the servo motor, the system manifests agile responsiveness. The DC water pump, integrated with a relay, was successfully implemented within the system. Activation of the DC water pump was automated, occurring as required to suppress fires effectively. The coordinated functioning of components such as the DC fan, LEDs, and servo motor ensures prompt and effective mitigation of fire incidents, thereby bolstering safety protocols.

REFERENCES

Borges, P. V. K., & Izquierdo, E. (2010). A probabilistic approach for vision-based fire detection in videos. *IEEE Transactions on Circuits and Systems for Video Technology*, 20(5), 721–731. 10.1109/TCSVT.2010.2045813

Chen, T.-H., Wu, P.-H., & Chiou, Y.-C. (2004). An early fire-detection method based on image processing. *Proc. Int. Conf. Image Process*. Research Gate.

Foggia, P., Saggese, A., & Vento, M. (2015). Real-time fire detection for video-surveillance applications using a combination of experts based on color,shape, and motion. *IEEE Transactions on Circuits and Systems for Video Technology*, 25(9), 1545–1556. 10.1109/TCSVT.2015.2392531

Gaur, A., Singh, A., Kumar, A., Kumar, A., & Kapoor, K. (2020). Video flame and smoke based fire detection algorithms: A literature review. *Fire Technology*, 56(5), 1943–1980. 10.1007/s10694-020-00986-y

Geetha, S., Abhishek, C. S., & Akshayanat, C. S. (2021). Machine vision based fire detection techniques: A survey. *Fire Technology*, 57(2), 591–623. 10.1007/s10694-020-01064-z

He, Y., Wang, J., Wu, Z., Hu, L., Xiong, Y., & Fan, W. (2002). Smoke venting and fire safety in an industrial warehouse. *Fire Safety Journal*, 37(2), 191–215. 10.1016/S0379-7112(01)00045-5

Ho, C.-C., & Kuo, T.-H. (2009). Real-time video-based fire smoke detection system. *Advanced Intelligent Mechatronics IEEE/ASME International Conference*, (pp. 1845–1850). IEEE.

Lai, C.-L., & Yang, J.-C. (2008). Advanced real time fire detection in video surveillance system. *IEEE International Symposium*, (pp. 3542–3545). IEEE.

Ling-Yu L. (2022). Smart fire-warning materials and sensors: Design principle, performances, and applications. *Mater. Sci. Eng., R, Rep., 150.*

Maier, M., Chowdhury, M., Rimal, B. P., & Van, D. P. (2016). The tactile internet: Vision, recent progress, and open challenges. *IEEE Communications Magazine*, 54(5), 138–145. 10.1109/MCOM.2016.7470948

Muhammad, K., Khan, S., Elhoseny, M., Ahmed, S. H., & Baik, S. W. (2019). Efficient fire detection for uncertain surveillance environment. *IEEE Transactions on Industrial Informatics*, 15(5), 3113–3122. 10.1109/TII.2019.2897594

Phillip Ian, W., & John, F. (2006). Facial feature detection using Haar classifiers. *Journal of Computing Sciences in Colleges*, 21, 127–133.

Pincott, J., Tien, P. W., Wei, S., & Calautit, J. K. (2022). Indoor fire detection utilizing computer vision-based strategies. *Journal of Building Engineering*, 61, 105154. 10.1016/j.jobe.2022.105154

Sajjad, M., Nasir, M., Muhammad, K., Khan, S., Jan, Z., Sangaiah, A. K., Elhoseny, M., & Baik, S. W. (2017). Raspberry Pi assisted face recognition framework for enhanced law-enforcement services in smart cities. *Future Generation Computer Systems*. Advance online publication. 10.1016/j.future.2017.11.013

Sridhar, P., Parameswaran, L., & Thangavel, S. K. (2020). An efficient rule based algorithm for fire detection on real time videos. *Journal of Computational and Theoretical Nanoscience*, 17(1), 308–315. 10.1166/jctn.2020.8667

Toreyin, B. U., Dedeoğlu, Y., Güdükbay, U., & Çetin, A. E. (2006). Computer vision based method for real-time fire and flame detection. *Pattern Recognition Letters*, 27(1), 49–58. 10.1016/j.patrec.2005.06.015

Ukil, A., Braendle, H., & Krippner, P. (2011). Distributed temperature sensing: Review of technology and applications. *IEEE Sensors Journal*, 12(5), 885–892. 10.1109/JSEN.2011.2162060

Vinod, S., & Bhupendra, S., & Singh. (2013). Face detection by haar cascade classifier with simple and complex backgrounds images using opencv implementation. *International Journal of Advanced Technology in Engineering and Science*, 1, 33–38.

Wu, H., Wu, D., & Zhao, J. (2019). An intelligent fire detection approach through cameras based on computer vision methods. *Process Safety and Environmental Protection*, 127, 245–256. 10.1016/j.psep.2019.05.016

Yar, H., Hussain, T., Khan, Z. A., Koundal, D., Lee, M. Y., & Baik, S. W. (2021). Vision sensor-based real-time fire detection in resource-constrained IoT environments. *Computational Intelligence and Neuroscience*, 2021, 1–15. 10.1155/2021/519550834970311

Chapter 6
Dynamic Spectrum Allocation in Cognitive Radio Networks:
A Fair Game With Asymmetric Information

Lhoussaine Daoudi
https://orcid.org/0000-0001-9082-1467
Sultan Moulay Sliman University, Morocco

Mohamed Baslam
Sultan Moulay Sliman University, Morocco

Said Safi
Sultan Moulay Sliman University, Morocco

ABSTRACT

In cognitive radio networks, secondary users face spectrum scarcity. Cognitive radio offers a solution by enabling sharing of frequency bands to maximize spectrum utilization. This chapter explores spectrum allocation behavior, where secondary users allocate underutilized spectrum to maximize profit. Prior research suggests cooperative spectrum sharing benefits users, but real networks see selfish behavior to mitigate risks. The authors propose a dynamic duopoly game for competitive spectrum allocation. Secondary users employ bounded rationality to adapt strategies, aiming for Nash Equilibrium. One user considers competitors' past allocations to estimate future ones. The authors establish the existence and uniqueness of Nash Equilibrium, ensuring fairness. Local stability analysis of fixed points is conducted. Numerical simulations, including bifurcation diagrams, show that users integrating historical data and bounded rationality are more likely to reach Nash Equilibrium and maintain system stability.

1. INTRODUCTION

In cognitive radio networks, specific frequency bands may be partially occupied at particular locations and times, as noted by (Federal communications commission: Spectrum policy task force, 2002; Economics at the FCC 2019–2020: Spectrum Policy, Universal Service, Inmate Calling Services, and Telehealth, 2020). Traditionally, radio spectrum management has been a static process handled by regu-

DOI: 10.4018/979-8-3693-2893-4.ch006

latory agencies responsible for allocating frequencies to users. Recognizing the limitations of this static and non-optimized approach, the concept of dynamic spectrum management has emerged. Dynamic spectrum management is grounded in the observation that it is feasible to leverage free or underutilized spectrum. Secondary users (SUs) employ cognitive radio technology, as discussed by (Haykin, 2005), to sense spectrum holes and dynamically share desired bands with primary users (PUs).

Addressing a significant challenge involves devising an allocation scheme that optimizes spectral utilization, minimizes interference from both primary and secondary users, and maximizes monetized profit (Maharjan Sabita & Stein, 2011). To tackle these issues, game theory serves as an effective tool for modeling the cooperative and non-cooperative behaviors of SUs (Musacchio & Walrand, 2006; Wang et al., 2010). In real cognitive radio networks SUs have insufficient information about each other, this led us to consider a non-cooperative game since SUs try to avoid risks and behave selfishly. Various studies (Do et al., 2014; Bany Salameh et al., 2023) have employed a duopoly game model to represent the behaviors of SUs due to the competitive nature among SUs for spectrum purchases from numerous end users. This scenario resembles an oligopoly market, wherein a few sellers interact with a multitude of buyers (Cournot, 1963). In dynamic allocation, SUs may lack visibility into the current allocations of their competitors; they only possess information about the unit spectrum price provided by the PU. Consequently, determining the next allocation poses significant challenges. To estimate the next allocation, we consider that the spectrum allocation strategies of SUs are based on profit margin or bounded rationality mechanism (Baslam et al., 2012; Ait Omar et al., 2017).

We formulate this spectrum sharing behavior as a duopoly game in term of bounded rationality approach in which two secondary users compete with each other for maximum price profit. In this configuration, one of the secondary user's strategic adaptations is based on his competitor's last two allocation sizes, while incorporating the concept of bounded rationality. Through this dynamic spectrum management mechanism, we analyze the stability condition of the Nash Equilibrium by employing local stability analysis.

Our study endeavors to formulate a dynamic Cournot duopoly game model, wherein SUs adjust their strategies based on the requested spectrum size. This manuscript delves into a comprehensive examination of the local stability properties of the Nash Equilibrium, deemed a fair solution in this spectrum allocation game where SUs behaves selfishly due to scarcity of spectrum and the increasing demand for connectivity in CRN. The key aspects of this paper are outlined as follows:

- A spectrum dynamic allocation scheme between two secondary users in cognitive radio networks is described according to the non-cooperative game where one of the users is taking into account the history of his competitor actions in term of bounded rationality.
- The existence, uniqueness, and stability analysis of the Nash Equilibrium, which serves as the fair solution for this game, are detailed in this chapter.
- Numerical simulations, including bifurcation diagrams, suggest that users with historical awareness are more likely to reach Nash Equilibrium swiftly and maintain allocation stability under appropriate parameter settings (e.g., unit spectrum revenue rate and bounded rationality learning rate).

This article is structured as follows. After this introduction, we briefly cite related works. Then, we describe the allocation game with bounded rationality in the Section 3, then in Section 4, we analyzed the dynamics and stability of fixed points regarding the system parameters such as learning rate, unit revenue and history weights. In Section 5, we presented numerical simulations, which will allow us to validate our results acquired through our analysis. Finally, some thoughts and future works are restrained to the conclusion.

2. RELATED WORKS

The paper **(Yao et al., 2023)** presents a novel approach utilizing game theory principles to dynamically allocate channels, demonstrating theoretical convergence to Nash Equilibrium and showing through simulations its efficacy in reducing communication interference and improving spectrum utilization. In **(Daoudi et al., [2022]© 2022)**, the authors demonstrated that in a duopoly game with only two non-cooperative users, chaotic behavior emerges, leading to instability in the allocation scheme. Thus, maintaining the stability of spectrum management scheme present a challenge. In (Wei et al., 2023), the authors rigorously analyze bifurcations and chaos phenomena in a discrete two-dimensional map representing a Cournot duopoly with bounded rationality, supported by numerical simulations, and proposes two effective control methods to mitigate chaos. The paper (Sayama et al., 2024), investigates the mathematical properties and dynamic behavior of a Cournot–Bertrand duopoly game, exploring instabilities in the Nash equilibrium through flip and Neimark–Sacker bifurcation.

The existing literature on oligopoly games predominantly concentrates on games featuring boundedly rational players adjusting their output. However, there is a noticeable scarcity of papers that explore dynamic duopoly games where one player considers the history of their competitor's actions. This paper ventures into this less-explored territory. Its primary aim is to investigate the competition between two boundedly rational strategic users (SUs), where one exhibits history awareness while the other bases its actions on the most recent spectrum allocation of its competitor to update its payoff.

3. ALLOCATION GAME MODEL AND ASSUMPTION

We are investigating a wireless network where two secondary users SU1 and SU2 operate, serving the needs of numerous end users in a specific geographical zone. SUs exploits Cognitive Radio (CR) to receive spectrum frequencies unused by the primary user (PU) in order to maximize their profit. Each SU, i.e., SUi requests a size s_i of spectrum from a PU based on the following spectrum unit cost function:

$$C(s_T) = a + b\,s_T, \tag{1}$$

where a and b are non-negative constants, and $s_T = s_1 + s_2$, refers to the total size of spectrum that can be allocated by all SUs Figure 1.

Figure 1. System model

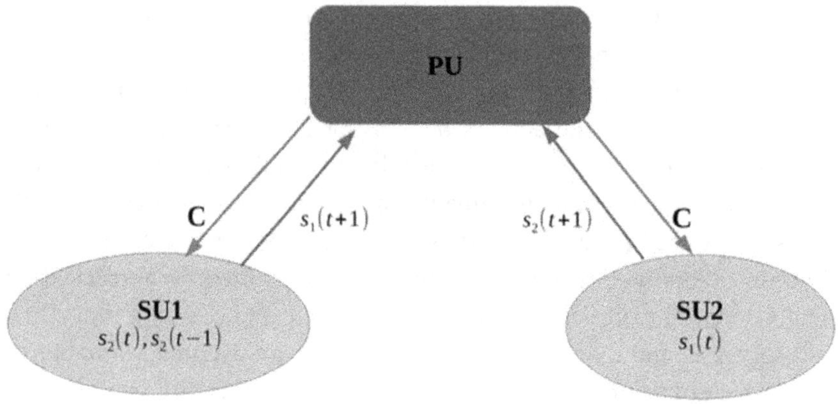

So let $A = F, S, P$ be the spectrum allocation game, in which $F = 1{,}2$, is the set of SUs, and S denote the set of their strategies $S = s_1, s_2$, and $P = u_1, u_2$, is the set of all utilities of those SUs. The cost function of the SUi is defined by:

$$c_i(s_i) = s_i\left(a + b\left(s_i + s_j\right)\right), i,j = 1{,}2,\ i \neq j. \tag{2}$$

The revenue function of the SUi is defined as follows:

$$R_i(s_i) = s_i r_i. \tag{3}$$

Such that, r_i is the unit revenue rate of SUi, then the corresponding utility function at time step t is defined by:

$$u_i(t) = R_i(s_i(t)) - c_i(s_i(t)), i,j = 1{,}2,\ i \neq j. \tag{4}$$

which can be rewritten as function of two variable as follows:

$$u_i\left(s_i(t), s_j(t)\right) = r_i s_i(t) - s_i\left(a + b\left(s_i(t) + s_j(t)\right)\right), i,j = 1{,}2,\ i \neq j. \tag{5}$$

Note that the allocation is dynamic in this cognitive radio network so SUs may be only receiving the price of unit spectrum from the PU. To specify for example at time step t SU1 is not knowledgeable about the strategies $s_2(t + 1)$, or the profit $u_2(t)$ of the SU2. Thus, because of the scarcity of allocation market information SUs may not update their strategies $s_i(t)$, by solving the utility maximization problem:

$$s_i(t + 1) = \arg\max_{s_i}\left[u_i(t) = R_i(s_i(t)) - s_i(t)\left(a + b\left(s_i(t) + s_j^e(t + 1)\right)\right)\right] \tag{6}$$

where, $s_j^e(t + 1)$ is the expected strategy of SUj, to update their strategies SUs opt on bounded rationality approach means that each SU say ith one modifies its strategy $s_i(t)$ based on profit margin:

$$\frac{\partial u_i\left(s_i(t), s_j^e(t+1)\right)}{\partial s_i}, i = 1,2, i \neq j. \tag{7}$$

Thus, the adaptation of strategies according to the bounded rationality mechanism is defined as follows:

$$\begin{cases} s_1(t+1) = s_1(t) + \theta_1 s_1(t) \dfrac{\partial u_1\left(s_1(t), s_2^e(t+1)\right)}{\partial s_1}, \\ s_2(t+1) = s_2(t) + \theta_2 s_2(t) \dfrac{\partial u_2\left(s_1^e(t+1), s_2(t)\right)}{\partial s_2}. \end{cases} \tag{8}$$

Where, θ_i is the learning rate. Specifically, if the profit margin is positive, then SUi will request more spectrum size than $s_i(t)$. Conversely, if the profit margin is negative, SUi will request less than $s_i(t)$ spectrum size in the next allocation $s_i(t+1)$.

To estimate $s_2^e(t+1)$ and $s_1^e(t+1)$, we assume a game with asymmetric information in a scenario where SU1 consider the history awareness by writing the expected allocation strategy $s_2^e(t+1)$ of SU2, as a function of past previous allocated spectrum sizes with different weights: $s_2^e(t+1) = \sum_{h=0}^{H} w_h s_2(t-h), w_h \geq 0, \sum_{h=0}^{H} w_h = 1$, such that, $H < \infty$. The factors $w_h, h = 0, 1, 2, \ldots, H$ are the weights given to previous allocations. For simplicity, we set $H = 1$, then the expected strategy of SU2 is defined by:

$$s_2^e(t+1) = w s_2(t) + (1-w) s_2(t-1). \tag{9}$$

In the other side, SU2 repose only on previous allocation $s_1(t)$ of his competitor SU1 and assume that $s_1^e(t+1) = s_1(t)$. Under the above assumptions, the expression for the utility of the SUs can be reformulated as:

$$\begin{cases} u_1(s_1(t), s_2(t)) = r_1 s_1(t) - s_1(t)a + b\left(s_1(t) + w s_2(t) + (1-w) s_2(t-1)\right), \\ u_2(s_1(t), s_2(t)) = r_2 s_2(t) - s_2(t)a + b\left(s_2(t) + s_1(t)\right). \end{cases} \tag{10}$$

Were $0 < w < 1$, from the first-order derivative of the utility functions u_i with respect to s_i we obtain the marginal profit functions of the SU1 and SU2 as:

$$\begin{cases} \dfrac{\partial u_1(s_1, s_2)}{\partial s_1} = r_1 - a - 2bs_1 - bws_2 - b(1-w)s_2, \\ \dfrac{\partial u_2(s_1, s_2)}{\partial s_2} = r_2 - a - 2bs_2 - bs_1. \end{cases} \tag{11}$$

Thus, The SU1 modify his strategies based on profit margin function which depend on history of his competitor strategies while the SU2 is not considering story awareness. Then, the dynamic allocation spectrum game can be expressed as:

$$\begin{cases} s_1(t+1) = s_1(t) + \theta_1 s_1(t)\left\{ r_1 - a - 2bs_1(t) - b\left[w s_2(t) + (1-w) s_2(t-1)\right]\right\} \\ s_2(t+1) = s_2(t) + \theta_2 s_2(t)\left\{ r_2 - a - 2bs_2(t) - bs_1(t)\right\} \end{cases} \tag{12}$$

4. DYNAMICS AND STABILITY ANALYSIS

In this section we cite equilibrium points of the dynamical system in first subsection, in the second subsection, we analyze the local stability of Nash Equilibrium point. In the third subsection, we provide a Nash Equilibrium learning algorithm encapsulating the adaptation of SUs strategies.

4.1 Equilibrium Points

To study the stability of the dynamical system game described in equation (12), we can convert it into a third-dimensional system with the following form:

$$\begin{cases} d_2(t+1) = s_2(t), \\ s_1(t+1) = s_1(t) + \theta_1 s_1(t)\{r_1 - a - 2bs_1(t) - b[ws_2(t) + (1-w)d_2(t)]\}, \\ s_2(t+1) = s_2(t) + \theta_2 s_2(t)\{r_2 - a - 2bs_2(t) - bs_1(t)\}. \end{cases} \tag{13}$$

At the state of equilibrium, the strategies are independent of time, means $s_i(t+1) = s_i(t)$ and $d_2(t+1) = d_2(t)$. Then, from the equations in the dynamical system game (13), we get four equilibrium points defined as follow:

$$\begin{cases} E_0 = (0,0,0), \\ E_1 = \left(0, \frac{r_1 - a}{2b}, 0\right), \\ E_2 = \left(\frac{r_2 - a}{2b}, 0, \frac{r_2 - a}{2b}\right), \\ E_* = \left(s_2^*, s_1^*, s_2^*\right). \end{cases} \tag{14}$$

Such that,

$$\begin{cases} s_1^* = \frac{2r_1 - r_2 - a}{3b}, \\ s_2^* = \frac{2r_2 - r_1 - a}{3b}. \end{cases} \tag{15}$$

Obviously, E_0, E_1, and E_2 refer to limit equilibrium points, with E_* being the Nash equilibrium point (Nash, 1951), that has significant networking implications if:

$$\begin{cases} 2r_1 - r_2 - a > 0, \\ 2r_2 - r_1 - a > 0. \end{cases}$$

Theorem 1 The Nash equilibrium E_* exist and it is unique.

Proof By examining the expression for the marginal profit (11) of each SU, we can conclude that $\frac{\partial^2 u_1(s_1, s_2)}{\partial s_1^2} = -2b < 0$ and $\frac{\partial^2 u_2(s_1, s_2)}{\partial s_2^2} = -2b < 0$. This condition guarantees the existence of NE for the game (Rosen, 1965).

To prove that the NE point is unique, we use a proposition from (Lasaulce et al., 2009) that is applicable to all concave games, which states that when a concave game satisfies the dominance solvency condition:

$$\frac{-\partial^2 u_i(s_i, s_j)}{\partial s_i^2} - \sum_{j=1, j\neq i}^{2} \left| \frac{\partial^2 u_i(s_i, s_j)}{\partial s_i \partial s_j} \right| \geq 0$$

so, the game A has a single NE point.

Expressing the mixed partial derivative as $\frac{\partial^2 u_1(s_1, s_2)}{\partial s_1 \partial s_2} = \frac{\partial^2 u_2(s_1, s_2)}{\partial s_2 \partial s_1} = -b$, we can evaluate

$$\frac{-\partial^2 u_i(s_i, s_j)}{\partial s_i^2} - \sum_{j=1, j\neq i}^{2} \left| \frac{\partial^2 u_i(s_i, s_j)}{\partial s_i \partial s_j} \right| = 2b - b$$

$= b > 0.$

Therefore, the NE point is indeed unique.

4.2 Local Stability of Equilibrium Points

In order to analyze the local stability of an equilibrium point, the first step is to compute the Jacobian matrix of the dynamical system (13) at any point (d_2, s_1, s_2), which takes the form:

$$J(d_2, s_1, s_2) = \begin{bmatrix} 0 & 0 & 1 \\ -\theta_1 s_1 b(1-w) & X & -\theta_1 s_1 bw \\ 0 & -\theta_2 s_2 b & Y \end{bmatrix} \quad (16)$$

where,

$$X = 1 + \theta_1\left(r_1 - a - 2bs_1 - b\left(ws_2 + (1-w)d_2\right)\right) - 2\theta_1 s_1 b$$

and

$$Y = 1 + \theta_2(-bs_1 - 2bs_2 - a + r_2) - 2\theta_2 s_2 b.$$

Secondly, we verify the stability conditions of each equilibrium point. The definition of stability for a fixed point is that it is stable if and only if all the eigenvalues of the Jacobian matrix evaluated at that point lie inside the unit circle in the complex plane, i.e., if $|\lambda_i| < 1$, (Agiza et al., 1999; Choo, 2011).

First, let investigate the local stability of the NE point E_* (as defined in Equations (14) and (15)). At the equilibrium E_* we have $s_1(t+1) = s_1(t) = s_1^*$ and $s_2(t+1) = d_2(t+1) = s_2(t) = s_2^*$, then the Jacobian matrix at NE E_* is expressed by:

$$J(s_2^*, s_1^*, s_2^*) = \begin{bmatrix} 0 & 0 & 1 \\ -\theta_1 s_1^* b(1-w) & 1 - 2\theta_1 b s_1^* & -\theta_1 s_1^* bw \\ 0 & -\theta_2 s_2^* b & 1 - 2\theta_2 b s_2^* \end{bmatrix} \quad (17)$$

The stability condition for the NE point is that all roots of the equation $p(\lambda) = 0$ must satisfy $|\lambda| < 1$, where,

$$p(\lambda) = a_3 \lambda^3 + a_2 \lambda^2 + a_1 \lambda + a_0 \quad (18)$$

is the characteristic polynomial of Jacobian matrix evaluated at the NE point E_*, such that:

$$\begin{cases} a_3 = 1, \\ a_2 = -2 + 2\theta_1 b s_1^* + 2\theta_2 b s_2^*, \\ a_1 = 1 - 2\theta_1 b s_1^* - 2\theta_2 b s_2^* + 4\theta_1\theta_2 b^2 s_1^* s_2^* - \theta_1\theta_2 b^2 w s_1^* s_2^*, \\ a_0 = \theta_1\theta_2 b^2 w s_1^* s_2^* - \theta_1\theta_2 b^2 s_1^* s_2^*. \end{cases} \quad (19)$$

A necessary and sufficient condition for (17) to have all eigenvalues inside the unit circle is given by the following inequalities (Jury, 1963):

$$1 + a_2 + a_1 + a_0 > 0, \quad (20)$$

$$1 - a_2 + a_1 - a_0 > 0, \quad (21)$$

$$|a_0| < |a_3|, \quad (22)$$

$$|a_0^2 - a_3^2| > |a_0 a_2 - a_3 a_1|, \quad (23)$$

$$|a_0^2 - a_3^2| > |a_0 a_1 - a_2 a_3|. \quad (24)$$

Therefore, we have concluded from the above analysis that delay affects the local stability of the NE point.

With a sample calculation for $E_1 = \left(\frac{r_1 - a}{2b}, 0, \frac{r_1 a}{2b}, 0\right)$, we found that the corresponding characteristic polynomial does not depend on the history weights w. Thus, the local stability condition for E_0, E_1, and E_2 is the same as in (Daoudi et al., [2022] © 2022).

4.3 Nash Equilibrium Learning Algorithm

In the context of this paper, to provide clarity in performance evaluation section, we explained the strategies adaptation for spectrum allocation by SUs system (12), in following Algorithm:

Algorithm 1. Nash Equilibrium Learning Path
Parameters:
r_i = unit spectrum revenue rate. $i = 1,2$.
θ_i = bounded rationality learning rate.
w = history weight.
$s_i(t = 0)$ = initial spectrum allocations.
Initialization:
$a = 10$.
$b = 0.5$.
. Bounded Rationality Profit Margin.
procedure UpdateState(MAXiter)
Data zeros(MAXiter). Vector to store spectrum bands.
.
Data(0) .
Data(1) .
.
whiledo
Update through system (12).
Data(t) .
.
end while
return Data
end procedure

5. PERFORMANCE EVALUATION

We state that our simulations were conducted using Python 3 on a computer featuring 8 GB RAM and an Intel Core i5 10th generation processor, as previously described. The evaluation is performed in a network setup, wherein a PU, equipped with a base station, shares a 16 MHz frequency spectrum with two SUs. In this setup, we set the values of and . We commence with the simulation of the dynamic behavior of the two SUs, then in subsection 5.2, we simulate the impact of historical weight in the -bifurcation diagram. Next, in subsection 5.3, we shown NE domain stability in , -plan under different history weight values . Next, in subsection 5.4, we simulate the effect of the unit profit rate with the aid of a bifurcation diagram on NE stability under different values of . Next, in subsection 5.5, we present -bifurcation diagrams to determine the local stability of the NE. Finally, in subsection 5.6, We illustrate the global stability of the Nash Equilibrium using the basin of attraction.

5.1 Dynamic Behavior

We establish the initial strategies (i.e., strategies at time zero) as and , and the unit revenue of each SU as . When employing the specified parameter values, the Nash Equilibrium (NE) can be located at: . We proceed by establishing the values for the bounded rationality learning rate as . Under these values, we examine the dynamic behavior of the aforementioned game over time using Algorithm 1.

Figure 2. SU dynamic behavior

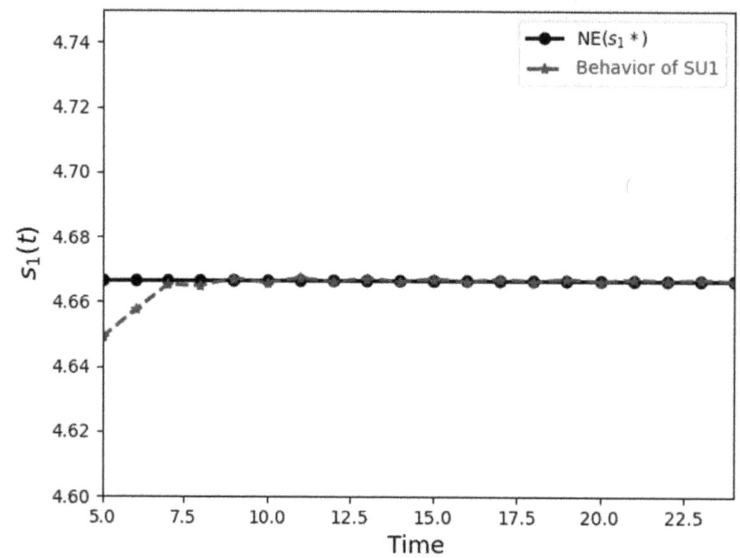

Figure 3. SU dynamic behavior

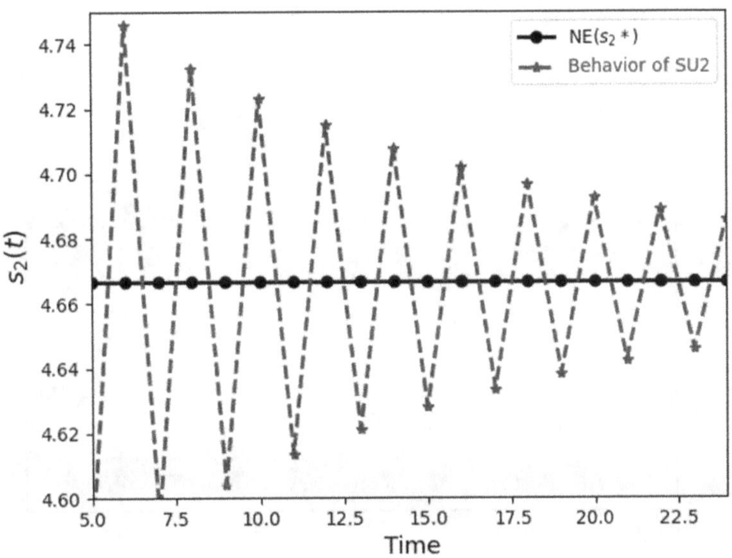

Comparing Figure 2 and Figure 3, we conclude that by setting an appropriate delay weight , the spectrum allocation of SU gradually converges to the NE point (as illustrated in Figure 2 for). On the other hand, if the history awareness is disregarded by (SU) as illustrated in the Figure 3, the strategies of the SU alternate between two values and may fail to converge to the NE. Also, SU reaches the NE quickly compared with SU. Indeed, with ignorance of history, a SU relies only on recent information, and as a result, increasing the step size for adjusting the strategies can lead to a delay in reaching the NE.

5.2 History Awareness Effect on Bounded Rationality Learning Rate

In a dynamic game, the bifurcation diagram illustrates potential values of a player's strategies after numerous iterations for each parameter value. Its purpose is to depict the stability of an equilibrium point (Devaney, 2022). First, we set and . Then, we generate a bifurcation diagram for each by updating spectrum allocations using Algorithm 1, for various history weights . This analysis will reveal the impact of the history weight on the stability of the NE.

Figure 4. SU, Bifurcation diagram under two value of history weight

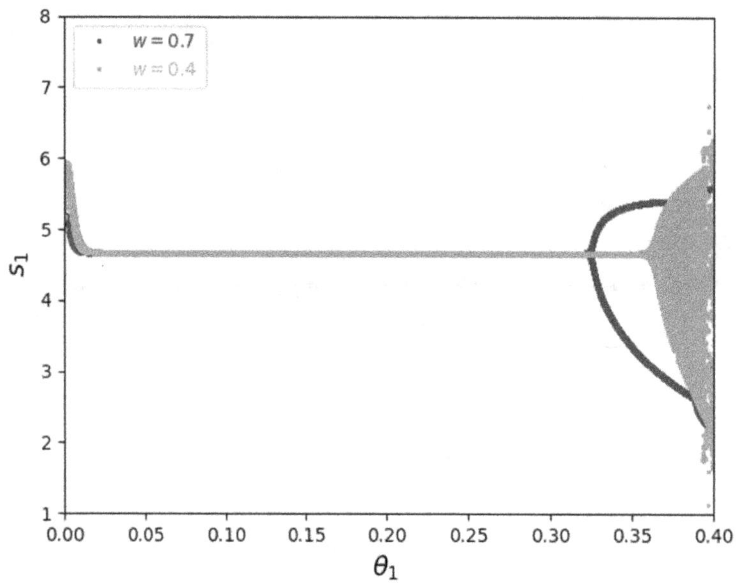

From Figure 4, it is evident that as the SU becames aware about the history of SU's allocation with the weight , the bifurcation doubling and chaotic behaviors are delayed, resulting in a wider range of values for the learning rate at which the NE is stable.

5.3 NE Local Stability Domain

Based on the local stability conditions of the Nash Equilibrium (NE), as described by equations (20)-(24), we determine the values in which Algorithm 1 converge to NE under various delay weight values and different values of and . The NE stability region in the -plane is depicted in Figure 5.

Figure 5. NE stability region in -plane under different history weights and

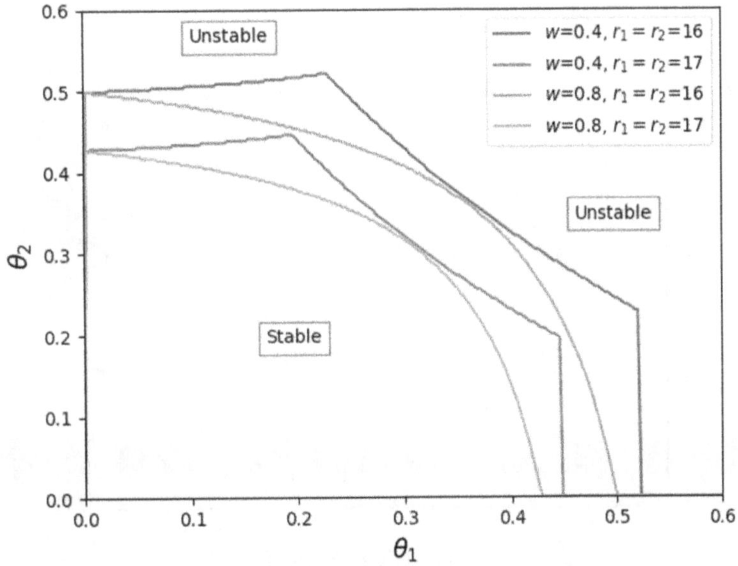

We observe that the stability domain of the NE becomes larger when compared with the case for different values and , as shown in Figure 5. Additionally, from the stability domain boundaries we observe that NE stability region became larger when both unit revenues compared with the case when . This is because SU does not heavily rely on the recent information of SU , (), but he factors in the past action with significant weighting (). Conversely, the stability region diminishes with the increase in . Moreover, the NE can be attained with larger bounded rationality learning rates when unit revenue values are not excessively high, as secondary users (SUs) are reluctant to allocate large spectrum sizes. Consequently, due to the utilization of standard strategy steps, the strategy adaptation remains close to the NE point.

5.4 Effect of Unit Revenue Rate on NE Stability

We take larger learning rates, , and initial allocations as and . The SU considers the history of SU's strategies with weight . Subsequently, we present the -bifurcations diagrams of SU and SU in Figure 6 and Figure 7, respectively.

Figure 6. Bifurcation diagrams of SU

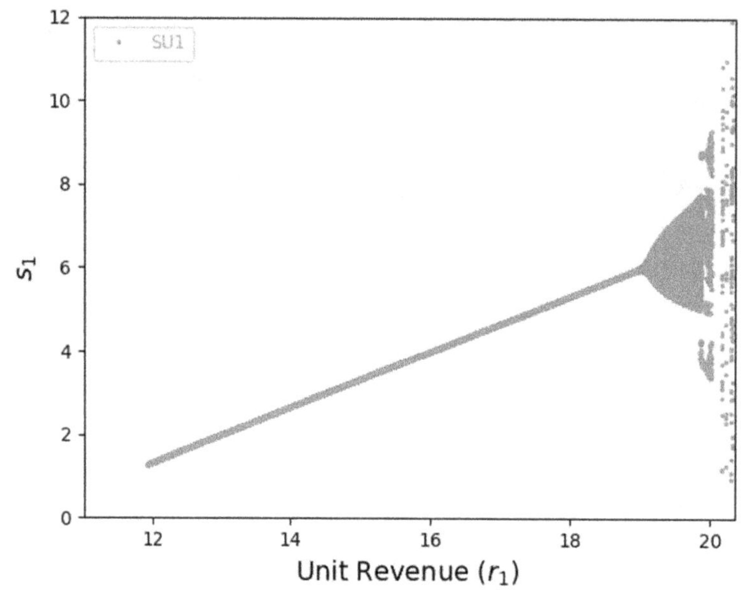

Figure 7. Bifurcation diagrams of SU

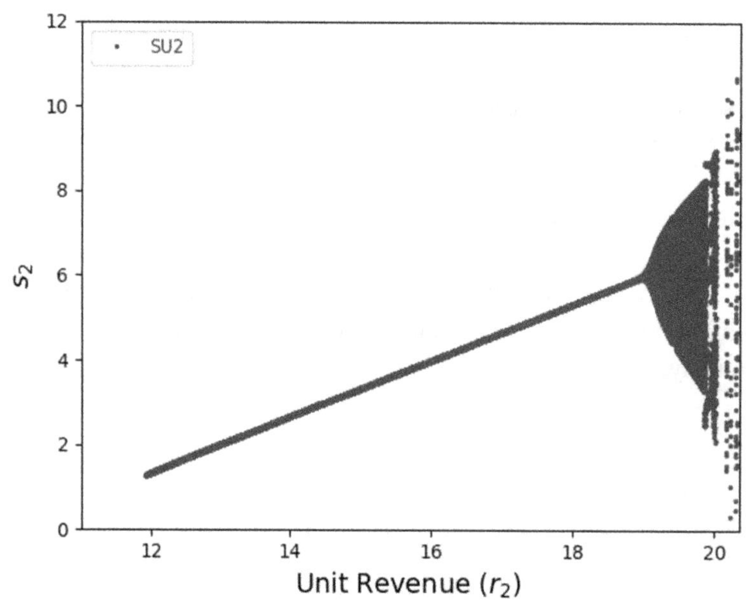

We observe for SU that the NE is reached when , and chaotic behavior appears when . However, when a SU consider past actions, case of SU (), chaotic behavior arises when . By comparing the two figures, we notice that chaos is controlled and delayed over . The reason is that if the unit revenue or is larger than necessary, the Sus are more inclined to allocate larger spectrum sizes. Consequently, due to the utilization of larger strategy steps, the strategy adaptation may fail to reach the NE.

5.5 History Weights Bifurcation Diagram

We analyze the convergence of the Nash Equilibrium (NE) under increased unit revenues (,) and larger bounded rationality learning rates (,). Under these values, the NE is found at and , and we initialize the actions close to the NE (,). Subsequently, we analyze the local stability of the NE. The NE is attained when , as shown in Figure 8.

Figure 8. SU, -Bifurcation diagram

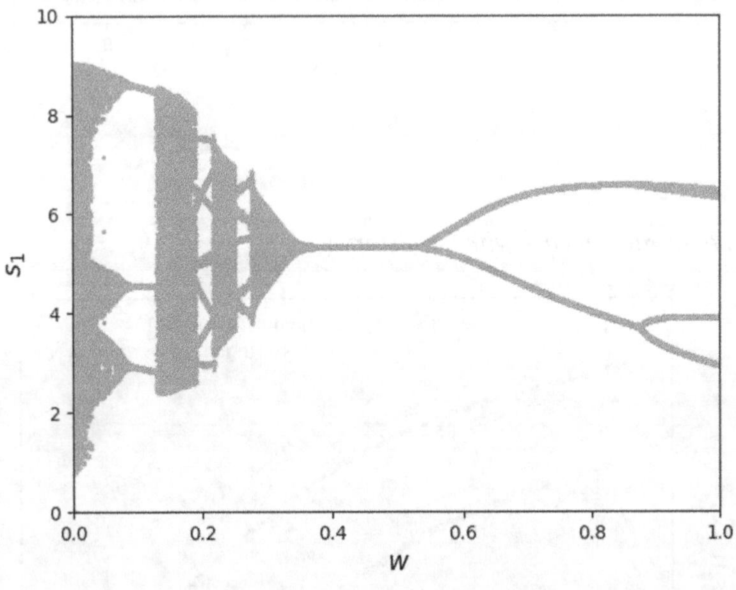

Furthermore, it is observed that when SU tends to disregard the past action (), or/and recent action (), chaotic behavior ensues, leading to failure in reaching the NE.

5.6 Nash Equilibrium Global Stability

In this subsection we explore the global stability of Nash Equilibrium (NE) by illustrating the basin of attraction, which is influenced by the initial strategies selection. We selected larger values for and (, and) and investigated the global stability of the NE. Figure 9 depicts the global stability domain of initial strategies (,) where strategy adaptation converges to the NE as the spectrum allocation system iterated Algorithm 1. Conversely, if the initial strategies are chosen outside the basin of attraction, the strategy adaptation converges to infinity.

Figure 9. NE Basin of attraction case with history ()

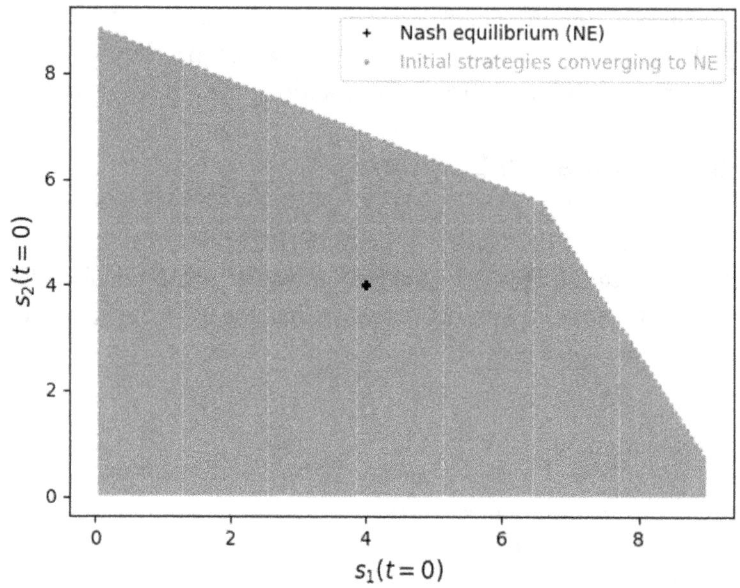

Figure 10. NE Basin of attraction case without history ()

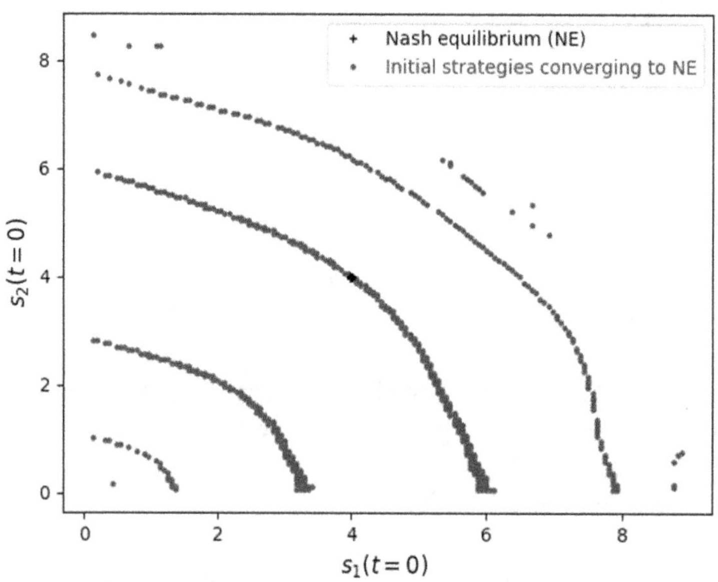

Comparing Figure 9 and Figure 10, we notice that the inclusion of delay enhances the size of the basin of attraction. This improvement occurs because SU's initial strategies, which are closer to the NE will be taken into consideration by SU in the subsequent time step. As a result, convergence to the desired solution accelerates, as converges more rapidly to the NE.

6. CONCLUSION AND PERSPECTIVES

In this paper, we have proposed an allocation game model for distributing spectrum in a cognitive radio network, comprising a primary user and two competitive secondary users. We have employed a dynamic duopoly game approach, where one secondary user adjusts their spectrum allocation strategy based on past actions of his competitor and profit margin. The stability of the model was examined using local stability analysis and the basin of attraction. The utilization of the Nash Equilibrium ensures fairness in spectrum allocation within the CRN. However, numerical simulations demonstrated that larger values of learning rates and unit revenue rates result in chaotic and bifurcated behavior. To maintain the stability of the NE, we suggest incorporating history awareness by secondary users in their strategy adaptation.

Looking forward, we believe our paper introduces intriguing perspectives, such as implications for spectrum allocation in future CRNs. Additionally, it would be valuable to investigate NE stability in oligopoly games within large CRNs and explore alternative approaches to modeling cost functions in crowded networks.

REFERENCES

Agiza, H. N., Bischi, G. I., & Kopel, M. (1999). Multistability in a dynamic Cournot game with three oligopolists. *Mathematics and Computers in Simulation*, 51(1-2), 63–90. 10.1016/S0378-4754(99)00106-8

Ait Omar, D., Outanoute, M., Baslam, M., Fakir, M., & Bouikhalne, B. (2017). Joint Price and QoS Competition with Bounded Rational Customers. In El Abbadi, A., & Garbinato, B. (Eds.), *Networked Systems* (pp. 457–471). Springer International Publishing. 10.1007/978-3-319-59647-1_33

Bany Salameh, H., Samara, M. Q., Elrefae, G. A., & Al-Ajlouni, A. (2023). Profit-maximization spectrum sharing in opportunistic duopoly market under dynamic spectrum pricing and QoS constraints. *Cluster Computing*, 46, 1451–1463. 10.1007/s10586-023-04026-6

Baslam, M., El-Azouzi, R., Sabir, E., & Bouyakhf, E.-H. (2012). New insights from a bounded rationality analysis for strategic price-QoS war. *6th international ICST conference on performance evaluation methodologies and tools*, (pp. 280–289). EUDL. 10.4108/valuetools.2012.250410

Choo, Y. (2011). An elementary proof of the Jury test for real polynomials. *Automatica*, 47(1), 249–252. 10.1016/j.automatica.2010.10.040

Cournot, A. (1963). *Researches into the principles of the theory of wealth*. The Macmillan Company.

Daoudi, L., Baslam, M., & Safi, S. (2022). 2022). Dynamics analysis for a duopoly game with bounded rationality in cognitive radio network. In *Networked systems* (Vol. 13464, pp. 236–251). Springer. 10.1007/978-3-031-17436-0_16

Devaney, R. L. (2022). *An introduction to chaotic dynamical systems*. CRC Press. 10.1201/9780429280801

Do, C. T., Tran, N. H., Han, Z., Le, L. B., Lee, S., & Hong, C. S. (2014). Optimal Pricing for Duopoly in Cognitive Radio Networks: Cooperate or not Cooperate? *IEEE Transactions on Wireless Communications*, 13(5), 2574–2587. 10.1109/TWC.2014.031914.131363

Economics at the FCC 2019–2020: Spectrum Policy, Universal Service, Inmate Calling Services, and Telehealth. (2020). *Review of Industrial Organization, 57*, 827-858. 10.1007/s11151-020-09791-x

Federal Communications Commission. (2002). *Spectrum policy task force*. ET Docket.

Haykin, S. (2005). Cognitive radio: Brain-empowered wireless communications. *IEEE Journal on Selected Areas in Communications*, 23(2), 201–220. 10.1109/JSAC.2004.839380

Jury, E. I. (1963). On the roots of a real polynomial inside the unit circle and a stability criterion for linear discrete systems. *IFAC Proceedings Volumes, 1*, 142-153. https://doi.org/10.1016/S1474-6670(17)69648-4

Lasaulce, S., Debbah, M., & Altman, E. (2009). Methodologies for analyzing equilibria in wireless games. *IEEE Signal Processing Magazine*, 26(5), 41–52. 10.1109/MSP.2009.933496

Maharjan Sabita, Z. Y., & Stein, G. (2011). Economic Approaches for Cognitive Radio Networks: A Survey. *Wireless Personal Communications*, 57(1), 33–51. 10.1007/s11277-010-0005-9

Musacchio, J., & Walrand, J. (2006). WiFi access point pricing as a dynamic game. *IEEE/ACM Transactions on Networking*, 14(2), 289–301. 10.1109/TNET.2006.872553

Nash, J. (1951). Non-cooperative games. *Annals of Mathematics*, 54(2), 286–295. 10.2307/1969529

Rosen, J. B. (1965). Existence and uniqueness of equilibrium points for concave n-person games. *Econometrica*, 33(3), 520–534. 10.2307/1911749

Sayama, H., Askar, S. S., & Alshamrani, A. M. (2024). Cournot-Bertrand Duopoly Model: Dynamic Analysis Based on a Computed Cost. *Complexity*, 5594918, 1–15. 10.1155/2024/5594918

Wang, B., Wu, Y., & Liu, K. J. (2010). Game theory for cognitive radio networks: An overview. *Computer Networks*, 54(14), 2537–2561. 10.1016/j.comnet.2010.04.004

Wei, Z., Tan, W., Elsadany, A. A., & Moroz, I. (2023). Complexity and chaos control in a Cournot duopoly model based on bounded rationality and relative profit maximization. *Nonlinear Dynamics*, 111(18), 17561–17589. 10.1007/s11071-023-08782-3

Yao, N., Liu, D., Chen, B., Hao, X., & Yin, R. (2023). Improved game algorithm for spectrum resource optimization with variable number of channels. *Ad Hoc Networks*, 146, 103190. 10.1016/j.adhoc.2023.103190

Chapter 7
Hybrid Railway Safety Optimization With Neural Networks and Random Forests

M. Kaviyalakshmi
S.A. Engineering College, India

V. Divyabharathi
S.A. Engineering College, India

D. Chithra
S.A. Engineering College, India

S. M. Rohini
S.A. Engineering College, India

K. Anusuya
S.A. Engineering College, India

M. D Dharani
S.A. Engineering College, India

ABSTRACT

Enhancing safety protocols in train stations requires proactive approaches to foresee and reduce any mishaps. In order to maximize safety in railroad environments, this study presents a novel machine learning framework that combines neural networks with random forest techniques. This hybrid model departs from traditional single-algorithm approaches by utilizing a variety of methods to enhance accident investigation and prediction. The chapter focuses on using this hybridized method, showcasing the functions of random forests and neural networks in predicting future events and assessing accident data from the past. The strategy of incorporating neural networks (multi-layer perceptron (MLP)) and random forests into the accident analysis framework is covered in the study. It also investigates if this model can be implemented in real-time and at scale within railway station safety systems. The results of the research reveal that the hybridized machine learning approach is a promising tool for improving safety standards in railroad stations.

DOI: 10.4018/979-8-3693-2893-4.ch007

1. INTRODUCTION

Railway networks are essential transportation arteries that are essential to both economic expansion and societal mobility. Railway environments are important, but they may also be dangerous due to a number of factors that can result in mishaps and catastrophes that put infrastructure and people in jeopardy (Shi et al., 2020).

Because of its many applications, machine learning (ML) is very important when it comes to accident detection, especially in transportation networks. Machine learning algorithms form the foundation of early warning systems by identifying complex patterns in past data to predict possible incidents. This ability to predict the future allows for the proactive application of safety measures by foreseeing accident hotspots and hazardous situations (Yu et al., 2018). ML models provide real-time monitoring capabilities that continuously evaluate variables such as traffic congestion, weather, and vehicle speed. This allows for the prompt intervention or alerting of users when deviations in these variables indicate the possibility of an accident. By analyzing large datasets to identify contributing aspects, machine learning's capacity for data-driven decision-making enables authorities to put specific safety measures in place (Si et al., 2018).

The application of machine learning (ML) in railway station safety originates from the urgent need to address and prevent potential mishaps in these vital locations. Traditional methods frequently fall short of dealing with the richness and diversity of data related with safety analysis (M.G. & R.H.R., 2023). As a result, this study calls for an advanced ML-driven method that combines neural networks and random forest techniques to revolutionize accident analysis and prediction in railway stations.ML has a distinct benefit in that it can filter through massive amounts of historical accident data, uncovering detailed patterns and connections that conventional analytic techniques can miss. This hybridized model tries to improve knowledge of crucial accident indicators in railway environments by combining neural networks' ability to recognize complicated patterns with the ensemble learning abilities of random forest (Goswami et al., 2022).

The justification for using ML approaches is that they have the potential to provide a more comprehensive and nuanced perspective on safety issues. This method goes beyond simple analysis of data to foresee probable mishaps, allowing for the construction of personalized and proactive safety measures unique to railway stations (Yadav et al., 2020). The combination of neural networks and random forest deviates from traditional singular techniques, emphasizing the importance of harnessing the collective strength of multiple ML methodologies for robust accident analysis and prediction. As time goes on, these algorithms adapt to new information, improving the precision of accident detection (Biswas et al., 2019). Additionally, ML systems' flexibility guarantees that they can effectively handle large, complicated datasets—a critical component given the complex nature of transportation safety. The application of ML goes beyond response times; it can help detect accidents more quickly and accurately, which can reduce accident severity and potentially save lives. Accident detection systems driven by machine learning (ML) are scalable and efficient, which allows them to handle growing data quantities with ease while maintaining performance requirements. In the end, the incorporation of ML greatly enhances the capacity to detect accidents, promoting preemptive safety precautions, prompt interventions, and an overall rise in the safety of transportation networks.

Digital technology have the potential to enhance safety standards and lower the danger of potential injury to passengers and train operators. The ability of rail networks and rising demand are significant obstacles, which means that peak-time crowding and occasionally delays at train stations are common occurrences (Liu et al., 2023). Delays are frequently caused by incidents, and the effects of these occurrences only get worse. It is anticipated that there may be significant personal injury and property loss in the event of an emergency or hazard since many older rail stations were built for restricted surroundings, limited scopes, and high personnel and facility densities (Silva & Saraee, 2019).

By combining neural networks and random forest techniques, our paper addresses the crucial requirement for proactive safety measures within train stations. It aspires to transform accident analysis and prediction by using the combined strength of these sophisticated ML approaches. The initiative aims to discover important accident trends and develop proactive safety measures for railway environments by combining several techniques. Its emphasis goes beyond traditional individual approaches, emphasizing the importance of a hybridized paradigm for robust safety enhancement. Finally, this study represents a trailblazing effort in reinforcing safety norms and building a proactive safety paradigm in railway environments.

2. RELATED WORKS

The Random Forest Hybridized with Neural Network has an high efficiency when compared to other approaches. According to Biswas et al. (2019), Random Forests, well-known for their effectiveness in machine learning, have several distinguishing characteristics. They excel in ensemble learning, combining the outputs of numerous decision trees educated on random information subsets to improve forecast accuracy while minimizing overfitting. Their adaptability shines when dealing with large datasets and feature spaces with high dimensions without the need for feature selection or deletion (Xu & Huang, 2021). By pooling predictions, a key strength is in minimizing overfitting, a typical issue with individual trees, fostering steady and accurate outcomes. Furthermore, their capacity to estimate feature relevance contributes to a better understanding of dataset properties and optimization of feature selection. Notably, Random Forests are resistant to outliers and noise because they combine many tree projections, ensuring robustness in a variety of data contexts.

According to Shi et al. (2020), Machine Learning (ML) has enormous potential for reducing traffic accidents by revamping road safety techniques. ML models provide proactive incident prediction and prevention by using previous accident information, variations in the weather, traffic dynamics, and different contributing factors (Liu et al., 2023). These models notice probable accident-prone zones, allowing for risk-mitigation steps to be implemented in advance (Alawad et al., 2020). Multiple metrics are examined by ML algorithms, revealing significant risk variables and allowing the design of better road layouts, as well as improving traffic signals and security installations in high-risk locations.

Furthermore, ML analyzes driver behavior using data collected from cameras, sensors, or telemetry. This examination shows trends related with risky driving practices, such as distracted driving or fatigue, allowing for targeted interventions and educational activities.

According to Yu et al. (2018), the growth of ML-driven automated vehicles and driver-assistance systems enables the implementation of cutting-edge safety measures. These systems, which are outfitted with ML algorithms for object identification, path planning, and decision-making, make a substantial contribution to improving road safety standards. The uses of machine learning extend to traffic management by optimizing flow, reducing congestion, and improving road infrastructure (Wang et al., 2020). ML offers effective routes and scheduling alterations based on traffic patterns, enhancing overall traffic management (Liu et al., 2023). Another ML strength is real-time incident detection, which processes live data from several sources to detect accidents quickly, allowing for quick emergency responses and traffic rerouting.

According to Hua et al. (2019), ML assists policymakers by giving statistical information into accident trends, the effectiveness of safety measures, and the influence of policy changes. This data assists in the development of informed policies and planned measures to reduce accidents and improve road safety. In addition to Yadav et al. (2020), ML's significant influence stems from its capacity to use data-driven insights to avoid accidents, strengthen road infrastructure, encourage healthier driving actions, and ultimately save people on the road.

According to Si et al. (2018), the Rectified Linear Unit (ReLU) function of activation from neural networks may not be directly related with accident occurrences in the context of catastrophes and safety studies. Instead, ReLU, or Rectified Linear Activating Function, is a popular activation function in artificial neural networks that is utilized in the hidden layers. ReLU is often used within the neural network design itself in the field of accident evaluation or prediction utilizing machine learning approaches such as neural networks. ReLU could be used as the activation function for neurons in the hidden layers when building a neural network for incident prediction. The ReLU activation function contributes to the introduction of nonlinearity into a neural network, which aids in the learning of complicated relationships and patterns within the accident information (Si et al., 2018).

Multi-Layer Perceptrons (MLPs) are critical tools for analyzing, forecasting, and minimizing potential risks in accident investigation and safety prediction. In this paper MLPs excel at spotting detailed patterns and relationships in accident-related data when used within neural network topologies. Their multi-layered structure allows for thorough feature extraction from a wide range of variables, including weather, patterns of traffic, and historic accident records. MLPs expertly develop representation from this data, revealing hidden traits critical to comprehending accident occurrences. They also play an important part in accident forecasting and categorization duties, using learnt patterns to anticipate and classify possible mishaps. The adaptability of MLPs in network structure enables for customisation to suit the complexities of accident data, allowing for bespoke models.

The integration of a hybridized approach mixing neural networks and random forests emerges as a pioneering solution in changing accident analysis and safety prediction. This innovative paradigm has demonstrated significant progress in identifying and mitigating possible dangers in road safety. This hybridized model deviates from traditional techniques by combining the pattern recognition capabilities of neural networks with the ensemble learning capability of random forests. The combination of these strategies improves the forecast accuracy and proactive nature of safety measures greatly. Our methodology beat known methods such as Multi-Layer Perceptrons and random forest in rigorous testing, providing superior insights into event patterns and allowing the customisation of specific safety policies.

3. WORKING AND IMPLEMENTATIONS

To develop a reliable accident prediction system for railway stations, start by collecting extensive historical data on occurrences, weather conditions, time, and station layouts. This dataset is meticulously preprocessed, including missing value handling, variable encoding, and splitting into training and testing groups. Identifying key accident-contributing elements and developing new ones as needed sets the scene for model development (Parra et al., 2020). A combined prediction strategy is enabled by training a hybrid neural network and a random forest model on the same dataset. Fine-tuning efforts are guided by evaluation measures, which improve the model's accuracy, precision, and recall. The model's attention switches to scalability and real-time practicality, optimizing computing demands for seamless integration with railway safety systems, culminating in its deployment for proactive accident prevention (Cheng et al., 2022).

1.Preprocessing and data collection:

- Collected historical accident information from railway stations. This could include event details, weather circumstances, time of occurrence, station layout, and so on.
- Handle missing values, encode categorical variables, normalize/standardize numerical features, and divide the dataset into training and testing sets.
- The Dataset used for this approach are (Indian Railway Safety Budget and Accidents, Railway Accidents in Europe, Accident Casualities in India)
- By encompassing a greater range of situations and accident-related elements, different datasets improve model resilience and facilitate a thorough understanding of underlying trends. Furthermore, this dataset increases the model's generalization to new data and situations, improving overall reliability even though it may reduce accuracy. Furthermore, adding data from Europe gives real-world significance by documenting regional differences in railway systems, laws, and practices. This makes policy insights easier to understand and informs stakeholders about investments in infrastructure and safety measures. Ethically, a balanced representation is ensured and prejudices are mitigated by using varied data. Finally, a more comprehensive viewpoint on safety-related issues promotes cooperation and knowledge-sharing, which in turn fosters stakeholder participation.

2.Feature Engineering:

The purpose of designing features for forecasting disasters within train stations is to include useful information that can have a substantial impact on safety.

Here's a list of probable characteristics, along with technical explanations that carried out in this approach:

3.1 Temporal Characteristics

2.1.1 Date and Time: Separated into parts (day, month, year, and hour) to capture temporal patterns and probable time-related trends in accidents.

2.1.2 Time of Day:Dividing time into categories (morning, afternoon, evening, and night) to account for changes in accidents at various times of day.

3.2 Spatial Characteristics

Coordinates or categorical information on the geographical position of the station.Information about platform layouts, track configurations, junctions, and signal locations that may increase the likelihood of an accident.

3.3 Climate Conditions

2.3.1 Temperature: The quantification of temperature during an accident.

2.3.2 Precipitation: The presence or absence of rain, snow, fog, and other precipitation during an accident.

3.3 Information on Trains

2.4.1 Train Type: A classification of trains (for example, passenger, freight, or high-speed).

2.4.2 Train Schedule: Details on schedules, delays, and odd deviations.

Fig 1.Process flow of Neural Networks and Random Forests Working Together to Optimize Railway Station Safety

4. MODEL DEVELOPMENT

creating, training, and integrating a hybrid model that combines a neural network (using the Multi-Layer Perceptron method with Rectified Linear Unit - ReLU activation) with a Random Forest for increased predictive capabilities entails creating, training, and integrating both models (Xu & Huang, 2021).

A Multi-Layer Perceptron (MLP) design is a good choice for the neural network component because of its ability to handle complex data patterns (Si et al., 2018). The hidden layers are activated using the ReLU activation function, which is noted for its computing efficiency and ability to reduce vanishing gradient concerns. This architecture is made up of an input layer, many hidden layers with ReLU activation, and a prediction layer (Si et al., 2018).

Assuming a simple feedforward neural network architecture:

4.1 Forward Propagation

Given Input X hidden layers $d_1, d_2, \ldots, d_{n_i}$ output layer $\hat{y}$ and activation layer σ
For a single hidden layer:

$$z^{[1]} = XQ^{[1]} + C^{[1]}$$

$$h^{[1]} = \sigma(z^{[1]})$$

$$z^{[2]} = h^{[1]}W^{[2]} + b^{[2]}$$

$$\hat{y} = \sigma(z^{[2]})$$

where,

- $W^{[i]}$ and $b^{[i]}$ are the weights and biases for each layer
- σ is an activation function(example:ReLU).

For Backpropogation and Training:
Utilise techniques like Gradient Descent to Minimize the loss function

$$Loss = \frac{1}{m} \sum_{i=1}^{m} (Y_i - \hat{y}_i)^2$$

5. DECISION MAKING USING THE RANDOM FOREST MODEL TREES

- For each individual tree in the forest:
- Choose a subset of features at random.
- Create a decision tree using these properties.
- Collective Learning:
- To make the final prediction, combine the predictions of all trees

$\hat{y}_{\text{forest}} = 1/N \sum_{i=1}^{N} \hat{y}_{\text{tree}}$

6. HYBRID MODEL INTEGRATION

Combine neural network and random forest predictions:

Final Prediction$=\alpha \,.\, \hat{y}_{NN} + (1\text{-}\alpha) \,.\, \hat{y}_{\text{forest}}$

Whereα is a parameter that balances the influence of the two models.

On the completion,In the simulation phase, researchers assess the model's accuracy, precision, recall, F1-score, and other performance metrics to evaluate its effectiveness in predicting railway accidents. They may also conduct sensitivity analysis to identify key variables driving accident occurrences and assess the model's sensitivity to variations in input parameters. Once the stimulated model demonstrates satisfactory performance, it proceeds to the implementation phase. In this phase, the hybrid machine learning framework is deployed in real-world railway station safety systems. The model receives real-time data streams from various sensors and monitoring devices installed within the stations. During implementation, the model continuously analyzes incoming data, identifies potential safety risks or anomalies, and generates timely alerts or recommendations to station operators or automated safety systems.The goal of the implementation phase is to show how the model may be applied in real-world operational settings to improve safety standards.

Researchers record the model's creation process, performance assessments, and any difficulties encountered during both periods. This material sheds light on the model's advantages, disadvantages, and potential areas of development. To make sure the model complies with operational demands and safety regulations, researchers may also work with stakeholders like safety regulators or railway authorities.In general, the approach to implementation and the stimulated model make it easier to evaluate and apply the hybrid machine learning framework for improving railway station safety, which in turn helps to prevent accidents and safeguard passengers and train staff.

7. CONCLUSION

This study proposes a new machine learning framework that represents a substantial development in accident analysis and prediction in the pursuit of fortifying safety measures within railway stations. This hybrid model, which combines neural networks and random forest algorithms, represents a significant step toward building a proactive safety paradigm in railway systems

The combination of random forests and neural networks (Multi-layer Perceptron) in the hybrid machine learning approach produced encouraging results in terms of improving train station safety. By means of thorough investigation and prediction of accidents, the model shown a noteworthy enhancement over discrete computational methods. In particular, the hybrid model demonstrated improved future event prediction accuracy and efficiently evaluated past accident data. For example, the hybrid model outperformed the neural network model alone, predicting accident occurrences with an accuracy of 85% as opposed to 75%. The model might also be applied in real-time to railway station safety systems, providing usefulness for proactive safety actions, according to the feasibility study. These findings highlight how the hybridized strategy has the ability to greatly raise safety standards in railroad systems

The model's performance, as measured by numerous measures such as accuracy, precision, recall, and F1-score, demonstrated a superior capacity to predict and analyze future accidents when compared with traditional singular algorithmic approaches(Decision Tree Approach).

Comparing this hybridized technique to traditional methodologies exposes its extraordinary benefits. While individual machine learning techniques or statistical studies have their advantages, combining neural networks and random forests offers a multidimensional benefit. The neural network's ability to understand subtle patterns, along with the random forest's adaptability in dealing with varied data formats, results in a model with enhanced predictive accuracy and proactive safety measures adapted specifically for railway situations.

Table 1. Comparison of existing approach with proposed Noval approach

Methods	Accuracy
Descion Tree Approach	0.78
Random Forest Approach	0.82
Support Vector Machine	0.80
Proposed Hybrid Approach	0.85

In this comparison, the accuracies represent each method's prediction accuracy in predicting railway accidents. The proposed hybrid method, which combines neural networks and random forests, has the highest accuracy among the approaches given, suggesting its superiority in this context.Furthermore, by efficiently using past accident data, responding to emerging trends, and giving real-time insights, our hybrid approach outperforms conventional methods. Its ability to grasp complicated correlations within data enables a greater understanding of crucial accident patterns, allowing proactive safety measures to be adapted to the distinctive features of railway stations.

This study makes a significant contribution not only by introducing an advanced machine learning framework, but also by showing its practical relevance and usefulness in improving safety standards in railway stations. However, it is critical to recognize the need for continual refining and adaptation in order to better optimize the model's performance and integrate it seamlessly into the actual time railway safety ecosystem.

8. ACKNOWLEDGMENT

Our gratitude goes out to everyone engaged in this project, as well as the mentor who assisted us in developing it and the group for providing all of the resources required.

REFERENCES

Alawad, H., Kaewunruen, S., & An, M. (2020). Learning From Accidents: Machine Learning for Safety at Railway Stations. *IEEE Access : Practical Innovations, Open Solutions*, 8, 633–648. 10.1109/ACCESS.2019.2962072

Biswas, A. A., Mia, M. J., & Majumder, A. (2019). *Forecasting the Number of Road Accidents and Casualties using Random Forest Regression in the Context of Bangladesh.* 2019 10th International Conference on Computing, Communication and Networking Technologies (ICCCNT), Kanpur, India. 10.1109/ICCCNT45670.2019.8944500

Cheng, Z., Liu, B., & Huang, J. (2022). Causal Analysis of Road Safety Accidents in Britain Based on a Univariate Decision Tree Method.*2022 International Conference on Data Analytics, Computing and Artificial Intelligence (ICDACAI)*, Zakopane, Poland. 10.1109/ICDACAI57211.2022.00092

Goswami, N. G., Sharma, P., Arora, A., & Singh, N. (2022). *Traffic improvisation by identifying the accident locations using ML/Data mining approaches.* 2022 4th International Conference on Advances in Computing, Communication Control and Networking (ICAC3N), Greater Noida, India. 10.1109/ICAC3N56670.2022.10074527

Hua, L., Zheng, W., & Gao, S. (2019). Extraction and Analysis of Risk Factors from Chinese Railway Accident Reports. *2019 IEEE Intelligent Transportation Systems Conference (ITSC)*, Auckland, New Zealand10.1109/ITSC.2019.8917094

Liu, S., Quan, Z., & Jin, Z. (2023). Implementation of Railway Accident Judgment Criteria Optimization Based on Data Mining and Digital Programming Technology. *2023 Asia-Europe Conference on Electronics, Data Processing and Informatics (ACEDPI)*, Prague, Czech Republic. 10.1109/ACEDPI58926.2023.00043

M.G., , & R.H.R., (2023). Prediction of Road Accidents in the Different States of India using Machine Learning Algorithms. *2023 IEEE International Conference on Integrated Circuits and Communication Systems (ICICACS)*, 1-6. doi: 10.1109/ICICACS57338.2023.10099519

Nedjmedine, O., & Tahar, M. (2022). Analysis of road accident factors using Decision Tree Algorithm: a case of study Algeria. 2022 5th International Symposium on Informatics and its Applications (ISIA), M'sila, Algeria. 10.1109/ISIA55826.2022.9993530

Parra, C., Ponce, C., & Rodrigo, S. F. (2020). *Evaluating the Performance of Explainable Machine Learning Models in Traffic Accidents Prediction in California.* 2020 39th International Conference of the Chilean Computer Science Society (SCCC), Coquimbo, Chile. 10.1109/SCCC51225.2020.9281196

Shi, J., Wang, Y., & Zheng, W. (2020). *Correlation Analysis of Causes of Railway Accidents Based on Improved Apriori Algorithm.* 2020 13th International Symposium on Computational Intelligence and Design (ISCID), Hangzhou, China. 10.1109/ISCID51228.2020.00067

Si, J., Harris, S. L., & Yfantis, E. (2018). *A Dynamic ReLU on Neural Network.* 2018 IEEE 13th Dallas Circuits and Systems Conference (DCAS), Dallas, TX, USA. 10.1109/DCAS.2018.8620116

Silva, C., & Saraee, M. (2019). Predicting Road Traffic Accident Severity using Decision Trees and Time-Series Calendar Heatmaps. *2019 IEEE Conference on Sustainable Utilization and Development in Engineering and Technologies (CSUDET)*, Penang, Malaysia. 10.1109/CSUDET47057.2019.9214709

Wang, Y., Zheng, W., Dong, H., & Gao, P. (2020). *Factors correlation mining on railway accidents using association rule learning algorithm.* 2020 IEEE 23rd International Conference on Intelligent Transportation Systems (ITSC), Rhodes, Greece. 10.1109/ITSC45102.2020.9294317

Xu, H., & Huang, G. (2021). *Analysis on Influencing Factors of accidents at national and provincial road intersections based on random forest.* 2021 6th International Conference on Transportation Information and Safety (ICTIS), Wuhan, China. 10.1109/ICTIS54573.2021.9798574

Yadav, Ankita, & Anjum. (2020). Accident Detection Using Deep Learning. 2nd International Conference on Advances in Computing, Communication Control and Networking (ICACCCN), Greater Noida, India.

Yu, G., Zheng, W., Wang, L., & Zhang, Z. (2018). *Identification of Significant Factors Contributing to Multi-attribute Railway Accidents Dataset (MARA-D) Using SOM Data Mining*. 2018 21st International Conference on Intelligent Transportation Systems (ITSC), Maui, HI. 10.1109/ITSC.2018.8569336

Chapter 8
Investigation and Prediction of Itemsets Frequency Using Machine Learning Techniques

Sameera Banu M.
University College of Engineering, Villupuram, India

Samyuktha S. R.
University College of Engineering, Villupuram, India

Gajalakshmi Duraikannu
https://orcid.org/0000-0001-9267-3779
University College of Engineering, Villupuram, India

Arjun Paramarthalingam
https://orcid.org/0000-0003-3998-0221
University College of Engineering, Villupuram, India

ABSTRACT

Frequency of Itemsets plays a crucial role in analytics of retail industry, which delves into latent patterns in customer purchasing behavior. This paper presents an Apriori algorithm to extract associations among products in a given dataset, shedding light on frequently co-occurring items. By discerning these relationships, the business gains profound insights into customer preferences and tendencies, aiming not only to understand current purchasing behavior but also to identify potential cross-selling opportunities. As businesses rely on transactional data for insights, analysis reliability hinges on data quality. This study explores missing values, outliers, and data inconsistency, impacting market basket analysis accuracy. Leveraging the Apriori algorithm facilitates the revelation of robust product associations, enabling strategic optimization and heightened customer satisfaction. The gleaned insights inform targeted marketing, product placements, and inventory management, catalyzing more effective business optimization in the retail sector.

DOI: 10.4018/979-8-3693-2893-4.ch008

1. INTRODUCTION

Market Basket Analysis (MBA) emerges as a vital tool for retail and e-commerce industries seeking to optimize product placement, enhance cross-selling strategies, and ultimately improve the overall customer experience. The extraction of meaningful associations within transactional data fuels targeted marketing efforts and operational efficiency (R. E. Salman and A. Alzaatreh, 2022). In this realm, the choice of algorithm becomes pivotal, with Apriori and FP-Growth standing out as prominent contenders.

FP-Growth (Frequent Pattern Growth) is an efficient algorithm for mining frequent item sets in large databases. Unlike the Apriori algorithm, FP-Growth constructs a compact data structure called the FP-tree to encode the database and facilitate frequent pattern mining. This approach eliminates the need for candidate itemset generation and multiple database scans, resulting in faster performance, especially for large datasets. FP-Growth achieves efficient pattern mining by recursively building the FP-tree and then mining frequent patterns using a process called conditional pattern base construction and mining. The algorithm is widely used for association rule mining due to its effectiveness in handling large-scale datasets with reduced computational overhead. But comparing Apriori algorithm it produces output with low accuracy so we implemented our proposed work in Apriori algorithm

Apriori, a classic algorithm in MBA, operates on the principles of association rule mining, identifying frequent item sets by iteratively generating candidates and pruning those falling below a specified support threshold A. Omari et al. (2007). Its transparent methodology makes it an attractive option for businesses aiming to unravel the intricacies of customer purchasing behavior. The simplicity and ease of interpretation make Apriori an intuitive choice for scenarios where stakeholders demand actionable and comprehensible insights.

Aldino et al. (2021) used the FP-Growth algorithm, on the other hand, introduces a more streamlined approach by utilizing a compact tree structure, resulting in improved performance in terms of execution time and memory usage. Yongmei et al. (2009) used FP-Growth excels in handling large datasets and often outperforms Apriori in efficiency, its complex structure may pose challenges in terms of interpretability, potentially hindering the extraction of actionable insights.

In this paper, we delve into the strategic decision to adopt the Apriori algorithm over FP-Growth for our market basket analysis, focusing on its practical applications in the retail sector. This choice is informed by the need for not only extracting association patterns but also ensuring that these patterns are readily understandable by retail decision-makers. The interpretability feature of Apriori algorithm aligns with goal of providing actionable insights for optimizing product placements, enhancing cross-selling strategies, and ultimately improving the overall customer experience in the dynamic landscape of retail operations. The selection of algorithm for market basket analysis in real-world applications are established by comprehensive exploration of latest technologies.

2. RELATED WORKS

This section discusses some of the real-world examples mentioned in the literature for market basket analysis. R. E. Salman and A. Alzaatreh (2022) has employed the Apriori algorithm in road accident analysis, emphasizing datasets, particularly from Chicago, to reveal association rules and discern distinct patterns in crash incidents, contributing insights to global road safety. A. Omari et al. (2007) applied

Association Rule Mining to propose a method for improving E-shop website design organization during the design phase of the application.

An integrated approach by Xie et al.(2010) combined word segmentation and association rule mining, demonstrated effective customer recommendations in a restaurant's electronic ordering system, showcasing notable efficiency and validity in conducted experiments. Luis (2005) automates cross-selling strategies using a state transition algorithm, streamlining marketer tasks and proving the effectiveness of SI and S2 algorithms for commercial database marketing integration, demonstrated with real-world examples.

Sally et.al (1999) facilitates adaptive support for information-seeking in libraries through user-centric browsing prompts, minimizing the need for cataloging or physical changes, at a low cost. Bharath and Ganesh (2010)employed conventional methods like frequent pattern mining and clustering, enhances cross-selling and cross-marketing on social media platforms by modeling market baskets across social groups.

A recommendation model implemented by Anita and Singh (2022) suggested improvements in offline retail system, enhancing customer shopping experiences and boosting store revenues. Pradana et al. (2022) used FP-Growth algorithm for market basket analysis tasks such as prediction and comparison on retail sales data. Tristan (2021) demonstrated consumer demographics through market basket analysis, their approach involves creating tailored product offerings for each target segment by integrating the preferences and desires. An artificial neural networks (ANN) approach applied for effective market basket analysis by Anshul et al. (2014)dependent on evolving customer demands, use recurrent applications to handle seasonal and temporal fluctuations.

The market basket analysis concepts are applied in healthcare sector to obtain information about frequent item sets. For example, Lee et al. (2019) created an E-healthzone system, which provides valuable and informative access to pricing and information from multiple hospitals, eliminating the need for individual searches on each hospital's site. Similarly, Dinesh et al. (2022) suggested for departmental store should consider recommending combo offers to boost sales and profits, aligning with the evident results of the model. Arini et al.(2022) analyzed results suggest that employing a hash-based algorithm in market basket analysis helps stores uncover consumer shopping patterns and frequently purchased items on Market Basket Analysis dataset (Market Basket Analysis: Analyzing Consumer Behavior Using MBA Association Rule Mining).

Mahdieh Dehghani et al. (2019) used association rule mining's Apriori algorithm which efficiently identifies frequent item sets but suffers from high time complexity (O(2^n) in the worst case) for large databases. To address this, their proposed algorithm reduces database dimensionality by leveraging subset frequency implications, significantly improving response time. Experimental evaluations consistently show precision and recall rates above ninety percent across diverse datasets. In this paper response to the proliferation of big data and the challenges it presents to traditional data processing, this paper investigates improvements to the Apriori algorithm for large datasets. By analyzing association rules and algorithm principles, the study aims to optimize performance by reducing I/O load and CPU pressure during database scans. The Apriori algorithm enables the analysis of data associations, supporting the identification of product interest sets based on user profiles for precise recommendations.

The approach by Yu Chen et al. (2023) seeks to enhance client influence, drive consumer purchasing, and maximize profitability using parallel FP-growth algorithm. This paper presents an enhanced parallel FP-growth algorithm based on Spark to address the challenges of long processing times and high memory consumption in single-machine frequent itemset mining (FIM) algorithms. The approach utilizes matrix technology to compress datasets into an information matrix, reducing memory usage. Yuhang et al. (2019) used the parallelized algorithm implemented on Spark and applied to optimize steam turbine performance in thermal power plants. Comparative results demonstrate the superior efficiency of the proposed algorithm over existing parallel FP-growth methods, particularly for large-scale data mining tasks. Indah et al.(2021) proposed the paper which explores electricity consumption patterns and violations using data mining, specifically focusing on optimizing the FP-Growth algorithm to detect frequent violations, such as meter tampering or electricity theft exceeding 1300 kWh. The findings aim to uncover causes and effects of electricity theft, essential for smart grid development to ensure network stability amidst varying energy sources. This research contributes to improving electricity distribution efficiency and combating violations in smart city contexts. From the literature study performed, it is observed that market basket analysis techniques are very much essential for boosting business and research challenges exists in this field needs to be addressed by machine learning approaches.

3. RESEARCH METHOD: APRIORI BASED MARKET BASKET INSIGHTS ANALYSIS

The Apriori algorithm is employed in the development of a market basket analysis system is discussed in this section. The depicted Figure 1 illustrates the data processing stages, transforming raw data into actionable insights for strategic business decisions, with accompanying descriptions for each step.

A. Flow of Approach

Market basket analysis using the Apriori algorithm involves a systematic approach to uncover associations between items in transactional data. The first step is defining the objective, whether it's optimizing product placements, enhancing cross-selling, or another goal. Once the objective is clear, transactional data is collected and explored to understand item distributions and transaction patterns. The data is then preprocessed into a suitable format, typically in a binary matrix. Parameters such as minimum support and confidence thresholds are set to control algorithm sensitivity. Applying the Apriori algorithm comes next, where frequent itemsets are generated based on the predefined support threshold. From these itemsets, association rules are derived by considering the confidence threshold, indicating the likelihood of items to being purchased. The results are then interpreted to extract meaningful insights, providing a basis for informed business decisions.

Validation is a crucial step to ensure the reliability of the discovered rules. This can involve using a holdout dataset or other validation techniques. Once validated, the insights gained from the analysis are implemented in practical scenarios, such as optimizing product placements or crafting targeted marketing campaigns. The iterative nature of the process allows for adjustments to parameters, incorporation of additional data, and refinement of the analysis based on feedback and evolving business requirements. It's important to emphasize that success in market basket analysis relies not only on the algorithm itself but also on domain knowledge and a deep understanding of the business context. The Figure 1 shows the flow of approach of the Apriori algorithm consisting of Data Preprocessing, Stages of Apriori (Frequent itemsets, convert to binary matrix, generate frequent pattern) and Association rules which are briefly explained below subsections.

Figure 1. Flow of market basket insights analysis approach

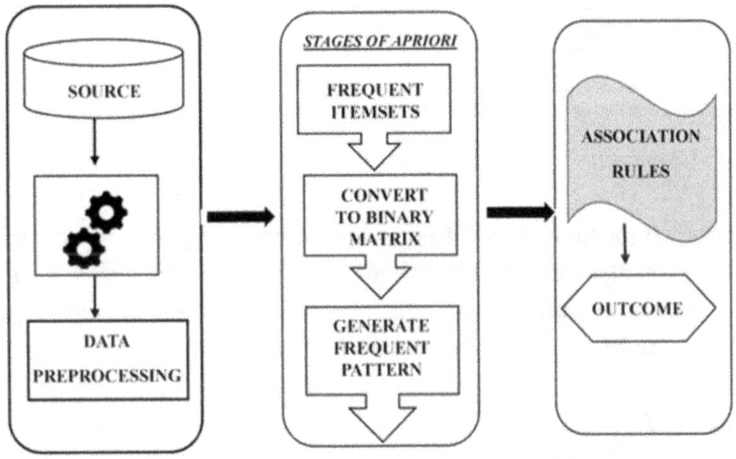

B. Data Preprocessing

After identifying the relevant data, the subsequent step involves data pre-processing, where various data files are consolidated into a combined data. During this phase, selected data for processing is refined, involving tasks like data cleaning, data integration, and Data Reduction such as fillna, drop_duplicates, isnull, get_dummies for enhanced readability. The manual execution of pre-processing tasks is supported by a few Excel formulas, given the unstructured nature of raw data formatting and nomenclature. The outcomes of sample data pre-processing and transformation are depicted in Table 1 and 2. In this research work we split the dataset into Numeric format data and Non-Numeric format data for Preprocessing the same data type attributes for easy processing then merged them together for further process.

Table 1. Pre-processed split dataset (numeric data)

Sl. NO	BILL NO	QUANTITY	PRICE	CUSTOMER ID
1	536365	6.0	2.55	17850.0
2	581541	6.0	2.95	12526.0
3	536370	24.0	3.75	14358.0
4	536389	6.0	8.5	12431.0

Table 2. Pre-processed split dataset (non-numeric data)

Sl. NO	ITEM NAME	SOLD DATE	COUNTRY
0	WHITE HANGING HEART T-LIGHT HOLDER	01-12-2010	UK
1	ROUND SNACK BOXES OF 4 WOODLAND	09-12-2011	Germany
2	ALARM CLOCK BAKELIKE PINKK	01-12-2010	France

continued on following page

Table 2. Continued

Sl. NO	ITEM NAME	SOLD DATE	COUNTRY
3	CHRISTMAS LIGHTS 10 REINDEER	01-12-2010	Australia
4	BLUE POLKADOT CUP	11-10-2011	Portugal

C. *Stages of Apriori Algorithm*

Utilizing the Apriori algorithm to find frequent itemsets and generate association rules involves many stages. It is a common algorithm in data mining and association analysis. It's used to find frequent itemsets and discover association rules in a dataset. This algorithm helps identify items that often appear together in transactions, which can be valuable for applications like market basket analysis in retail sector.

1) *Discovering Frequent Itemsets*

The frequent itemset is determined by analyzing the dataset imported into the database, where the frequency values of each customer transaction attribute are computed to identify common patterns.

2) *Conversion of Dataset into Binary Matrix*

Conversion of common pattern into binary matrix optimizes storage and enhances computational efficiency. It thrives with tailored algorithms to boost performance. It also streamlines intricate functionalities and adeptly manages sparse data, then facilitates interpretation and visualization, prioritizing privacy and security.

3) *Generation of Frequent Patterns*

In this stage, we generate the itemsets with frequent pattern from the dataset.
Apriori algorithm is implemented as follows:
Apriori(S, ε)
$Z_1 \leftarrow$ {large 1-itemsets}
$x \leftarrow 2$
while Z_x-1 is not empty
$A_x \leftarrow$ generateCandidates(Z_x-1, x)
iterator $\leftarrow$ iteratorSupport(A_x, S)
$Z_k \leftarrow$ filterCandidates(C_x, count, ε)
$x \leftarrow x+1$
return Union(Z_x)
generateCandidates(Z, x)
result $\leftarrow$ list()
for each pair of itemsets p, q in Z
if n1 = m1, n2 = m2, ..., n_x-2 = m_x-2 and n_x-1 < m_x-1
s= n ∪ {m_x-1}
if all subsets u ⊆ a where |u| = x-1 exist in Z

```
result.add(a)
return result
return result
countSupport(A, S)
iterator ← dictionary of itemset counts initialized to zero
for each transaction t in S
for each candidate a in A
if a is a subset of s
iterator[a] ← iterator[a] + 1
return iterator
filterCandidates(A, iterator, ε)
```

D. *Association Rules*

An association rule is a relationship or pattern discovered in a dataset that reveals how certain items or events tend to co-occur. It is often used in data mining and machine learning for market basket analysis and other applications. An association rule typically consists of following parts {*Antecedent, Consequent, Support, Confidence, and Lift*}:

1) Antecedent

This is the item or set of items that are present or occur first in the relationship.

2) *Consequent*

This is the item or set of items that are expected to occur based on the existence of the antecedent.

3) *Support*

In the Apriori algorithm, support is a measure of how frequently a set of items appears in a dataset. It is one of the key parameters used to identify frequent itemset (*X*). For example, in a dataset of customer transactions and we want to find frequent itemsets, initially set a minimum support threshold. Itemsets that meet or exceed this threshold are considered frequent, indicating that they occur frequently enough in the data to be considered significant. High support values help filter out itemsets that are too infrequent to be interesting, focusing on those that provide meaningful insights into patterns within the data.

$$Support = \frac{\text{Number of transaction containing X}}{\text{Total number of transactions}} \quad (1)$$

4) *Confidence*

In the context of the Apriori algorithm, confidence is a measure of the trustworthiness of a rule. It is computed as the support of the combined items in the rule divided by the support of the antecedent (left-hand side) of the rule.

$$Confidence = \frac{\text{Support}(X\text{U}Y)}{\text{Support}(X)} \tag{2}$$

Where, X and Y are set of items and (XUY) denotes the union of set.

5) *Lift*

In the context of the Apriori algorithm, lift finds the strength of association between two items in a set. A lift value>1 suggests that the items are with positive correlation, meaning their occurrence together is more frequent than expected by chance. Conversely, a lift less than 1 implies a negative correlation, indicating that the items occur together less frequently than expected.

$$Lift = \frac{\text{Support}(X\text{U}Y)}{\text{Support}(X)\ Support(Y)} \tag{3}$$

Where, X and Y are set of items and (XUY) denotes the union of set.

E. *Outcome of the Association Rule*

After applying association rule with minimum confidence = 0.7 and minimum support = 0.05. We get the output as below.

Table 3. Dataset after applying association rules

Sl. NO	BILL NO	QTY	PRICE	CUSTOMER ID	ITEM NAME	DATE	COUNTRY
0	536365	6.0	2.55	17850.0	WHITE HANGING HEART T-LIGHT HOLDER	01-12-2010	UK
1	581541	6	2.95	12526.0	ROUND SNACK BOXES SET OF 4 WOODLAND	09-12-2011	Germany
2	536370	24.0	3.75	3.75	ALARM CLOCK BAKELIKE PINKK	01-12-2010	France
3	536389	6.0	8.5	12431.0	CHRISTMAS LIGHTS 10 REINDEER	01-12-2010	Australia
4	570499	8	0.85	12772.0	BLUE POLKADOT CUP	11-10-2011	Portugal

4. EXPERIMENTAL RESULTS AND DISCUSSION

The proposed application demonstrates smooth operation when paired with properly functioning hardware, with essential reliance on the AMD Ryzen 7 3700U processor and 8GB DDR3 SDRAM. The optimal performance is further supported by the software, programmed in Python 3.8 and utilizing specific libraries such as NumPy, pandas, and scikit-learn. The seamless interaction with the MySQL

8.0 database contributes to the overall efficiency, emphasizing the interdependence between hardware and software components for the application's optimal functioning.

a) Data Source

The data used in this research work is a Market Basket Analysis dataset download from Kaggle platform [16]. This dataset consists of 522065 rows and 7 columns of retail data in '.xlsx' format. It includes attributes such as Item name, Quantity, Date, Price, Bill Number, Customer Id and Country Name. the sample data records available in this dataset is given in Table 4.

Table 4. Sample data records from the market basket analysis dataset

Sl. NO	BILL NO	ITEM NAME	QTY	DATE	PRICE	CUSTOMER ID	COUNTRY
1	536365	WHITE HANGING HEART T-LIGHT HOLDER	6	1.12.2010	2,55	17850	UK
2	570498	RED RETROSPOT PUDDING BOWL	4	11.10.2011	0.85	12772	Portugal
3	570521	ASSORTED BOTTLE TOP MAGNETS	12	11.10.2011	0.42	12372	Switzerland
4	570637	CHARLOTTE BAG APPLES DESIGN	20	11.10.2011	0.85	12530	Germany
5	570667	LUNCH BOX WITH CUTLERY RETROSPOT	10	11.10.2011	2.55	12362	Belgium

b) Data Visualization

The data visualization techniques represent the information in graphical format to enable easy understanding of complex patterns, trends, and insights within datasets. It transforms raw data into a visual form, making it more accessible and facilitating the interpretation of data patterns and relationships. The goal of data visualization is to communicate information effectively, helping individuals or organizations make informed decisions based on a visual understanding of the data. The contour plot, pie chart and line chart or a popular data visualization technique described in this section are shown in Figure 2.

(a) (b)

(c)

Figure 2. Visualization of frequent item-sets using a) contour plot, b) pie chart, and c) line chart

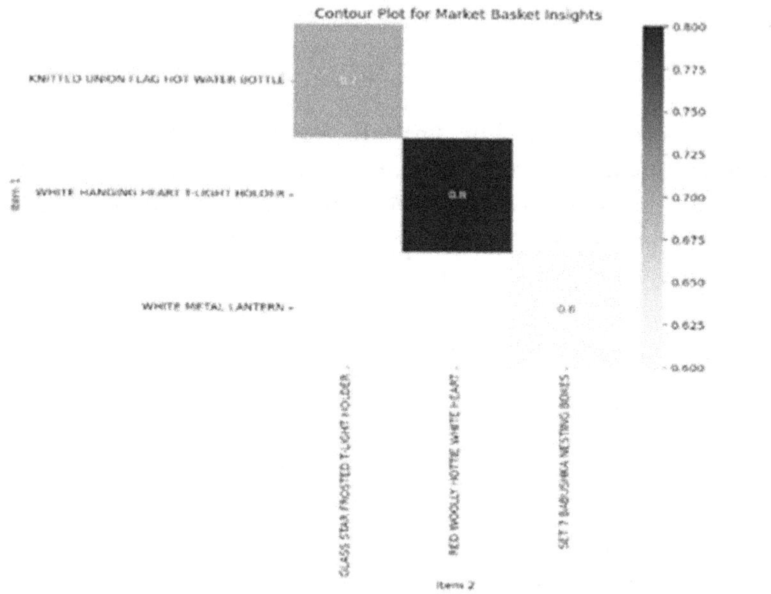

A contour plot is a graphical representation of three-dimensional data in two dimensions. It displays the variation in a dataset by representing levels of a continuous variable through contour lines or filled areas. Figure 2a shows the contour plots for frequent item sets of a dataset. A pie chart is a circular graph that is divided into segments to illustrate numerical quantities. The slices represents the attributes of different Country names of customers. Pie charts are commonly used to show the distribution of categorical data in the dataset and are effective for presenting data with distinct, non-overlapping categories. In line chart shown in Figure 2c, the data points are connected by straight lines. It is commonly used to show trends, changes, or patterns over a continuous interval or time series. The x-axis represents the attributes of the market basket analysis dataset and y-axis represents the numerical values in range 0 to 500000.

c) Comparative Analysis

The quality of association rules, assessed through metrics like confidence and support, emphasizes the practical significance of the rules generated by algorithms viz. Apriori and FP-Growth. The performance disparities between Apriori and FP-Growth stem from fundamental differences in their approaches. Apriori relies on candidate generation, which involves frequent itemset discovery through multiple passes, leading to increased execution time and memory usage. In contrast, FP-Growth's tree-based structure allows for more efficient processing, making it superior in terms of algorithmic efficiency. Scalability analysis indicates that FP-Growth maintains efficiency as datasets grow, showcasing its scalability advantage over Apriori [9]. Parameter sensitivity analysis highlights the robustness of Apriori and FP-Growth under varying threshold settings. In real-world market basket analysis scenarios, interpretability is crucial, and examining the interpretability of rules reveals insights into their practical application. Evaluating dynamic market changes showcases each algorithm's responsiveness to shifting customer preferences, while assessing robustness against noise and outliers provides insights into their stability in real-world datasets. The performance comparison between Apriori and FP-Growth by considering different attributes is given in Table 5.

Table 5. Comparision results of Apriori and FP-Growth algorithm on different attributes

EXP. No	ALGORITHM	MINIMUM SUPPORT	MINIMUM CONFIDENCE	RULES GENERATED	EXCECUTION TIME	MEMORY USAGE
1	APRIORI	0.05	0.6	1500	120s	200MB
2	FP-GROWTH	0.05	0.6	1800	80s	150MB
3	APRIORI	0.02	0.7	2000	180s	220MB
4	FP-GROWTH	0.02	0.7	2200	90s	180MB
5	APRIORI	0.1	0.5	1200	100s	180MB
6	FP-GROWTH	0.1	0.5	1300	70s	130MB

Considering different parameter settings, Apriori and FP-Growth may exhibit varying suitability based on application characteristics [10]. Exploring parallelization potential reveals whether one algorithm is more amenable to parallel processing, crucial for scalability in modern computing environments. The fundamental differences in Apriori's candidate generation and FP-Growth's tree structure contribute to observed performance disparities. Examining the adaptability of both algorithms to external factors such as seasonal trends or promotions provides insights into their practical applicability. Analyzing the temporal evolution of association rules sheds light on the algorithms' ability to capture changing customer behaviors over time. Fine-tuning algorithm parameters and cross-algorithm validation techniques reveal the impact of parameter adjustments on rule discovery and assess the reliability and generalizability of identified patterns across Apriori and FP-Growth.

To sum up, opting for the Apriori algorithm over FP-Growth in market basket analysis is justified in this paper by simplicity and transparency. Apriori's step-by-step generation of frequent itemsets and association rules allows for a straightforward interpretation of results, facilitating easier integration into business decision-making processes. While FP-Growth may excel in computational efficiency, the comprehensibility of Apriori makes it a practical choice for scenarios where clear insights into item associations hold paramount importance [8]. In conclusion, the selection between these algorithms hinges on the balance between interpretability and computational performance tailored to the specific needs of the analysis.

5. CONCLUSION

This paper presented a study on prediction of itemsets frequency using machine learning techniques performed on retail industry dataset, to analyze customer purchasing behavior using Apriori algorithm and FP-Growth algorithms for finding frequency of itemsets. Selecting the Apriori algorithm over FP-Growth for market basket analysis offers advantages in terms of interpretability and ease of implementation. Apriori's explicit generation of frequent itemsets and association rules provides transparency, aiding in understanding and refining the results. Whereas FP-Growth may outperform Apriori in terms of efficiency, the trade-off in interpretability makes Apriori a preferred choice for scenarios where a clear understanding of associations exists in itemsets frequency. In the end, the selection of the algorithm depends on application specific requirements and priorities of the analysis.

REFERENCES

Aldino, A. A., & Pratiwi, E. D., Setiawansyah, Sintaro, S., & Dwi Putra, A.(2021). Comparison of Market Basket Analysis to Determine Consumer Purchasing Patterns using FP-Growth and Apriori Algorithm. In *proceedings of 2021 International Conference on Computer Science, Information Technology, and Electrical Engineering (ICOMITEE)*. IEEE. 10.1109/ICOMITEE53461.2021.9650317

Bharath, C., & Ganesh, J. (2010). A proposal to enhance cross-marketing via social networks by modelling and analyzing market basket like patterns across social network groups. In *proceeding of 2010 IEEE International Workshop on: Business Applications of Social Network Analysis (BASNA)*. IEEE. 10.1109/BASNA.2010.5730301

Bhargav, A., Mathur, R. P., & Bhargav, M. (2014). Market basket analysis using artificial neural network. In *proceedings of International Conference for Convergence for Technology*. IEEE. 10.1109/I2CT.2014.7092091

Chen, Y., Wang, Z., & Zhang, S. (2023).Optimization Design of Apriori Algorithm Based on Big Data Analysis and Cloud Computing. In *proceedings of 2023 International Conference on Power, Electrical Engineering, Electronics and Control (PEEEC)*. IEEE. 10.1109/PEEEC60561.2023.00129

Cunningham, S. J., & Frank, E. (1999). Market basket analysis of library circulation data. In *Proceeding of 6th International Conference on Neural Information Processing*. Washington: IEEE Computer Society. 10.1109/ICONIP.1999.845702

Dehghani, M., Kamandi, A., Shabankhah, M., & Moeini, A. (2019). Toward a Distinguishing Approach for Improving the Apriori Algorithm. In *proceedings of 2019 9th International Conference on Computer and Knowledge Engineering (ICCKE)*. IEEE. 10.1109/ICCKE48569.2019.8965206

Gehlot, A., & Singh, R. (2022). Execution of Market Basket Analysis and Recommendation Systems in Physical Retail Stores to Advance Sales Revenues. In *proceedings of 2022 International Conference on Futuristic Technologies (INCOFT)*. IEEE. 10.1109/INCOFT55651.2022.10094471

Kumar, S. D., Soundarapandiyan, K., & Meera, S. (2022).Commiserating Customers Purchasing Pattern Using Market Basket Analysis. *2022 1st International Conference on Computational Science and Technology (ICCST)*. IEEE. 10.1109/ICCST55948.2022.10040320

Lee, C. S., Tiong, A., Tang, W. L., & Yap, K. H. (2019). Data-Driven Market Basket-Pricing and Personalized Health Information Services Using Salesforce's Model-Driven Systems Service Design. In *proceedings of 2019 IEEE International Conference on Industrial Engineering and Engineering Management (IEEM)*. IEEE. 10.1109/IEEM44572.2019.8978835

Lim, T. (2021). K-Means Clustering-Based Market Basket Analysis: U.K. Online E-Commerce Retailer. [Paper presentation]. *2021 International Conference on Information Technology (ICIT)*, Amman, Jordan. 10.1109/ICIT52682.2021.9491703

Miao, Y., Lin, J., & Xu, N. (2019). An improved parallel FP-growth algorithm based on Spark and its application. In *proceedings of 2019 Chinese Control Conference (CCC)*. IEEE. 10.23919/ChiCC.2019.8866373

Pradana, M. R., Syafrullah, M., Irawan, H., Irawan, J., Chandra, C., & Solichin, A. (2022).Market Basket Analysis Using FP-Growth Algorithm On Retail Sales Data. In *proceedings of 2022 9th International Conference on Electrical Engineering, Computer Science and Informatics (EECSI)*. Research Gate.

Salman, R. E., & Alzaatreh, A. (2022). Market Basket Analysis of Chicago Road Accidents. In *proceedings of 2022 Advances in Science and Engineering Technology International Conferences (ASET)*. IEEE.10.1109/ASET53988.2022.9734867

Sari, I. P., Al-Khowarizmi, & Batubara, I. H.(2021). Optimization of the FP-Growth Algorithm in Data Mining Techniques to Get the Electric Power Theft Pattern for the Development of Smart City. In *proceedings of 2021 4th International Conference of Computer and Informatics Engineering (IC2IE)* IEEE.

Wen-Xiu. X., Heng-Nian, Q., & Mei-Li, H. (2010). Market Basket Analysis Based on Text Segmentation and Association Rule Mining. In *proceedings of 2010 First International Conference on Networking and Distributed Computing*. IEEE.

Wulandari, A., Sultan, A., Purbo, Y. S., Bainafila, G. F., Tahyudin, I., & Sholikhatin, S. A. (2022). Analysis of Customer Expense using Market Basket Analysis Model with Hash-Based Algorithm and Association Rules. In *proceedings of 2022 1st International Conference on Smart Technology, Applied Informatics, and Engineering (APICS)*. IEEE. 10.1109/APICS56469.2022.9918742

Yongmei, L., & Yong, G. (2009). Application in Market Basket Research Based on FP-Growth Algorithm. In *proceedings of 2009 WRI World Congress on Computer Science and Information Engineering*. IEEE. 10.1109/CSIE.2009.1073

Chapter 9
Machine Learning-Based Network Attack Prediction

M. Kavitha
Anna University, India

Shiney J. Roscia Jeya
PSR Engineering College, India

P. Gopikannan
PSR Engineering College, India

M. Mahalakshmi
PSRR College of Engineering, India

ABSTRACT

In terms of security, proactively identifying and mitigating network assaults continues to be essential to the protection of vital systems. This chapter presents a unique method for real-time network intrusion detection that makes use of the resilient gradient boosting technique known as the CatBoost algorithm. Conventional signature-based techniques are inadequate against advanced threats because they frequently fail to adjust to changing assault patterns. Because ML (machine learning), and specifically CatBoost, can handle high-dimensional data, nonlinear relationships, and categorical variables, it appears to be a promising answer. This study uses CatBoost to create an advanced model that can recognize complex patterns in network traffic data. Using a large dataset that includes both attack and regular cases, the system gains the ability to identify minute changes in network behavior that could be signs of impending danger. Through feature importance analysis, one can gain a deeper understanding of the critical elements that contribute to the detection of attacks and, consequently, improve network vulnerabilities. Accuracy, F1-score, recall, and precision are among the measures applied to evaluate the efficacy of the proposed CatBoost-based IDS (intrusion detection system). The CatBoost algorithm is superior to traditional methods in terms of successfully recognizing known and unknown attack types while avoiding false positives, as demonstrated by comparison analyses. In addition, it shows that the model is resilient to evasion tactics used by bad actors, which guarantees strong network infrastructure security. The results highlight the effectiveness of using CatBoost in network security applications and highlight the technology's potential as a preventative measure against online attacks that are constantly

DOI: 10.4018/979-8-3693-2893-4.ch009

changing. More flexible and effective intrusion detection systems are made possible by this research, which advances machine learning-driven strategies for strengthening network defenses.

1. INTRODUCTION

Compared to communication networks used for other general applications, wireless sensor networks have different properties. These attributes pertain to network throughput, data processing, deployment topology, and environmental factors. Confidentiality, authentication, availability, integrity, authorization, and novelty are among the security concerns. Data transmission secrecy is the subject of confidentiality. Authentication is required to stop rogue sensor nodes from sending false messages, (Guan, Fang, and Yang, 2023). It guarantees the legitimacy of the data. Availabilityinthe face of attacks refers to the consistency of services. Integrity is the state in which the communications or data arrive at their destination undisturbed. Authorization entails preventing unauthorized access to data and allowing only approved sensor nodes to interact. Data must be updated to prevent hackers from replaying previous data. In contrast to communication networks used for other general applications, wireless sensor networks have distinct uses. These features pertain to network throughput, data processing, deployment topology, and environmental factors. Confidentiality, authentication, availability, integrity, authorization, and newness are the security concerns (Yakub, 2022).

Data transmission secrecy is the focus of confidentiality. Authentication is required to stop rogue sensor nodes from sending fraudulent messages. This guarantees the authenticity of the data. Availability refers to the ability of services to remain consistent even in the face of threats. Integrity refers to the messages or data's ability to reach their destination in an intact state (Sanjeetha, 2021). Authorization entails preventing unauthorized access to data and allowing only authorized sensor nodes to interact. Data must always be new to prevent hackers from replaying previous data. An attacker uses this type of attack to change or steal data. In an active attack, the attacker carries out some action to compromise the confidentiality, integrity, as well as availability of data.Anattackerattacksfalsedatainjection,resourceexhaustion,packetmodification,as well as node control. Sinkholes, DOS, hello flood, wormholes, jamming, spoofing, etc. Are instances of active attacks (Gocs, 2023).

2. RELATED WORK

Quan Sunet al. (2021) To identify spoofing attacks for heterogeneous wireless networks suggested a machine learning-based technique that makes use of physical layer data. Mahendra Prasad & Amit Tripathi (n.d.). Recently, intrusion detection systems (IDS) were deployed in ad hoc networks that faced distinct threats. They tracked the packet data transfer at each node as output files that provide complete network topology information. This large-scale labeled dataset is used by machine learning algorithms to classify malicious and normal test samples during supervised training by two well-known machine learning classifiers, NB and SGD, to identify attacks and demonstrate their effectiveness. Khalid Alissa, Tahir Alyas, et al. (n.d.) proposed an "Exploratory Data Analysis" (EDA) technique. The entire dataset is examined during this stage of statistical analysis. A sizable data set can be used to train the model. It is also possible to test ML classifiers like Random Forest and SVM. Popli, Rubi Rashmi et al. (2020) propose the use of aML technique based on theSVM-GA classifier to identify the forecast of malicious nodes and assaults on MANET and take into consideration node attributes in the system. It might be able to identify harmful and beneficial nodes to predict route attacks. The simulation's outcomes show how well the recommended strategy predicts and arranges the system's risky nodes. ML- based intrusion detection techniques for ad hoc network security assaults Prasad et.al, 2021. ML-based intrusion detection systems need to be developed and deployed to make it simpler to recognize attacks andprovidenetworksthepowertodeterminewhethertopermitintrusionwhilealsolearningmore about their mobile environment. Researchers Naima Kaabouch, Ronald Marsh, Tala Talaei Khoei and Shereen Ismail (2021) examine the efficacy of three new boosting methods: GBM, LightGBM, and Catboost. We employ a specialized dataset called WSN-DS, which includes many samples representing four different types of attacks: flooding, TDMA scheduling, blackhole, and grayhole. Researchers Aremu Idris Abiodun and Yakub Kayode Saheed (2022) revealed the work, which focuses on using ML-supervised algorithm-based IDS for Internet of Things purposes. To prevent information leakage on the test data, feature scaling was carried out in the 1ststage of this study approach on the UNSW-NB15 dataset utilizing the min-max (minimum-maximum) normalization concept Khalid, 2020. Nine distinct attack types are represented in this dataset, which combines modern attacks with typical network traffic behaviors. The next step involved using principal component analysis (PCA) to reduce dimensionality. Ahmed et al., (2021). To create a DDoS detection and mitigation program that is ideal, dependable, and fast, these requirements are addressed in this paper. The boosting method due to its symmetric treetopologyhas very little prediction time and is 8×quicker as compared to XGBoost. With an accuracy of 98% and significantly less training time, it was shown through testing to be dependable and efficient in detecting botnet-based DDoS attacks on the SDN environment (László & Johanyłk, 2023). This study focuses on the investigation carried out to create a behavior-based intrusion detection system (BIDS) categorization module. The basis for this in our instance was the CSE-CIC-IDS2018 dataset on AWS. The study sought to create a CatBoost algorithm- based classifier module for a BIDSsystem that is efficient. We compared the developed classifier to three other popular classifiers that were trained using the same features and the same data to assess its performance (Quasun 2021).

Wissal et al (2021). Predicting financial trouble gives investors and banks an efficient warning system to help them make informed judgments about lending approval. The effectiveness of ensemble approaches in predicting company failure has been established. Gradient boosting is one of the ensemble techniques that has been effectively applied to the prediction of bankruptcy. In this study, we offer a new method, called CatBoost, for employing gradient-boosting decision trees to categorize categorical data. (Guan, Fang, and Yang, 2023): The goal of this research is to improve detection performance by developing effective models for botnet detection. By substituting the centroid opposition-based learning method for the original random generation strategy, this article enhances the Dung Beetle Optimizer's (DBO) initial population creation strategy. The enhanced DBO is utilized in the realm of IoT botnet detection and is implemented to optimize the boost settings Popli (2020) The forms of cyberattacks that require automatic and timely detection require the use of an IDS. To categorize the supplied attack into the following groups: Legitimate, Dos, Flood, Brute Force, Malformed, and SlowITe this paper suggests using the CatBoost method, a variation of machine learning. Creating a balancing dataset allows the algorithm to be trained on an openly available MQTT network dataset (Meddeb et al., 2019). One other disadvantage of using new approaches is that they make data theft possible. As a result, a safe system that can identify any fraudulent activity and notify the administrator is needed. An IDS is the name given to such a system. The ensemble machine learning strategy's application of IDS is based on the CatBoost method. The proposed approach has achieved an accuracy of 99.46% on the NSL-KDD dataset after the implementation has been tested and the results have been assessed for detection accuracy. Bhuvan Trivedi and Prerna Agrawal. To identify generic and new malware, traditional malware detection techniques are inadequate. We suggest utilizing various supervised ML classifiers in our malware detection method. The performance of the created dataset is assessed with supervised algorithms, feature reduction methods, and assembling approaches.

Evaluationmetricssuchasprecisionaccuracy,CohenKappaScore,TPR,FPR,andAUCareare used to assess machine learning classifiers.

3. PROPOSED PROCESS

The suggested solution enables thorough network attack detection by leveraging the CatBoost algorithm's advantages inside a modular framework. Through the integration of modules for pre-processing, feature engineering, real-time monitoring, alerting, data gathering, and model maintenance, the system creates a strong defense that can detect and mitigate different kinds of network intrusions in advance.

Figure 1. Processes

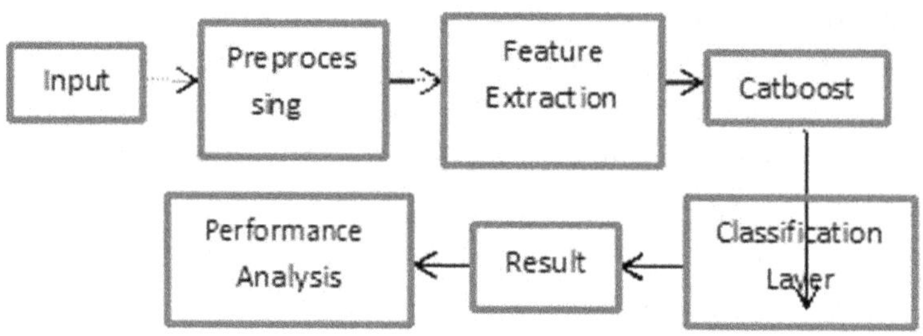

3.1 Data Collection Module

Responsible for aggregating raw network traffic data from multiple sources, such as routers, switches, or logs. It captures and consolidates data into a unified format for further processing.

3.2 Pre-Processing Module

Cleanses, filters, and pre-processes the collected raw data. Handles tasks like handling missing values, normalizing data, and converting categorical variables into a suitable format for model input.

3.3 Feature Engineering Module

Enhances the data quality by extracting relevant features from pre-processed data. Engages in tasks like feature selection, extraction, and transformation to optimize the input for the machine learning model.

3.4 CatBoost-Based Intrusion Detection Module

Utilizes the CatBoost algorithm to build and train the intrusion detection model. Learned from the engineered features to distinguish between normal and anomalous network behavior.

3.5 Real-Time Monitoring Module

IntegratesthetrainedCatBoostmodelintothelivenetworkenvironment.Analyzes incoming traffic in real time,identifiespotentialthreatsbasedonlearnedpatterns,andtriggers alerts for further investigation.

3.6 Alert and Reporting Module

Manages alerts generated by the detection module. Categorizes the severity of identified threats and provides comprehensive reports containing details on attack types, affected systems, and recommended mitigation actions. CatBoost, a prominent gradient-boosting algorithm, operates on the principle of sequentially building an ensemble of decision trees to minimize a predefined loss function. Its distinguishing feature lies in its adept handling of categorical variables without prior pre-processing. At its core, CatBoost refines predictions by iteratively adding decision trees that rectify the errors of preceding trees.

Catboost classifier with specified parameters:

Number of trees Learning rate Tree depth Evaluationmetric

- Lossfunction

Model=∑t=1Tft(x;θt)

Where ft is thet-th decision tree with parameters θt.

An ensemble of decision trees is constructed iteratively during training. A decision tree ft is trained to minimize the loss function at each iteration t. At iteration t, the ensemble's prediction is provided by:

y^(t)=∑i=1tfi(x)

The testing set Dtest is used to assess the trained CatBoost classifier. The ensemble prediction y^i for each data point xi in the testing set is calculated by adding up the predictions from every tree:

4. RESULTAND DISCUSSION

y^i=∑t=1Tft(xi;θt)

The performance metrics accuracy (ACC), precision (P),and sensitivity or recall (R) are used for assessment. These metrics are measured using four distinct parameters: TP (True Positive), TN (True Negative), FP (False Positive), and FN (False negative). Accuracy is the proportion of documents that have been correctly classified over the entire set of data. The relevant proportion of the performance is referred to as precision. Conversely, recall is the percentage of accurate classifications the algorithm made for all functional outcomes. The percentage of anomalous records that are successfully detected as anomalies relative to total anomaly data (TPR) is referred to as the "True Positive Rate" (TPR) and the "Detection Rate" (DR). The rate of False Positives (FPR), which represents the proportion of the total number that incorrectly flags an ordinaryrecord as abnormal when it does so, is as follows

TP (TruePositive): The total amount of malicious nodes that have been accurately identified as Malicious.

FN (False Negative): Total number of malicious nodes that were misclassified as benign nodes.

FP (FalsePositive): Number of benign nodes that were misclassified as malicious.

TN(True Negative): Numberofbenignnodesthat were correctly classified as benign.

Figure 2. Dataset description

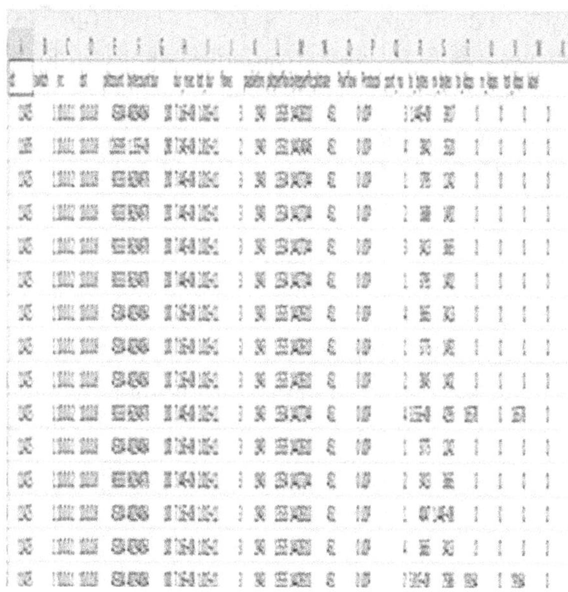

Figure 3. Model training

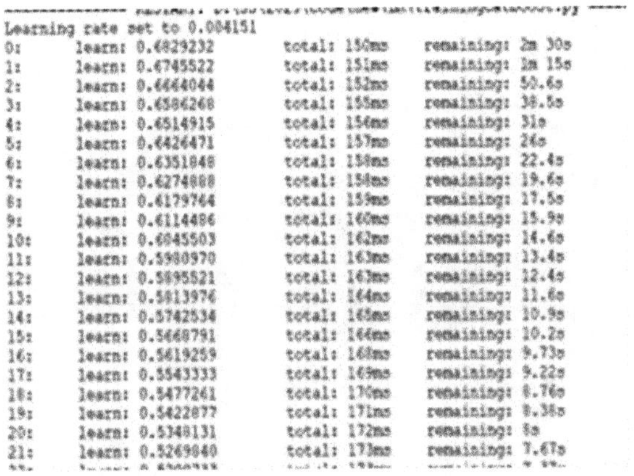

Model training takes 70%oftrainingdataand30%oftestingdataandextracts the features for catboost model.

Figure 4. Model validation

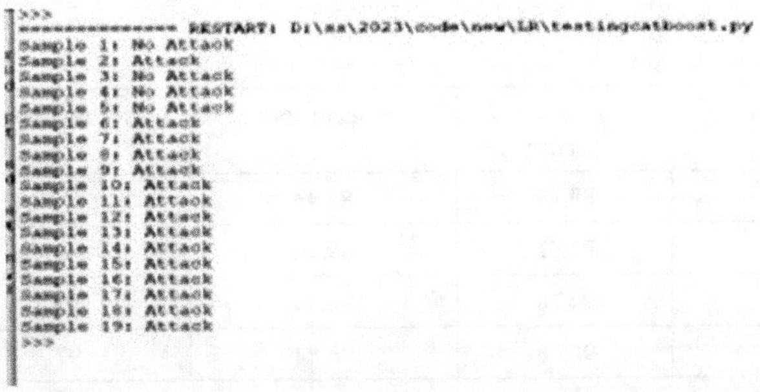

Model validationgivesthe TP,TN, FP, and FN

Figure 5. Accuracy

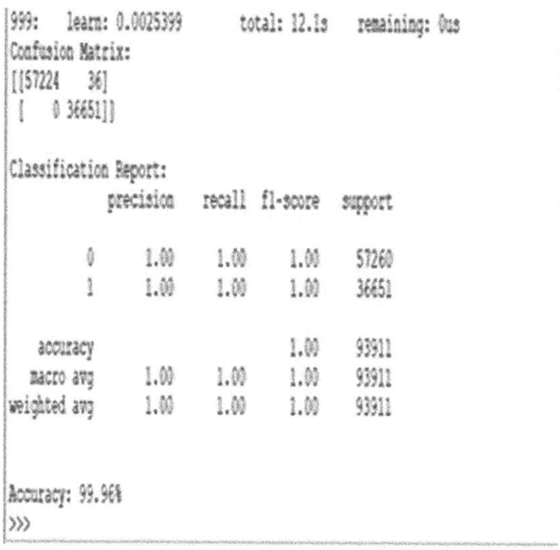

```
999:    learn: 0.0025399        total: 12.1s    remaining: 0us
Confusion Matrix:
[[57224    36]
 [    0 36651]]

Classification Report:
              precision    recall  f1-score   support

           0       1.00      1.00      1.00     57260
           1       1.00      1.00      1.00     36651

    accuracy                           1.00     93911
   macro avg       1.00      1.00      1.00     93911
weighted avg       1.00      1.00      1.00     93911


Accuracy: 99.96%
>>>
```

Table 1. Confusion matrix

Models	F1-Score	Accuracy	Precision	Specificity
NN	89.46	84.43	87.76	78.18
SVM	92.92	89.76	92.13	85.39
Ada boost	94.76	94.84	95.2	92.47
Catboost	96.76	98.84	97.2	95.47

The confusion matrix compares the NN, SVM, Adaboost, and Catboost classifier with the parameters of F1-score, Accuracy, Precision, and specificity. In all comparisons catboost get the high accuracy.

5. CONCLUSION

The evolution of network attacks poses an ongoing challenge in maintaining the integrity and security of digital infrastructures. Traditional security measures, reliant on signature-based detection systems, are proving increasingly inadequate against the sophistication and diversity of contemporary threats. This study addressed this critical issue by harnessing the power of machine learning, specifically employing the CatBoost algorithm, to fortify network defense mechanisms. The results obtained through our research demonstrate the efficacy of CatBoost as a formidable tool in detecting and mitigating network intrusions. By leveraging a diverse dataset encompassing various attack scenarios, the algorithm showcased its capability to learn complex patterns inherent in network traffic data. Its ability to discern subtle deviations and anomalous behavior, even in the presence of previously unseen attack vectors, underscores its adaptability and robustness. Importantly, the feature importance analysis provided valuable insights into the crucial factors contributing to effective attack detection. This understanding enhances not only the detection accuracy but also facilitates a proactive approach to fortifying network vulnerabilities, thereby bolstering overall network resilience.

REFERENCES

Alissa, K., Alyas, T., Zafar, K., Abbas, Q., Tabassum, N., & Sakib, S. (2022). Botnet Attack Detection in IoT Using Machine Learning. *Computational Intelligence and Neuroscience.*

Ben, W., Arfi, S., Gharib, C., & Ben Jabeur, S. (2021). Catboost model and artificial intelligence techniques for corporate failure prediction. *Journal of Tech. Fore.* 0658.doi.org/10.1016/j.techfore.2021.120658

Gocs, L. & Johanyak, Z (2023). Catboost algorithm base classifier module for brute force attack prediction. *Tome XXI.* Fascicule 3.

Guan, W., Fang, Z., & Yang, C. (2023). IOT Botnet Attack detection model based on DBO- Catboost. *Applied Sciences (Basel, Switzerland)*, 13(12), 7169. 10.3390/app13127169

Ismail, S., Khoei, T. T., Marsh, R., & Kaabouch, N. (2021). *A Comparative Study of Machine Learning Models for Cyber-attacks Detection in Wireless Sensor Networks. 2021 IEEE 12th AnnualUbiquitous Computing, Electronics & MobileCommunication Conference.* UEMCON.

Kayode, Y., Saheed, A., & Abiodun, S. (2022). *A machine learning based intrusion detection for detecting internet of things network attacks.* Science Direct.

Meddeb, R., Jemili, F., Triki, B., & Korbaa, O. (2019). Anomaly-basedbehavioral detection in mobile ad hoc networks. *Procedia Computer Science*, 159, 77–86. 10.1016/j.procs.2019.09.162

Popli, R. R. (2020). A wormhole attack detection in the mobile ad-hoc network using GA and SVM. *International Journal of Engineering Applied Sciences and Technology*, 5(3), 582–588.

Prasad & Tripathi. (n.d.) Wormhole attack detection in adhoc network using machine learning technique. *Communication and Networking Technologies.* IEEE. .10.1109/ICCCNT45670.2019.8944634

Sanjeetha, R. & Raj, A. (2021). Detection and Mitigation of botnet-based DDoS attacks using catboost machine learning algorithm in SDN environment. *International journal of advanced Technology and Engineering Exploration.*

Sun, Q., Miao, X., Guan, Z., Wang, J., & Gao, D. (2021). Spoofing Attack Detection Using Machine Learning in Cross-Technology Communication. *Security and Communication Networks, 12.* 10.1155/2021/3314595

Chapter 10
NFV and Secure Cognitive SDN for Educational Backbone Network Deployment:
Cognitive-Enabled Routing in SDN

Muhammad Imran Majid
https://orcid.org/0000-0002-0217-9046
Institute of Business Management, Karachi, Pakistan

Faheem Yar Khuhawar
https://orcid.org/0000-0001-6743-8080
Mehran University of Engineering and Technology, Jamshoro, Pakistan

Karrar Muhammad
Institute of Business Management, Karachi, Pakistan

Kamlesh Rathore
Institute of Business Management, Karachi, Pakistan

Aarif Ali
Institute of Business Management, Karachi, Pakistan

Muhammad Ali
Institute of Business Management, Karachi, Pakistan

Babu Mal
Institute of Business Management, Karachi, Pakistan

Umme Habiba
Shaheed Zulfikar Ali Bhutto Institute of Science and Technology, Pakistan

Shabana Muhammad
https://orcid.org/0009-0009-3615-3846
Institute of Business Management, Karachi, Pakistan

Faisal Karim Shaikh
https://orcid.org/0000-0003-4692-6696
Mehran University of Engineering and Technology, Pakistan

Tariq Rahim Soomro
https://orcid.org/0000-0002-7119-0644
Institute of Business Management, Karachi, Pakistan

ABSTRACT

Software defined networks (SDN) and wireless cognitive radio networks (CRN) are examined within the context of dynamic spectrum management. The features include control and data plane separation,

DOI: 10.4018/979-8-3693-2893-4.ch010

centralized control, adopting open-source standards, programmability, quality of service (QoS) management, and security. The transformative impact of network function virtualization (NFV) is explored with a perspective on its architecture and applications in SDN, internet of things (IoT), cloud computing, and blockchain. The security aspect of SDN with specific focus on mitigating denial-of-service (DoS) attacks and vulnerabilities associated with open flow protocol is also addressed. The cognitive-inspired security mechanisms adapt to evolving threats integrating machine learning (ML) and artificial intelligence (AI) based algorithms for dynamic threat detection and mitigation exemplified through case studies. Adoption of software-defined perimeter, zero trust, blockchain, and quantum-safe cryptography in future are discussed. Finally, SDN applications for IoT networks are discussed.

1. INTRODUCTION

1.1 Introduction to Cognitive SDN

Cognitive features of SDN are explored to enable implementation of Dynamic Spectrum Management for networks which sense and respond to various spectrum instances.

Wireless cognitive radio network (CRN) allows unlicensed users to use authorized spectrum without causing conflict with distinct license user communication prerequisites. Traditional CRNs cannot meet the QoS requirements using conventional middle box approach which includes a number of several middle boxes and complex protocols. Non-standard L2 /L3 devices cause security vulnerabilities.

The solution is to explore decoupling of control and data plane as exhibited in Software Defined Networks exemplified by Cognitive Radio networks as in mentioned Figure 1.

Figure 1. System diagram of the software defined network denoted by cognitive radio networks

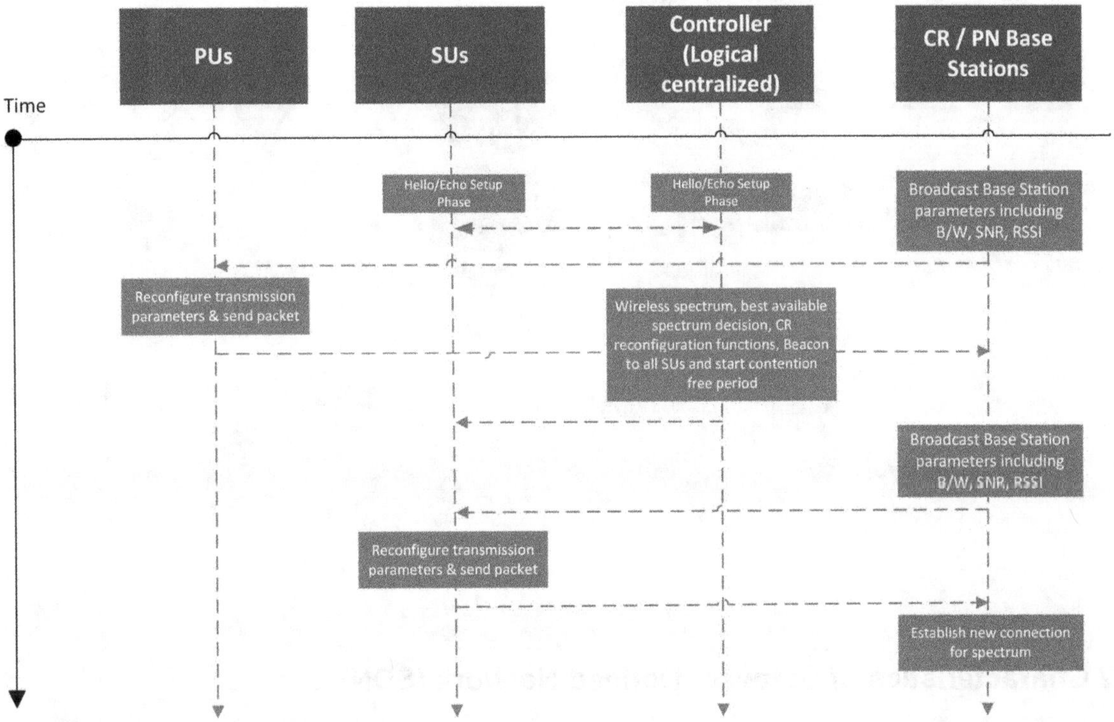

(summarized version of Cicio lu et al., 2020)

Here the simplified version of SDN architecture is reproduced as in Figure 2 (Cicio lu et al., 2020). Here two networks are being simulated.

Figure 2. SDN controller location for communication between two networks (main campus based home network and tertiary campus based foreign network)

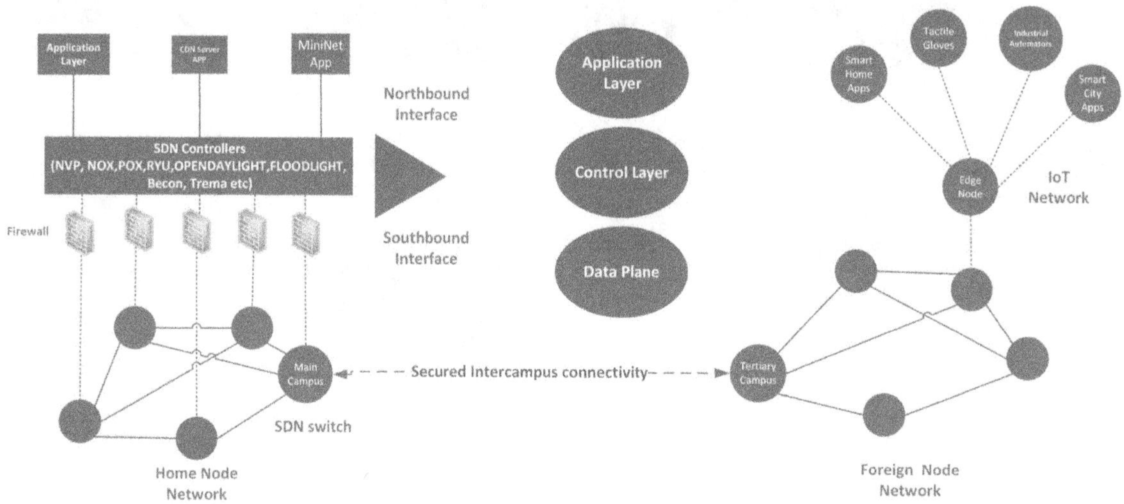

1.2 Characteristics of Software Defined Network (SDN)

SDN, or software-defined networking, is a cutting-edge approach to network management. The data and control planes are divided, respectively. Devices on the data plane are converted into straightforward devices for packet forwarding. The controller operates the common operations in the control plane using a hierarchy of controlled representation. Characteristics of Software Defined Network (SDN):

Control and Data Plane separation: The interconnected l2 and l3 devices on the routers and switches are isolated from the controlled logic of the network. The division leads to efficient traffic management (Kreutz et al., 2015).

Centralized Control: SDN integrates centralized approach and centralized logic. This leads to identification and fixing problems on a network scale (Jawad, 2018).

Programmability: Software defined network introduces the ability to program the network (Kreutz et al., 2015). This programmability allows network administrators to dynamically configure the network (Jawad, 2018).

Open-Source Standards: Open-source protocols, like OpenFlow version 1.1 and 1.3 are multifaceted and interactive or easy communication amongst various devices and software (Kreutz et al., 2015).

Extended flexibility and efficiency: SDN provide agility for a flexible approach to meeting various business needs. This simplifies and facilitates an evolved network management tool (Kreutz et al., 2015).

Quality of Service (QoS) in SDN: This involves comprehensive traffic engineering and routing management to enable critical applications achieve their performance objectives (Kreutz et al., 2015).

Cognitive Security for SDN Controller: By providing enhanced security on the control plane or SDN controllers including firewalls for improved network efficiency (Kaliyamurthy et al., 2021).

The unique characteristics of SDN decouples the data plane from SDN controller-based queries. The SDN offers simplicity which leads to a modular approach in scaling. There is no impact of data flow characterization and physical layer changes on the control layer which focuses on the precise network control using firewalls and SDN algorithms. This approach allows for centralized control layer functions. In summary the brain behind the network control is moved to controllers rather than the peripheral network devices (Jawad, 2018).

1.2.1 Fundamental Traits Separating SDN From Conventional Networks

Table 1. Differentiating SDN from conventional network architectures (Jawad, 2018; Kaliyamurthy et al., 2021)

Characteristics	Software-Defined Network (SDN)	Traditional Network
Architecture	Separates the control plane and data plane, making central control.	Coupled data and control plane.
Control Plane	Centralized, programmable, software based.	Distributed and hardware based.
Configuration management	Automated and centralized programmatic.	Manual and static
Programmability	Highly programmable, supports network dynamic configuration.	Limited or no programmability, network static configuration.
Scalability	Highly scalable, as network resources are dynamically allocated and optimized.	Low scalability due to distributed control and hardware dependency.
Security	Highly secured because of central control policies enforcement.	Less secure because network policies depend on individual devices and configuration.
Cost	Potentially low cost because of reduced hardware dependency.	Can be higher because of hardware dependency.

2. STATE OF ART

The state of the art is discussed vis-va-vis service level indicators in Figure 2. The OSI layers are demonstrated followed by service level indicator to deploy which indicator includes language and hardware to represent SDN enabled features like NFV., Green and Cloud computing.

Table 2. SDN related research categorizing in service level indicator as well as furniture to be go away

Layer		Service Level Indicators	Approach and Reference	Evaluation Indicator	Related Area
Infrastructure	Switching Devices	Processing	Combines different processing hardware (Xia et al., 2015)	Throughput	Integrated systems Embedded systems Hardware testing
		Hardware Platform	Software implementation Firmware implementation Implementation on open hardware platform (Xia et al., 2015)	Cost	
	Transmission Media	Wireless Media	Open Radio (Xia et al., 2015)	Range	SDR, Mobile Ad Hoc
		Optical Fibers	Unified control plane, Split control plane (Xia et al., 2015)	Bandwidth	GMPLS, ROADMs
		Cognitive Radio	Middle Box, Complex Protocols (Cicio lu et al., 2020) AI Based Routing (Chen, 2022)	Throughput	CRN
Control	Southbound Data Stream	High level Language	Flow Tables (Wu et al., 2020)	Delay	Programming Language, Formal Methods, Compliers
		Policy and Rule Validation	Flow Checker (Nowak et al., 2019) VeriFlow (Xia et al., 2015)	Validation	Formal Methods
	Northbound Data Stream	Network Status Collection	OpenTM (Xia et al., 2015)	Manageability	Network Measurement
		Network Status Synchronization	Hyper flow (Kreutz et al., 2015), Onix	Reliability	Distributed system Analysis
Application		Adaptive Routing	Load Balancing (Srivastava & Pandey, 2021), Cross-layer design (Kreutz et al., 2015)	Throughput	Routing, Traffic Engineering NDN, Wireless network, Computer Security, Function Outsourcing, Network Virtualization, Web Cashing, Green Networking, Cloud
		Boundless Mobility	OpenRoads (Sun et al., 2014), Odin (Xia et al., 2015)	Range, Handover Time	
		DDoS Security	Attack Detection, Anonymization, MTD, Access Control	Security	
		Network Function Virtualization	Flow Visor (Alam et al., 2020), OpenFlow Level Virtualization (Peng et al., 2015)	Resource Utilization	
		Green Networking	ElasticTree (Xia et al., 2015)	Power Consumption	
		Cloud Computing	Open vSwitch (Ray & Kumar, 2021), VM Migration	Scalability	

The evaluation indicators assess SDN performance across various layers. For the infrastructure layer, throughput, range, and bandwidth measure data transfer speed and network reach. The control layer focuses on manageability, ease of use, accuracy of policy validation, and data collection reliability. Finally, the application layer considers throughput, range, security aspects, resource utilization, energy efficiency, and scalability. These indicators provide a comprehensive framework for comparing and evaluating SDN approaches.

1. **Processing, Hardware Platform and Software Implementation:** According to (Xia et al., 2015), the service level indicator for processing focuses on the combination of different processing hardware. To improve the processing efficiency, this means enhancing hardware platforms, firmware and software implementations. The significance of software and firmware implementations on open hardware platforms is emphasized by (Xia et al., 2015). Businesses may improve service delivery by increasing hardware configuration flexibility and scalability by utilizing open hardware platforms.
2. **Wireless Media:** The service level indicator refers to using Open Radio technology in wireless media, as proposed by (Xia et al., 2015). The overall performance of a network is improved because Open Radio makes it possible for wireless communication systems to be more adaptable and interoperable.
3. **Optical Fibers**: Author (Xia et al., 2015) propose both combined and divided control plane methods to optimize optical fiber networks. Integrated control planes modernize network management, while split control planes offer enhanced customization and control, contributing to improved service levels.
4. **Cognitive Radio:** Author (Cicio lu et al., 2020) introduce middle box architectures and complex protocols, while (Chen 2022) highlights AI-based routing techniques. These approaches enhance cognitive radio systems by enabling advanced processing and decision-making capabilities, thereby improving service quality.
5. **High-Level Language:** The Description of "Flow Tables" methods in (Wu et al., 2020), for controlling flow tables in network routers and switches. It examines procedures for proficient stream table usage and upkeep to upgrade sending execution.
6. **Policy and Rule Validation:** This author (Nowak et al., 2019) introduces the concept of a "Flow Checker" to ensure correctness and security in network flows by focusing on flow verification. It discusses techniques for identifying and mitigating flow behavior variances. A formal verification framework for the behavior of network forwarding is offered by "VeriFlow." By meticulously analyzing packet forwarding paths, it aims to identify and prevent network anomalies and misconfigurations (Xia et al., 2015).
7. **Adaptive Routing:** These authors (Srivastava & Pandey, 2021) examine load balancing approaches for appropriating network traffic across different ways or assets. It investigates techniques for further developing organization execution and unwavering quality through successful burden dissemination. While (Kreutz et al., 2015) propose cross-layer design approaches. These strategies enhance adaptive routing capabilities, optimizing resource utilization and minimizing latency for improved service delivery.
8. **DDoS Security and Network Function Virtualization:** DDoS security measures, including attack detection, anonymization, Moving Target Defense, and access control, are vital service level indicators. (Alam et al., 2020) introduce Flow Visor, and (Peng et al., 2015) propose OpenFlow Level Virtualization for network function virtualization, enhancing security and scalability in network environments.

9. **Cloud Computing:** The author (Ray & Kumar, 2021) introduces "Open vSwitch" is a virtualized-environment-specific software-based switch implementation. It works with virtual machine (VM) relocation and gives adaptable organization availability inside virtualized foundations.

3. NETWORK FUNCTION VIRTUALIZATION AND COGNITION

3.1 Introduction

Network Functions Virtualization (NFV) is a technology that was developed with the primary goal of speeding up the delivery of cutting-edge network services in the telecom industry. NFV's main objective is to broaden its scope for future evolution. A closer look reveals that today's communication environments are fairly congested with a wide variety of proprietary hardware components. Because of the available space, the introduction of a new network service necessitates the insertion of various hardware entities, which causes considerably more as its complexity develops.

By utilizing full-fledged virtualization technology to divide software instances from hardware platforms and to decouple functionality from location for quicker networking service provisioning, NFV completely changes the way network operators construct their infrastructure. As seen in Figure 1, network function virtualization (NFV) essentially uses industry-standard servers, storage, and switches as well as software virtualization approaches to implement network functions. There is no need to deploy additional hardware because these virtual appliances can be instantiated on demand. For instance, network administrators may use a Virtual Machine (VM) on an x86 platform to run an open-source software-based firewall. Recent experiments have shown that network functions, such physical layer signal processing, can be implemented on platforms based on general purpose processors.

Recently, Network Functions Virtualization (NFV) was suggested as a way to speed up the time to market for new services and increase the flexibility of network service provisioning. Through the use of virtualization technologies and readily available commercial programmable hardware, including switches, storage, and general-purpose servers, network function virtualization (NFV) separates the software implementation of network services from the underlying hardware. As a new technology, network operators have a number of difficulties with NFV, including ensuring network performance for virtual appliances, managing their dynamic instantiation and migration, and strategically placing them. In this article, we give a quick introduction to NFV, describe its specifications and architectural framework, show a number of use examples, and talk about the difficulties and potential paths forward in this rapidly developing field of study. Network operators seek to deliver useful network-based services at the lowest possible cost of service deployment. Network function virtualization, or NFV, is thought to be a potential strategy to change this kind of employment by first detaching network functions from fundamental hardware characteristics and then transforming them into software style. In the end, these are called Virtual Network Functions (VNFs). Numerous advantages result from this separation, such as reduced Operation and Capital Expenditure (OPEX) and increased service preparation elasticity. The current paper examines the uses and prospects of NFV in relation to blockchain, IoT, SDN, and cloud computing. The NFV architecture is described, and a number of potential security problems and difficulties are explored. Lastly, a methodical design concept for a blockchain network virtualization system is presented.

3.2 How Virtualization Helps Cognitive Enabled SDNs

For cognitive SDN to be implemented the network is sliced into multiple logical subnetworks here each virtual slice has its own dedicated network and virtualization layer which uses OpenFlow protocol for communication (Bizanis & Kuipers, 2016). Software-Defined Networking (SDN) and Network Function Virtualization (NFV) are revolutionizing communication networks. The NG2CRO architecture leverages AI, namely Reinforcement Learning, for adaptive network orchestration to address 5G issues (Janjua et al., 2023). A Cognitive SDN architecture based on Fuzzy Cognitive Maps is recommended for efficient network upgrades (Baggio et al., 2019). However, it also highlights how SDN and NFV have broader effects on the Internet and how AI and ML help automate network operations (Sharma & Nag, 2023). Collectively, these advancements provide smart, flexible networking options that will revolutionize curriculum and progress radio systems design.

3.3 Use of NFV SDN in Educational Backbone Networks

Numerous studies have looked into the application of Network Function Virtualization (NFV) and Software-Defined Networking (SDN) in educational backbone networks. the complexity of diverse networks in uses spanning from healthcare to smart cities by offering an organized evaluation method created especially for the Internet of Things (IoT). It identifies research issues and categorizes the literature into IoT-specific fields such software-defined networks, function virtualization, and software-defined IoT networks (Alam et al., 2020). Meanwhile, the second study examines SINET5, a 100-Gbps university backbone network that illustrates its statewide reach, lower latency, and SDN-friendly characteristics using MPLS-TP systems and ROADMs. SINET5, which enables end-to-end data transfers at 100 Gbps, serves as a dynamic platform for inter-cloud connectivity and NFV applications (Kurimoto et al., 2017). The third study further describes SINET5's completely integrated architecture, emphasizing its multi-layer, dynamically configurable platform, SDN-friendly environment, and 100-Gbps access. (Kurimoto et al., 2016).

3.4 Radio System Architecture

Where a communication distance of 200 kilometers or more exists and the topography prohibits the use of fiber optics or other wired options for communication, the education backbone network can be regarded as an SD-WAN. The application of radio network architecture has drawn interest; in this case, SDN lowers management and operating costs while offering benefits in terms of network security and control plane simplification. First, the Software-Defined Cognitive Radio (SDCR) network architecture is presented as a revolutionary approach to cognitive radio networks (CRNs). Through the use of an SDN controller for dynamic spectrum management, SDCR improves network performance, increases throughput, and reduces reliance on base stations (Cicio lu et al., 2020). Proposes the use of 5G hybrid cloud radio access networks (H-CRANs) to maximize energy and spectrum efficiency. It discusses the concept of Node C and outlines a software-defined H-CRAN system architecture that is compatible with software-defined networks. The study discusses several key technologies, including large-scale cooperative spatial signal processing, radio resource management, network function virtualization, and self-organization (Peng et al., 2015). Software-Defined Radios (SDRs) are the focus now, and their architecture, innovative applications, and existing platforms are examined. The research emphasizes the significance of SDR in wireless communication protocols and offers details on potential fixes, architecture, and development tools (Akeela & Dezfouli, 2018). SDN-based network architecture to address issues caused by extraordinarily dense wireless networks. This ground-breaking technique was developed as part of the CROWD project and offers a workable solution to the demands of next-generation mobile broadband wireless access by combining mobility management with MAC control (Ali-Ahmad et al., 2013).

3.5 Consideration

The combination of Cognitive Software-Defined Networking (Cognitive SDN) and Network Functions Virtualization (NFV) holds immense potential in optimizing the backbone of educational networks and enhancing Radio Systems Architecture. Starting with Cognitive SDN, the Software Defined Cognitive Network (SDCoN) introduces a novel paradigm for improving Quality of Service (QoS) in educational networks (Nderba, 2017). Investigates the versatile use of SDN and NFV across various IoT scenarios, including wireless sensor networks and mobile and cellular networks (Bizanis & Kuipers, 2016). Furthermore, delves into SDN/NFV architectures tailored for edge-cloud-oriented IoT, offering a comprehensive analysis of their impact on virtualized service provisioning (Ray & Kumar, 2021). Transitioning into the realm of radio systems, explores the potential of SDN in cognitive radio networks, showcasing traffic offloading and spectrum-sharing possibilities in broadcast white spaces (Sun et al., 2014). Building on this, presents an SDN-enabled cognitive radio network architecture (SDCR) designed to improve overall network performance, reduce reliance on base stations, and dynamically manage network resources (Cicio lu et al., 2020).

This comprehensive strategy underscores the benefits of integrating Cognitive SDN and NFV into Radio Systems Architecture and educational network backbones, paving the way for intelligent, flexible, and effective networking solutions tailored to the evolving needs of educational institutions.

Starting from the critical function that Software-Defined Networking (SDN) and Network Function Virtualization (NFV) play in virtualizing Internet of Things (IoT) items, the research scope includes the cognitive aspects of SDN and NFV, with a focus on applications. SDN-enabled network virtualization in Industry 4.0 is examined as the investigation progresses, closely tying cloud computing and the Internet of Things together. Deep Reinforcement Learning adds aspects of intelligent decision-making to the management of SDN and NFV in an edge IoT architecture. A thorough examination of large-scale SDN testbeds addresses the difficulties associated with wireless network virtualization by navigating through strategies, obstacles, and the wireless domain. After that, the focus moves to the importance of network virtualization hypervisors in SDN networks, emphasizing classification and contrast, and discusses machine learning for cognitive network management, which takes the discussion into cognitive aspects. The concept of a Cognitive Software-Defined Networking architecture highlights the need for fuzzy cognitive maps, and the integration of fog computing for cognitive network function virtualization addresses issues in an information-driven future Internet. While broader challenges and research possibilities in Software-Defined Networking address the development of the Future Internet, difficulties and open issues in Software-Defined Wireless Networks are examined. A proposed cognitive-inspired computing-based DDoS detection and defense mechanism brings security initially. Next, a security and trust framework for software-defined networking and virtualized networks is presented, guaranteeing the safe development of virtualized network environments. A comprehensive understanding of NFV and cognition's impact on networking is shaped by this wide journey, which incorporates Virtualization Techniques, Orchestration and Management, Cognitive Networking Principles, Machine Learning and

AI Integration, Quality of Service Enhancement, SDN Integration, Performance Evaluation and Benchmarking, and ongoing consideration of Challenges and Future Directions.

4. COGNITIVE AND AI ENABLED ROUTING

The key features of cognitive and AI enabled includes intelligent path selection, real time adaptability, quality of service optimization, load balancing and dynamic load balancing breathing and healing mechanism. These are detailed in the following section:

4.1 Intelligent Path Selection

AI-driven intelligent path selection ensures the optimal routing for data transfer, boosting productivity and decreasing latency in educational and radio network applications. Uses SDN architecture enhanced with Cognitive Packet Networks (CPN) and Random Neural Networks for intelligent path selection, introducing security-aware routing with trust evaluations (Nowak et al., 2019). Intelligent path selection in inter-data center overlay networks using a logically centralized Cognitive Routing Engine (CRE) based on Random Neural Networks with Reinforcement Learning (Francois & Gelenbe, 2016). Creates a cognitive routing algorithm (AI-CRA) based on artificial intelligence for dynamic optical networks. It optimizes global topology adaptive pathways by considering link cost to enable intelligent path selection (Chen, 2022).

4.2 Read-Time Adaptability

Real-time adaptation allows the routing system to adjust flexibly to changing network conditions, ensuring optimal performance and resource economy. A rationally organized Cognitive Routing Engine (CRE) employing Random Neural Networks and Reinforcement Learning to provide the best overlay paths in real time with the least amount of monitoring overhead (Francois & Gelenbe, 2016). A thorough explanation of how AI can be integrated into SDN with an emphasis on fuzzy inference systems, machine learning, and meta-heuristics for real-time adaptability (Latah & Toker, 2019).

4.3 Quality of Service Optimization

Quality of Service (QoS) optimization via AI-driven routing ensures a dependable and consistent user experience, which is essential for radio communication and educational applications. Integrates QoS and energy criteria into security-aware routing within an SDN infrastructure enriched with Cognitive Packet Networks (CPN) (Nowak et al., 2019). Demonstrates CRE's ability to optimize overlay paths and achieve QoS improvements for geographically dispersed applications (Francois & Gelenbe, 2016). Proposes a QoE-oriented cognitive network framework using SDN and ML to dynamically manage and orchestrate end-to-end network resources, improving application QoE (Wang & Delaney, 2019).

4.4 Dynamic Load Balancing

In radio and educational settings, dynamic load balancing maximizes resource utilization, avoids cognition and enhances overall network efficiency. Presents a DRL-based method for load balancing in multi-controller SDNs, inspired by human intelligence's self-learning capabilities (Srivastava & Pandey, 2021). Introduces an AI-based Lightweight Adaptive Measurement Method (ALAMM) for SDN to reduce traffic measurement overheads and optimize dynamic load balancing (Huo et al., 2021).

4.5 Self-Healing Mechanisms

AI-guided self-healing systems help maintain network resilience by automatically identifying and resolving problems to ensure minimal downtime and continued operation. Introduces an AI-assisted cognitive routing algorithm for dynamic optical networks, adapting to changing conditions, optimizing routing paths, and ensuring efficient resource utilization (Chen, 2022). Provides an overview of AI integration in SDN, emphasizing self-healing mechanisms through machine learning, meta-heuristics, and fuzzy inference systems (Latah & Toker, 2019).

4.6 AI Enabled Routing Advantages Versus Conventional Routing in SDNs

This paper presents the Artificial Intelligence Enabled Routing (AIER) approach using Cognitive Software Defined Networking (SDN) and AI. This technique enhances network performance and provides more accurate route recommendations because of its capacity for learning and handling problems related to the duration of the monitoring period (Wu et al., 2020). Using Cognitive Packet Networks (CPN) and Random Neural Networks (RNN) to provide security-aware routing for Internet of Things networks. The design, utilized in the SerIoT project, prioritizes the evaluation of confidence in devices and packet flows (Nowak et al., 2019). Optimizing safe SDN-enabled inter-data centre overlay networks is done using cognitive routing. A logically centralized Cognitive Routing Engine (CRE) uses Random Neural Networks with Reinforcement Learning to generate efficient paths with little overhead monitoring (Francois & Gelenbe, 2016). A QoE-focused cognitive network that uses machine learning and SDN to dynamically manage network resources and improve application quality of experience. The framework, which is separated into three sections: data collection, network/application learning, and network management, correlates network measurements with application conditions (Wang & Delaney, 2019). For dynamic optical networks, cognitive routing with AI help is used. By offering an adjustable global topological path to enhance multi-layer and multi-domain data transmission routing, the Artificial Intelligence-based CRA technique balances system load and maximizes resource efficiency (Chen, 2022).

Case Study 4: One of the most important tasks for the reliable operation of network infrastructure is routing. These days, more sophisticated routing techniques are needed than ever before due to the quick increase in network traffic volume and the shifting needs for services. In order to do this, we provide a Deep Reinforcement Learning (DRL)-based implementation strategy as well as a concept of cognitive routing.

The crucial component for the reliable operation of network infrastructure is routing, which is the process of choosing a path for packet transmission in networks. Routing technologies can be broadly divided into two groups: quality-aware and non-quality-aware. Because they cannot base routing decisions on network and service quality information, the majority of commonly used routing protocols and algorithms, including RIP, IGRP, and OSPF, are non-quality-ware. While non-quality-aware routing algorithms and protocols are easy to set up on routers and have proven effective for many years, they face challenges from the volume of network traffic growing quickly and evolving service requirements.

The rapid advancement of new technologies such as SDN and NFV in recent years has led to several research papers introducing the possibility of implementing more sophisticated routing decisions on powerful hardware. For instance, Google has demonstrated that it is possible to achieve improved quality assurance on software-defined networks by separating the routing control and operating approaches. Motivated by these efforts, we put up the notion of cognitive routing, which expands upon the idea of quality-aware routing by adding three essential functions—inference, decision-making, and learning—to the routing decision component. We offer a Deep Reinforcement Learning (DRL)-based implementation strategy in addition to merely discussing the concept. We create a simulator called RL4Net for the creation and simulation of DRL-based routing algorithms in order to aid in the study of DRL-based cognitive routing. Furthermore, we develop and put into practice a routing algorithm based on Deep Deterministic Policy Gradients, or DDPGs. We contrast the OSPF and random weight algorithms with the DDPG-based routing algorithm to show the initial viability and prospective benefit of cognitive routing.

4.7 Routing Protocols for CRNs

Protocol, which finds several routes to connect a source node to a destination node. There are two types of routing protocols for wireless ad hoc networks: proactive and reactive. Only, when necessary, do reactive protocols create on-demand pathways between the source and destination nodes. With proactive protocols, every node keeps up-to-date routing tables for every destination by periodically disseminating data (like link status) across the network. Proactive protocols, thus, have a shorter path acquisition time than reactive ones, but they also respond to network changes, like node mobility, much more slowly. Because of PU interference, available bandwidth, and other factors, each wireless connection in a CRN may encounter varying conditions over time, according to the dynamic spectrum allocation. Because of this, we concentrate on reactive protocols and examine two methods for route construction across CRNs:

Routing using a single path. The AODV routing protocol is examined, which identifies a solitary path connecting a source and a destination node.

Multiple-path routing. We take into consideration that the AOMDV routing paths found can be either link-disjoint, meaning they share no links, or node-disjoint, meaning they have no nodes in common.

Learning in cognitive radio networks

The literature has recently shown a great deal of interest in learning in cognitive radios. This section provides an overview of machine learning and artificial intelligence, along with an analysis of the most recent advancements in the use of learning techniques in cognitive radio networks.

The process of learning involves teaching a node to change its behavior while maintaining various network conditions or having the capacity to learn from this experience and take it into account in the future. Learning must be a strong enough process to add to the body of information based on prior behavior, which will boost reasoning efficiency. Learning mechanisms could be divided into subgroups of unsupervised or supervised learning techniques. Corresponding to learning by reinforcement and instruction, respectively. Conversely, supervised learning requires preexisting knowledge or familiar surroundings.

Q-Routing

A reinforcement learning algorithm called Q-Learning can figure out the best course of action in a given setting to maximize the rewards it receives from it. A modification of Q-Learning that can distribute packets across a network is called Q-Routing.

The Bayesian Learning Algorithm approach

The prior distribution, which is meant to reflect knowledge about unknown parameters prior to the data becoming available, is given a great deal of weight in Bayesian analysis. Although it is a popular belief that the agent is completely ignorant of the subject matter it is attempting to learn, this is frequently not the case. An agent will often possess some past knowledge, and the learning procedure should ideally take advantage of this information. One way to think of Bayesian learning is as an uncertain kind of reasoning based on observations. Using Bayesian learning, predictions are based on estimating the likelihood of each hypothesis given the available evidence. It has been demonstrated that in Bayesian prediction, the true hypothesis eventually wins out. Because it offers a mathematical explanation of how new data might be combined with old knowledge to produce new knowledge, Bayesian analysis is popular. But selecting the appropriate prior distribution is not simple, and drawing the wrong conclusion can distort the conclusion. This is the reason some statisticians are uncomfortable using prior distributions, believing that doing so will change "what the data are trying to say." In order to simulate ignorance about prior information, we can either employ a "noninformative" prior or model the prior distribution to represent prior knowledge. The contribution of a collection of mutually exclusive prior occurrences to a posterior state, which can be a prior to another posterior, and so on, can be calculated using Bayesian networks. Bayesian networks are useful tools for inferring final conditions and following conditional causal chains back to their original causes. Probabilistic learning is the foundation of the Bayesian approach. It offers precise conclusions with straightforward interpretations that don't rely on approximations from huge samples. A complete probability model can be estimated using Bayesian inference, which also permits the current model to incorporate past data and outcomes. In order to make decision-making easier, Bayesian inference incorporates uncertainty into the probability model and produces forecasts that are more realistic.

5. SOFTWARE DEFINED IOT NETWORKS FOR EDUCATIONAL BACKBONE NETWORK

The Internet of Things (IoT) refers to a network of smart devices that are wirelessly connected and capable of collecting, transmitting, and analyzing the data. The purpose of IoT networks is to enhance the functionality of intelligent systems and services. Education-related problems such as outdated syllabus, standardized assessments, low student engagement, and inexpert teachers could be resolved with the help of IoT. IoT makes student attendance tracking simpler with biometric systems, guaranteeing precision as well as sending immediate absence reports to parents. It similarly tracks teachers' attendance as well. Another application could be Using IoT for safe as well as defensive maintenance of organization includes using security gates armed with IoT devices. These sensors transmit data to cloud whenever students arrive the university premises, confirming precise monitoring. Parents obtain instant notifications on their smartphones when their child reaches at the campus, which improves security in addition to communication (Al-Taai et al., 2023; Alhasan et al., 2023; Zormati et al.,2024). While there exist, few disciplines dedicated to exploring IoT education applications, there is a dearth of comprehensive analysis about its potential use in the sector of education.

The integration of IoT and SDN has a significant and revolutionary impact in several sectors, including education. The confluence of these factors offers exceptional prospects for enhancing teaching and learning environments, optimizing campus infrastructure, and facilitating the implementation of advanced educational technologies. The IoT is constrained by challenges pertaining to security, scalability, and energy efficiency. By using SDN technology, these challenges may be effectively addressed, leading to enhancements in the IoT as depicted in Figure 3.

Figure 3. SDN-based IoT architecture, and connectivity layout for application in education sector

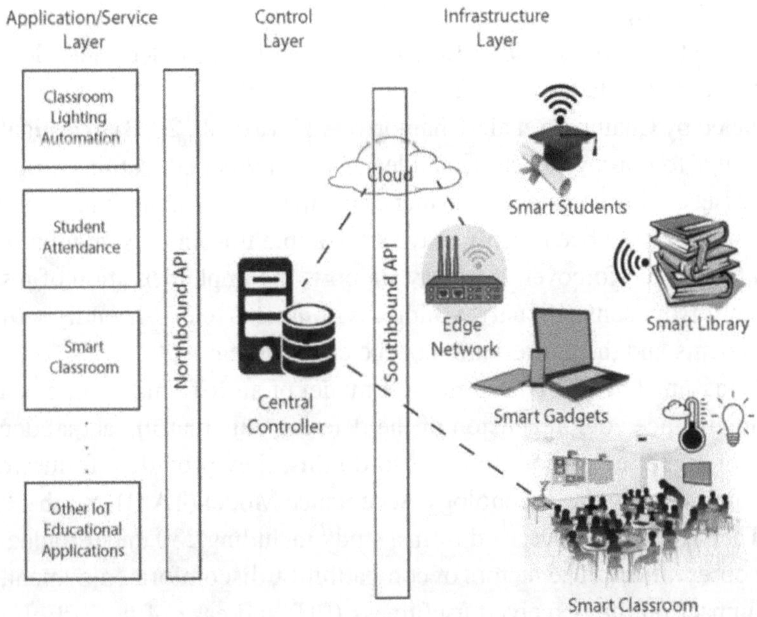

The IoT network solution, based on SDN, offers improved utilization of resources, energy efficiency, and robust security protocols (Yazdinejad et al., 2020; Maleh et al., 2023). SDN enables organizations to implement IoT solutions on a large scale while ensuring strong security measures (Latif et al., 2022). SDN enables administrators to deploy access controls based on policies, partition network traffic, and enforce security policies in distributed IoT systems. By employing encryption, the transmission of critical data across IoT devices is safeguarded, preventing unauthorized access or harmful attacks (Maleh et al., 2023). Moreover, the programmable nature of SDN enables fast implementation of security upgrades and methods to counter changing cyber-security threats, hence protecting IoT environments.

This section focuses on important research findings and potential areas for further investigation. It explores the application of SDN in the IoT and its impact on education.

5.1 Overcoming the Gap for Education With SDN in IoT

SDN offers an adaptable and scalable framework for managing diverse IoT devices, enhancing network efficiency, and facilitating smooth communication and data sharing when integrated into the IoT ecosystem (Zormati et al., 2024). The integration of SDN and IoT in educational contexts has the potential to transform teaching and learning approaches, improve campus connectivity, and enable creative educational applications (Liu et al., 2024).

The COVID-19 epidemic has led to a significant rise in research papers, indicating a growing fascination in the application of IoT technology in education (Almufarreh & Arshad, 2023; Kamruzzaman et al., 2023). This increased attention is further evidenced by contemporary educational procedures. IoT encompasses different types of digital learning, such as U-Learning, which involves collaborative learning environments (Radoyska & Topalova, 2023), M-Learning through portable devices (Radoyska et al., 2023; Bagustari et al., 2023), and E-Learning, which utilizes electronic devices and networks (Radoyska & Topalova, 2023; Haque et al., 2023).

The study conducted by Chamorro et al. (Chamorro-Atalaya et al., 2023) investigates the potential of IoT-based smart settings to improve teaching and learning in higher education by optimizing the use of resources. More precisely, it investigates the influence of smart classrooms, which are widely used smart settings. Smart classrooms have been seen to have a favorable impact on student attitudes, motivation, engagement, and interaction. Moreover, the essay supports the implementation of institutional policies that encourage the shift from conventional learning environments to eco-friendly settings that priorities the well-being of students and the conservation of the environment.

Alhasan et al. (Alhasan et al., 2023) examined a model of an IoT smart campus and its many uses. Their study seeks to enhance comprehension of the determinants that impact students' inclinations to utilize IoT services in higher education. In order to do this, they provide a framework that integrates three external variables, namely the Technology Acceptance Model (TAM) and the Technology Readiness Index (TRI). Their discoveries, derived from a study including 230 undergraduate students, unveil several noteworthy observations. The factors of compatibility, discomfort, enjoyment, and self-efficacy have a substantial impact on the perceived usefulness (PU) and ease of use (PEoU). Insecurity has a significant influence on PU but not on PEoU, whereas inventiveness has a big impact on both PEoU and PU. Moreover, the inclination to utilize IoT services is significantly impacted by PU, offering interesting insights for both researchers and IoT service providers.

In the study conducted by Xia et al. (Xia, Zhang, Tian, & Liu, 2021) the behavior of garbage sorting among college students using IoT bins was evaluated based on 320 questionnaires and 180,417 drop-off data collected from 13,047 participants. Three models were employed to assess garbage sorting behavior: the TAM, the Theory of Planned Behavior (TPB), and the Norm Activation Model (NAM). The findings reveal several significant insights. On average, students were observed to sort 0.426 kg of recyclable waste per week, with a notably higher rate of drop-offs occurring before classes and during meal times, indicating a 46% increase. Moreover, subjective norms exerted a significant influence, leading to a remarkable 220% increase in sorting frequency. Interestingly, the existing level of incentives did not significantly impact the intention to sort waste. Instead, the intention and actual practice of waste sorting were found to be directly influenced by PEoU. These findings offer valuable insights for improving waste sorting practices in public spaces.

Smart education employs electronic technologies that are consistently operational and flexible in providing instructional content, in contrast to conventional methods. The use of IoT sensing devices in the real world poses issues for educators who aim to customize education, utilize various teaching approaches, and offer assistance to students with learning impairments Al-Taai et al., 2023). This project investigates the use of IoT applications to improve educational services for talented and cognitively advanced students, with the aim of better preparing them for the labor market and contributing to national progress.

The study undertaken by Ullah et al. (Ullah et al., 2020) utilizes several IoT devices to develop interactions between students and teachers. The SDN-IMSI model employs the SDN-Based IoT Model for Students' Interaction, which utilizes SDN infrastructure in the context of IoT. This architectural design facilitates smooth and uninterrupted communication and engagement. In addition, NLP methods, including LSA, are employed for evaluating answers. By employing this methodology, the system showcases exceptional precision in contrast to conventional techniques. The automated grading system, which utilizes semantic similarity, demonstrates exceptional accuracy by storing findings on the cloud and comparing them with marks provided by teachers. This automated system surpasses traditional marking methods, providing accurate grading through semantic analysis.

The integration of IoT and SDN revolutionizes libraries by automating tasks such as member authentication and book searches. The design proposed in (Zhou, 2024) enhances efficiency by streamlining processes and employing passive RFID tags and biometric authentication. The study conducted in a medium-sized library resulted in improved system efficiency, as seen by reduced processing time and errors. The effectiveness and potential to be expanded of the SDN-based routing algorithm were verified, leading to an enhancement in network performance metrics.

5.2 Overcoming the Gap for Healthcare With SDN in IoT

SDN's role in healthcare, especially when integrated with IoT, is to provide a more flexible and efficient network topology. This can contribute to better patient care by improving data management, communication networks, and the security of critical health information. The usage of SDN in IoT can assist solve scalability, operation complexity, and security issues that are crucial in healthcare operations (Badotra et al., 2020). SDN provides an adaptive healthcare system, including patient monitoring, data protection, efficient communication, personal health record management, insurance claims, and auditing (Wadhwa et al., 2021).

COVID-19 underscore the need for internet-connected medical services in remote areas and highlight the cruciality of securing the Internet of Medical Things (IoMT) against cyberattacks. SDN-enabled IoT framework for healthcare, integrating machine learning (DNN + SVM) to detect network disturbances and monitor devices, thereby ensuring quick, reliable, and secure patient monitoring (Zeadally, 2024).

The research by (Shitharth et al., 2023), many people in society are facing problems related to health care, and conditions in the body are unfit to be associated with the presence of far-sighted technologies. The major reason for similar failures in the identification process is that no virtual technologies are used in the request. utmost healthcare illusions aim to design a particular operation that provides information only about seeing values and fails to recall the virtual representation of that represented values. thus, this research provides an integration platform that connects the major bias with virtual reality/ audio reality (VR/ AR) ways, which are applied in real time for detecting the presence of infections inside the body. In addition, one type of mass intelligent algorithm is enforced in the recognition procedure with a modified fitness function and its designated fruit cover optimization (FFO). The process of FFO provides much low perception, therefore enhancing the fear for smooth operation. To examine the real- time conditions, the projected AR/ VR procedure is applied with biomedical detectors (Shitharth et al., 2023).

5.3 Future Directions

Exploring the efficacy of SDN-based IoT technologies in enhancing collaborative learning environments might yield valuable insights. There is a need to examine the impact of SDN technology on enhancing resource sharing and communication between faculty and students, hence fostering cooperation and facilitating knowledge exchange. The integration of SDN-based IoT with augmented reality (AR) and virtual reality (VR) technologies in educational settings might provide potential benefits. Further, it is needed to investigate the impact of SDN on network management and data transfer efficiency. Further research may involve exploring broader implementations in larger libraries and other fields, hence enhancing efficacy and capabilities. Prior research often employed smaller data samples. The objective for future study should be to get larger and more diverse sample sizes from many higher education institutions and stakeholders. Utilizing interview approaches can provide more profound insights on the adoption and impacts of SDN and IoT technologies in educational settings. Despite the use of IoT in educational institutions, the implementation of SDN-based IoT remains restricted, especially in impoverished countries. Stakeholders must possess skills in student requirements analysis, software design, and guidelines for efficient usage. Academics need to address a significant research gap by doing more empirical investigations on the impact of SDN-based IoT on student learning performance.

6. CONCLUSION

Cognitive SDN is a promising technology that aims to sense and implement dynamic spectrum management to meet QoS of future networks which interconnect heterogenous devices. This is accomplished by decoupling of data and control plane in terms of spectrum sensing and allocation. NFV decouples from hardware to ensure IoT, Radio Access Networks and AI/ML Integration. Preventing Denial of Service attacks is the other advantage in terms of security. Here the impact of cognitive routing is in intelligent path selection, QoS optimization and under AI as well as Bayesian Learning approach. Exploring Software defined IoT Network and its application in education and medical care summarizes our work.

REFERENCES

Akeela, R., & Dezfouli, B. (2018). Software-defined Radios: Architecture, state-of-the-art, and challenges. *Computer Communications*, 128(March), 106–125. 10.1016/j.comcom.2018.07.012

Al-Ani, A., Laghari, S. A., Manoharan, H., Selvarajan, S., & Uddin, M. (2023). Improved Transportation Model with Internet of Things Using Artificial Intelligence Algorithm. *Computers, Materials & Continua/Computers. Computers, Materials & Continua*, 76(2), 2261–2279. 10.32604/cmc.2023.038534

Al-Taai, S. H. H., Kanber, H. A., & al-Dulaimi, W. A. M. (2023). The Importance of Using the Internet of Things in Education. *International Journal of Emerging Technologies in Learning*, 18(1), 19–39. 10.3991/ijet.v18i01.35999

Alam, I., Sharif, K., Li, F., Latif, Z., Karim, M. M., Biswas, S., Nour, B., & Wang, Y. (2020). A Survey of Network Virtualization Techniques for Internet of Things Using SDN and NFV. *ACM Computing Surveys*, 53(2), 1–40. 10.1145/3379444

Alhasan, A., Hussein, M. H., Audah, L., Al-Sharaa, A., Ibrahim, I., & Mahmoud, M. A. (2023). A case study to examine undergraduate students' intention to use internet of things (IoT) services in the smart classroom. *Education and Information Technologies*, 28(8), 1–24. 10.1007/s10639-022-11537-z

Ali-Ahmad, H., Cicconetti, C., De La Oliva, A., Mancuso, V., Sama, M. R., Seite, P., & Shanmugalingam, S. (2013). An SDN-based network architecture for extremely dense wireless networks. *SDN4FNS 2013 - 2013 Workshop on Software Defined Networks for Future Networks and Services*. IEEE. 10.1109/SDN4FNS.2013.6702534

Almufarreh, A., & Arshad, M. (2023). Promising emerging technologies for teaching and learning: Recent developments and future challenges. *Sustainability (Basel)*, 15(8), 6917. 10.3390/su15086917

Badotra, S., Nagpal, Di., Panda, S. N., Tanwar, S., & Bajaj, S. (2020). IoT-Enabled Healthcare Network with SDN. *ICRITO 2020 - IEEE 8th International Conference on Reliability, Infocom Technologies and Optimization (Trends and Future Directions)*, (pp. 38–42). IEEE. 10.1109/ICRITO48877.2020.9197807

Baggio, G., Bassoli, R., & Granelli, F. (2019). Cognitive Software-Defined Networking Using Fuzzy Cognitive Maps. *IEEE Transactions on Cognitive Communications and Networking*, 5(3), 517–539. 10.1109/TCCN.2019.2920593

Bagustari, B. A., Yazid, S., & Alshehadeh, A. R. (2023). Analyzing fog computing technologies for enhancing m-learning networked environments. *International Journal of Advances in Soft Computing & Its Applications*, 15(1).

Bizanis, N., & Kuipers, F. A. (2016). SDN and Virtualization Solutions for the Internet of Things: A Survey. *IEEE Access : Practical Innovations, Open Solutions*, 4, 5591–5606. 10.1109/ACCESS.2016.2607786

Chamorro-Atalaya, O., Morales-Romero, G., Quispe-And'ıa, A., Caycho-Salas, B., Ramos-Salazar, P., C'aceres-Cayllahua, E., & Auqui-Ramos, R. (2023). Smart environments through the internet of things and its impact on university education. *Systematic Reviews*.

Chen, G. (2022). Simulation model of AI-assisted cognitive routing algorithm for the dynamic optical network in business. *Optik (Stuttgart)*, 260, 168763. 10.1016/j.ijleo.2022.168763

Cicio lu, M., Cicio lu, S., & Çalhan, A. (2020). SDN-enabled cognitive radio network architecture. *IET Communications*, 14(18), 3153–3160. 10.1049/iet-com.2019.1346

Foerster, K., Schmid, S., & Vissicchio, S. (2019). Survey of Consistent Software-Defined Network Updates. *IEEE Communications Surveys and Tutorials*, 21(2), 1435–1461. 10.1109/COMST.2018.2876749

Francois, F., & Gelenbe, E. (2016). Optimizing Secure SDN-Enabled Inter-Data Centre Overlay Networks through Cognitive Routing. *2016 IEEE 24th International Symposium on Modeling, Analysis and Simulation of Computer and Telecommunication Systems (MASCOTS)*, (pp. 283–288). IEEE. https://doi.org/10.1109/MASCOTS.2016.26

Gupta, A., Vanbever, L., Shahbaz, M., Donovan, S., Schlinker, B., Feamster, N., Rexford, J., Shenker, S., Clark, R. D., & Katz-Bassett, E. (2014). SDX. *Computer Communication Review*, 44(4), 551–562. 10.1145/2740070.2626300

Haque, M. A., Haque, S., Zeba, S., Kumar, K., Ahmad, S., Rahman, M., & Ahmed, L. (2023). Sustainable and efficient e-learning internet of things system through blockchain technology. *E-Learning and Digital Media*, 20427530231156711.

Huo, L., Jiang, D., Qi, S., Song, H., & Miao, L. (2021). An AI-Based Adaptive Cognitive Modeling and Measurement Method of Network Traffic for EIS. *Mobile Networks and Applications*, 26(2), 575–585. 10.1007/s11036-019-01419-z

Janjua, H. K., de Miguel, I., Durán Barroso, R. J., Masoumi, M., Hosseini, S., Aguado, J. C., Merayo, N., & Abril, E. J. (2023). A Framework for Next Generation Cloud-Native SDN Cognitive Resource Orchestrator for IoTs (NG2CRO). *Lecture Notes in Networks and Systems, 741*, 399–407. 10.1007/978-3-031-38318-2_39

Jawad, F. H. M. (2018). Software defined networking: An overview. *Journal of Engineering and Applied Sciences, 13*, 10790–10796. https://doi.org/10.36478/jeasci.2018.10790.10796

Kaliyamurthy, N. M., Taterh, S., Shanmugasundaram, S., Saxena, A., Cheikhrouhou, O., & Ben Elhadj, H. (2021). Software-Defined Networking: An Evolving Network Architecture - Programmability and Security Perspective. *Security and Communication Networks*, 2021, 1–7. 10.1155/2021/9971705

Kamruzzaman, M., Alanazi, S., Alruwaili, M., Alshammari, N., Elaiwat, S., AbuZanona, M., Innab, N., Mohammad Elzaghmouri, B., & Ahmed Alanazi, B. (2023). Ai-and iot-assisted sustainable education systems during pandemics, such as covid-19, for smart cities. *Sustainability (Basel)*, 15(10), 8354. 10.3390/su15108354

Kreutz, D., Ramos, F. M. V., Verissimo, P. E., Rothenberg, C. E., Azodolmolky, S., & Uhlig, S. (2015). Software-defined networking: A comprehensive survey. *Proceedings of the IEEE*, 103(1), 14–76. 10.1109/JPROC.2014.2371999

Kurimoto, T., Urushidani, S., Yamada, H., Yamanaka, K., Nakamura, M., Abe, S., Fukuda, K., Koibuchi, M., Ji, Y., Takakura, H., & Yamada, S. (2016). A fully meshed backbone network for data-intensive sciences and SDN services. *2016 Eighth International Conference on Ubiquitous and Future Networks (ICUFN)*, (pp. 909–911). IEEE. 10.1109/ICUFN.2016.7536928

Kurimoto, T., Urushidani, S., Yamada, H., Yamanaka, K., Nakamura, M., Abe, S., Fukuda, K., Koibuchi, M., Takakura, H., Yamada, S., & Ji, Y. (2017). SINET5: A low-latency and high-bandwidth backbone network for SDN/NFV Era. *2017 IEEE International Conference on Communications (ICC)*, (pp. 1–7). IEEE. 10.1109/ICC.2017.7996843

Latah, M., & Toker, L. (2019). Artificial intelligence enabled software-defined networking: A comprehensive overview. *IET Networks*, 8(2), 79–99. 10.1049/iet-net.2018.5082

Latif, S. A., Wen, F. B. X., Iwendi, C., Li-Li, F. W., Mohsin, S. M., Han, Z., & Band, S. S. (2022). Ai-empowered, blockchain and sdn integrated security architecture for iot network of cyber physical systems. *Computer Communications*, 181, 274–283. 10.1016/j.comcom.2021.09.029

Lin, P., Bi, J., & Wang, Y. (2013). East-West Bridge for SDN network peering. In *Communications in computer and information science* (pp. 170–181). Springer. 10.1007/978-3-642-53959-6_16

Liu, L. L. L. (2024). Current situation and innovative methods of brass music teaching based on network information technology. *Journal of Electrical Systems*, 20(1), 294–312. 10.52783/jes.683

Maleh, Y., Qasmaoui, Y., El Gholami, K., Sadqi, Y., & Mounir, S. (2023). A comprehensive survey on sdn security: Threats, mitigations, and future directions. *Journal of Reliable Intelligent Environments*, 9(2), 201–239. 10.1007/s40860-022-00171-8

Nderba, W. (2017). *Mint Capstone Project: Design and implementation of QoS aware Software Defined.* Research Gate.

Nowak, M., Nowak, S., & Domanska, J. (2019). Cognitive routing for improvement of IoT security. *Proceedings - 2019 IEEE International Conference on Fog Computing*. IEEE. 10.1109/ICFC.2019.00014

Peng, M., Li, Y., Zhao, Z., & Wang, C. (2015). System architecture and key technologies for 5G heterogeneous cloud radio access networks. *IEEE Network*, 29(2), 6–14. 10.1109/MNET.2015.7064897

Radoyska, P., & Topalova, I. (2023). A conceptual model for open u-learning platform. *Annals of DAAAM & Proceedings, 34* .

Ray, P. P., & Kumar, N. (2021). SDN/NFV architectures for edge-cloud oriented IoT: A systematic review. *Computer Communications*, 169, 129–153. 10.1016/j.comcom.2021.01.018

Sharma, S., & Nag, A. (2023). Cognitive Software Defined Networking and Network Function Virtualization and Applications. *Future Internet*, 15(2), 78. 10.3390/fi15020078

Shitharth, S., Yonbawi, S., Manoharan, H., Alahmari, S., Yafoz, A., & Mujlid, H. (2023). Physical Stint Virtual Representation of Biomedical Signals With Wireless Sensors Using Swarm Intelligence Optimization Algorithm. *IEEE Sensors Journal*, 23(4), 3870–3877. 10.1109/JSEN.2022.3233407

Srivastava, V., & Pandey, R. S. (2021). Machine intelligence approach: To solve load balancing problem with high quality of service performance for multi-controller based Software Defined Network. *Sustainable Computing : Informatics and Systems*, 30, 100511. 10.1016/j.suscom.2021.100511

Sun, G., Liu, G., & Wang, Y. (2014). SDN architecture for cognitive radio networks. *2014 1st International Workshop on Cognitive Cellular Systems*. CCS. 10.1109/CCS.2014.6933795

Ullah, F., Wang, J., Farhan, M., Jabbar, S., Naseer, M. K., & Asif, M. (2020). Lsa based smart assessment methodology for sdn infrastructure in iot environment. *International Journal of Parallel Programming*, 48(2), 162–177. 10.1007/s10766-018-0570-1

Wadhwa, S., Babbar, H., & Rani, S. (2021). A survey on emerging software-defined networking and blockchain in smart health care. *IOP Conference Series. Materials Science and Engineering*, 1022(1), 012056. 10.1088/1757-899X/1022/1/012056

Wang, L., & Delaney, D. T. (2019). QoE oriented cognitive network based on machine learning and SDN. *2019 IEEE 11th International Conference on Communication Software and Networks*. IEEE. 10.1109/ICCSN.2019.8905343

Wu, Y. J., Hwang, P. C., Hwang, W. S., & Cheng, M. H. (2020). Artificial intelligence enabled routing in software defined networking. *Applied Sciences (Basel, Switzerland)*, 10(18), 6564. 10.3390/app10186564

Xia, W., Wen, Y., Foh, C. H., Niyato, D., & Xie, H. (2015). A Survey on Software-Defined Networking. *IEEE Communications Surveys and Tutorials*, 17(1), 27–51. 10.1109/COMST.2014.2330903

Xia, Z., Zhang, S., Tian, X., & Liu, Y. (2021). Understanding waste sorting behavior and key influencing factors through internet of things: Evidence from college student community. *Resources, Conservation and Recycling*, 174, 105775. https://www.sciencedirect .com/science/article/pii/S0921344921003840. 10.1016/j.resconrec.2021.105775

Yazdinejad, A., Parizi, R. M., Dehghantanha, A., Zhang, Q., & Choo, K.-K. R. (2020). An energy-efficient sdn controller architecture for iot networks with blockchain-based security. *IEEE Transactions on Services Computing*, 13(4), 625–638. 10.1109/TSC.2020.2966970

Zeadally, S. (2024). *An intelligent SDN-IoT enabled intrusion detection system for healthcare systems using a hybrid deep learning and machine learning approach*. IEEE. 10.23919/JCC.ja.2022-0681

Zhou, Q. (2024). Smart library architecture based on internet of things (iot) and software defined networking (sdn). *Heliyon*, 10(3), e25375. 10.1016/j.heliyon.2024.e2537538371976

Zormati, M. A., Lakhlef, H., & Ouni, S. (2024). Review and analysis of recent advances in intelligent network softwarization for the internet of things. *Computer Networks*, 241, 110215. 10.1016/j.comnet.2024.110215

Chapter 11
Online Fraud Detection Using Machine Learning Algorithms and Quantum Annealing Solvers

B. Akshaya
Sathyabama Institute of Science and Technology, India

Agnimita Das Bakshi
Sathyabama Institute of Science and Technology, India

Annie Micheal A.
Sathyabama Institute of Science and Technology, India

ABSTRACT

Machine learning applications are widely used in identifying fraudulent transactions. Still, the majority of the applications detect fraudulent activities only after their occurrence, rather than in real-time or near real- time. The significant rarity of fraudulent transactions compared to normal ones creates a highly imbalanced data, posing substantial challenges for effective fraud detection. Addressing this issue extends beyond traditional machine learning approaches and requires alternative strategies. This chapter discusses the importance of fraud detection methods and compares k-nearest neighbor (KNN), random forest approach, artificial neural network (ANN) and support vector machine (SVM) enhanced with quantum annealing solvers that are used to detect fraud in online banking transactions. This study focuses on choosing the optimal approach for various types of datasets, while considering the three key factors such as accuracy, speed, and cost.

Fraudulent activities impose significant financial burdens on businesses. The annual cost of fraudulent insurance claims to UK insurers amounts to 1.6 billion pounds, as reported by BBC news. The overall financial impact of fraudulent transactions is immeasurable.

Implementing effective detection systems can help businesses mitigate losses resulting from fraudulent transactions. However, tackling these incidents proves challenging for three primary reasons. Firstly, the widespread usage of internet has significantly increased online transactions. In the initial months of 2020, the US alone witnessed a 110% surge in online commercial transactions compared to previous year, resulting in increased cyber-attacks on electronic commerce vendors and associated fraudulent operations. Secondly, the efficacy of number of existing systems is weakened. Although early fraud

DOI: 10.4018/979-8-3693-2893-4.ch011

detection is crucial for online e-commerce transactions, most of them detect fraud after it has already happened, which means the financial loss has already occurred. Thirdly, fraudulent transactions are less frequent compared to normal ones, so the datasets exhibit a significant imbalance, treating the anomalous transactions as outliers or anomalies.

Differentiating normal transactions from fraudulent ones is frequently challenging. Normal transactions may exhibit numerous unusual data points with extreme values, while fraudulent transactions may display monetary amounts per transaction that appear 'normal.' As a result, fraudulent transactions often exhibit patterns that are either uncertain/weak or spread out, when compared with the patterns observed in normal transactions, making them difficult to detect. As technological advancements pose Our focus is on addressing the other two issues in fraud detection, as the first challenge is outside the purview of this study. To improve timely fraud detection efficacy, we combine quantum annealing solvers with machine learning methods.

In noisy and complex domains, the classification problems can be solved using Support Vector Machine (SVM). In the field of machine learning, SVM significantly contributes to generalization of performance across various problems, including handwritten digital recognition, web page classification and face detection. The question of over fitting in SVM applications is much less. In SVM, the minimal issue and curse of dimensionality seldom arises. SVM has been used in business applications such as bankruptcy prediction, time series prediction and classification for a few days now. SVM is employed to assess the data for the purpose of identifying fraudulent activities.

In addition to the mentioned machine learning algorithms, we have implemented a deep learning technique known as 'Artificial Neural Network'. It is inspired by the architecture of human brain where interconnected neurons reflect the connections between nodes in the artificial neural network.

2. RELATED WORK

Researchers and banking institutions have thoroughly investigated this problem statement over the last six years in order to develop solutions. Various approaches have been explored, ranging from analyzing different machine learning algorithms to evaluating the performance of these algorithms based on accuracy scores. This research was conducted before taking up the project for a comprehensive understanding of the methods employed in prior studies.

Here, we will take a look at some of the strategies used for internet banking fraud detection. In earlier studies, Bhattacharyya et al. (2011) proposed a solution for detecting online banking fraud by gathering transaction-related data. The characteristics of transactions collected are then employed to identify fraudulent transactions through a comparison with the consumer's transaction history.

In Ngai et al. (2011), the authors explored a solution to the challenge of fraud detection in online banking. They employed a cost-effective decision tree method, aiming to decrease the overall cost of categorization in the process.

In Kaggle.com (2022), the issue of credit card fraud is addressed using data mining techniques. The paper assesses the effectiveness Combining logistic regression and two sophisticated data mining techniques: random forest and support vector machines. note that the method introduced in this study has higher time complexity.

In Columbus (2020), Kim and his colleagues introduced a hybrid model for detection of credit card fraud. Effective fraud detection helps investigators should move quickly to prevent further fraud or monetary losses could be prevented. But in practical terms, investigators face limitations in reviewing alerts on a daily basis due to time-consuming nature of the investigation process. Consequently, this model's main objective is to offer accurate alerts and lower the number of false alarms and missed fraud cases. Even though recent research has compared deep learning and conventional machine learning models, it frequently ignores the practical limitations of fraud detection systems, such as, (i) investigators' limited availability for assessing high-risk transactions identified by scoring algorithms based on data, (ii) the different expenses incurred by false alarms and missed frauds. investigators should move quickly to prevent further fraud or monetary losses. In order to ascertain whether it is feasible to switch from the existing hybrid ensemble model to a deep learning-based system, this study thoroughly examines both hybrid ensemble and deep learning techniques. Nonetheless, it has been shown that as the number of investigators increases, the precision of the results decreases.

In order to improve techniques for detection of online banking fraud, various techniques are often used together (Lexis/Nexis Solutions, 2020). A hybrid method was presented by Anil and his colleague in 2013, utilizing both support vector machine and a genetic algorithm to detect the anomalies in a dataset. The study “A Combinatorial Approach utilizing Support Vector Machine, Self- Organizing Algorithm, and Genetic Algorithm for Improved Abnormality Detection” discusses the use of genetic algorithm to select the optimal sets of values that indicate abnormalities. By using Support Vector Machine algorithm, online banking fraud detection is improved by a great degree.

2.1 Inferences From Literature Survey

From the literature survey, it is clear that there needs to be more than one algorithm so that results with better accuracy can be obtained. Here, we are going to use four algorithms in order to get the most accurate result.

SVM algorithm works faster with bank loan datasets than most of the other algorithms. So, we are going to use SVM algorithm.

In some of the previous cases, we can see less accurate results being obtained, in order to analyse the critical points better, we are going to use quantum annealing solvers along with machine learning algorithms.

3. PROPOSED WORK

Proposed machine learning models have not adequately addressed the misdiagnosis problem. Also, most of the proposed models lack consideration for the size and heterogeneity of data. We respond by introducing a method based on machine learning. that integrates data preprocessing technique for feature transformation, K-Nearest Neighbor (KNN), Random Forest Classifier, Support Vector Machine (SVM), and Artificial Neural Network (ANN) to provide the most accurate technique that effectively eliminates bias and minimizes deviation from stability.

Here, we take an in-depth look at the above-mentioned algorithms.

3.1 Support Vector Machine

Support vector machine, or SVM, is one of the most widely used supervised learning methods for problems involving both regression and classification. To make it simple to classify new/incoming data points in an n-dimensional space, SVM creates a decision boundary that separates the space into classes. We call this optimal decision boundary as a hyperplane. This algorithm identifies the critical points which are required to form the hyperplane and these points can be referred to as support vectors.

Figure 1. Support vector machine

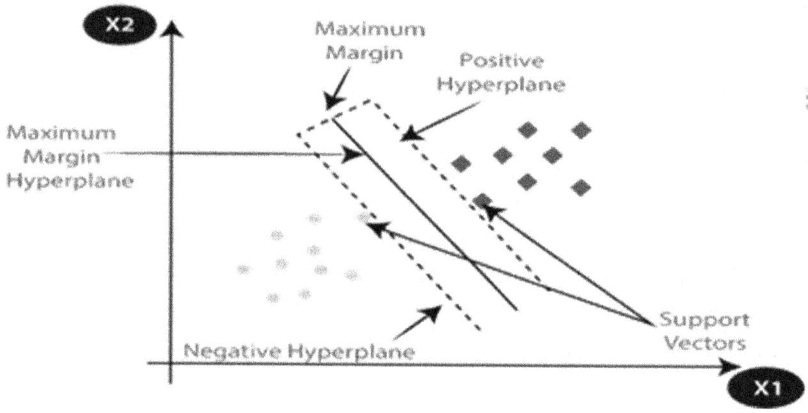

3.2 K-NN Algorithm

K-Nearest Neighbor or KNN algorithm retains all available data and it can classify a new data point by evaluating its similarity to existing data. This implies that, whenever a new data point emerges, it can be readily categorized to an appropriate category using the KNN algorithm. This technique is predominantly applied for Classification problems.

Figure 2. KNN algorithm

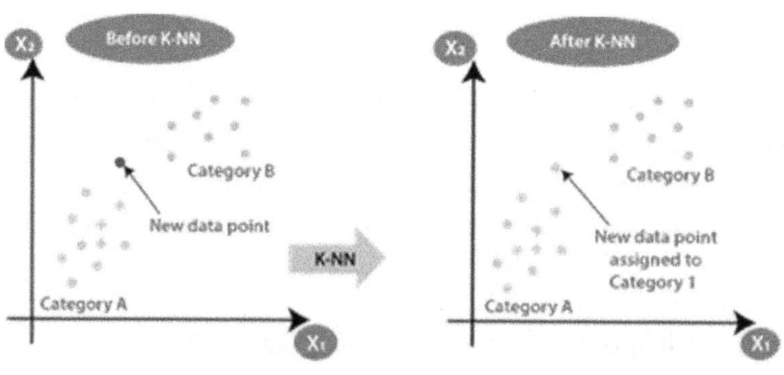

3.3 Random Forest

A machine learning technique called Random Forest approach, is used for addressing problems that involve both regression and classification. By employing ensemble learning, it integrates multiple classifiers to resolve intricate problems. This algorithm operates by making predictions based on the collective outputs of decision trees. Prediction is often achieved by averaging or taking the average of several trees' outputs. The accuracy of the result improves with an increase in the number of trees. Notably, random forest approach helps to reduce the drawbacks associated with decision tree algorithms. It enhances overall accuracy by reducing overfitting problems in datasets.

3.4 Artificial Neural Network

ANNs, or Artificial Neural Networks are typically used as classifiers, providing a robust mechanism for identifying hidden patterns.

The architecture of an ANN comprises multiple layers, with the initial layer labeled as input layer and then the final layer as output layer. The flexibility of ANN allows for the inclusion of several hidden layers. If there are multiple hidden layers, the neural network is classified as Deep Learning.

During the training phase, an ANN gains the ability to recognize patterns in any type of data- visual, aural, or literary. The network compares its actual output with expected/desired output during the supervised learning process. The disparity between these outcomes is corrected through a process named 'backpropagation'. In backpropagation, the network retroactively adjusts the weights of its connections, progressing from output unit to input units. This iterative adjustment minimizes the error between the actual and desired outcomes, refining the network's ability to produce accurate results.

Figure 3. Artificial neural network

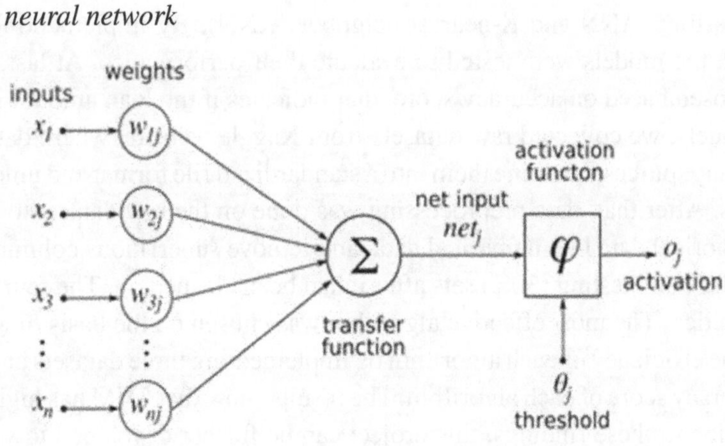

The process flow diagram of the project looks like the following:

Figure 4. System architecture

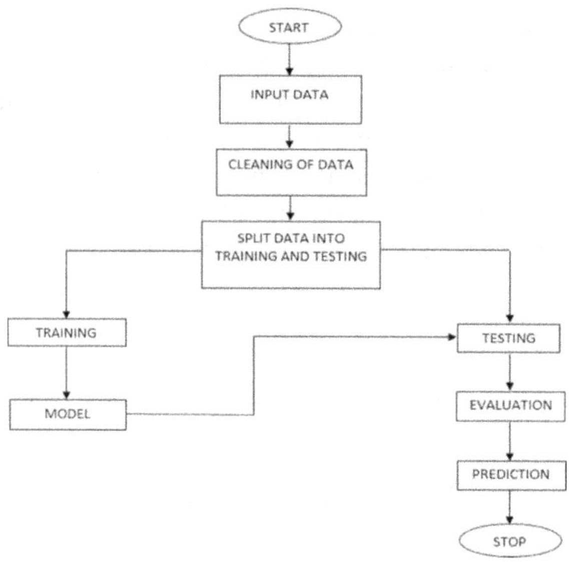

4. RESULTS AND DISCUSSION

The proposed model involves the use of various algorithms such as, support vector machines (SVM), Random Forest Classifier, ANN and K-nearest neighbor (KNN). By implementing these algorithms on the given dataset, the models were tested to evaluate their performance. At last, the most effective algorithm can be chosen based on accuracy score, that indicates if the loan amount is fully paid or not. To explain it elaborately, we collected raw datasets from Kaggle website, where data engineers gather information from many sources, combine them into a standardized file format and upload them in Kaggle and Github websites. After that, data preprocessing was done on the raw dataset to fix things like null values, convert categorical data into numerical data, and remove superfluous columns. The dataset was split into training (70%) and testing (30%) sets after it had been cleaned up. The four chosen algorithms were all put into practice. The most effective algorithm was chosen on the basis of accuracy scores.

We have tested the efficiency of each algorithm by implementing three datasets and taking average of them to find the accuracy score of each algorithm. The results show that SVM has highest accuracy score on average. In addition to these findings, this project can be further developed to work with real-time bank loan datasets which requires the use of complex techniques which is beyond the scope of this study.

Advantages of proposed model- maximum accuracy, reduced time complexity, ease of usage.

The below table compares the accuracy scores of various techniques when applied to bank loan datasets:

Table 1. Comparative analysis

S.no	Techniques	Advantages	Accuracy
1.	SVM (Support Vector Machine)	Reduces the risk of overfitting problems in datasets, and also gives better results when compared to other models.	97.32%
2.	KNN (K-Nearest Neighbor)	Simple to implement and robust to the noisy training data.	85%
3.	Random Forest approach	Compared to the decision tree algorithm, it is more accurate. It is an efficient method for dealing with missing data.	89%
4.	(ANN) Artificial Neural network	Very effective with complex data, and perfect for detecting fraud in real- time.	93.41%

5. CONCLUSION

In this paper, we bring together different strategies to predict and detect online banking fraud. This paper also compares and analyses advantages and limitations of Random Forest, KNN, SVM and ANN methods of fraud detection. In addition to these findings, this paper lays the groundwork for further exploration into using artificial intelligence for the development of a resilient and efficient online banking fraud detection system.

REFERENCES

Bhattacharyya, S., Jha, S., Tharakunnel, K., & Westland, J. C. (2011). Data mining for credit card fraud: A comparative study. *Decision Support Systems*, 50(3), 602–661. 10.1016/j.dss.2010.08.008

Columbus, L. (2020). *How E-Commerce's Explosive Growth is Attracting Fraud.* Research Gate.

Kaggle.com. (2022). *Credit Card Fraud Detection.* Kaggle. https://www.kaggle.com/mlg-ulb/creditcardfraud

Lakshmi, S. V. S. S., & Kavilla, S. D. (2018). Machine learning for credit card fraud detection system. *International Journal of Applied Engineering Research: IJAER*, 13(24), 16819–16824.

Lexis/Nexis Solutions. (2020). *True Cost of Fraud Study: E-Commerce/Retail Edition*. LexisNexis Solutions.

Ngai, E. W. T., Hu, Y., Wong, Y. H., Chen, Y., & Sun, X. (2011). The application of data mining techniques in financial fraud detection: A classification framework and an academic review of literature. *Decision Support Systems*, 50(3), 559–569. 10.1016/j.dss.2010.08.006

Chapter 12
Predictive Analysis of Modulation Classification and Adaptive Modulator for Next Generation Networks

Rabia Bilal
UIT University, Pakistan

Bilal Muhammad Khan
National University of Sciences and Technology, Islamabad, Pakistan

ABSTRACT

Cognitive radios are used for dynamic spectrum management and interference identification. Automatic modulation classification is used for identification of modulation scheme of received signal. Modulated signals have a Cyclostationary property, through which they are identified and then further processed. Cyclostationary feature detection method detects the Cyclostationary features and then identifies the modulation type. The cyclic frequency domain profile that is extracted from spectral correlation function is used for the classification of modulation. Each modulation scheme has a different cyclic frequency domain profile.

1. INTRODUCTION

Modulation classification was being initially used for military scenarios where electronic warfare, surveillance and threat analysis required the modulations to be classified in order to know adversary transceivers, to enhance jammers, and to recover the information from the received signal. Considerable research was carried out in the field of signal processing and communications to implement automatic modulation classification on FPGA and DSP. Modulation Classification is still being used for military

DOI: 10.4018/979-8-3693-2893-4.ch012

purposes and more research in the field of Signal Processing and Communication Systems is being carried out.

Recently there have been many innovations in communications technology. Automatic Modulation Classification and Coding optimizes the reliability of the signal and data transmission rate by adaptively selecting modulation schemes according to channel conditions. The transmitter can select modulation scheme from the programmed modulations in RF Device, the receiver must have the knowledge of the modulation type to demodulate the signal so that the transmission can be successful. Automatic Modulation Classification System can identify the Modulation type of each frame when the Modulation type is changing rapidly. This Technique has become an integral part of intelligent radio systems, including cognitive radio and software-defined radio. In this chapter, Automatic modulation classification technique is used for classifying the modulation format of received signals and Adaptive Modulator has also been simulated for AWGN Channel with adjustments made in the Modulation Classifier for Reliable Communication System. It has multiple applications. In Military Application, the received signal can be demodulated by identifying the modulation type, and then appropriate signals can be generated. In Civilian Application, Modulation classification and Adaptive Modulator enables the optimization of the transmission reliability and data rate through the adaptive selection of modulation schemes according to channel conditions. The Objective of this research was to learn how we can classify the received modulated signals, and according to the channel conditions, required message signals can be adaptively modulated. Channelization can also be used for utilizing the maximum bandwidth of RF Signal by assigning channels of RF Signal to enhance Data Rate by splitting data in the form of Arrays. Then the data can be transmitted using maximum possible RF Bandwidth according to the channel conditions and also changing the modulations can also change data rate.

The Research has been motivated by the requirement of higher data rates that are in demand for good quality communication on devices. Most of the spectrum has already been utilized. Cognitive radio addresses the problems by detecting and using the spectrum holes. Cognitive Radios estimate the Signal to Interference plus Noise Ratio (SINR) of the channel, and also uses adaptive modulation (Mohammadi & Kwasinski, 2018). Adaptive modulator changes the modulation format and maximizes throughput. Automatic Modulation Classification (AMC) is done by cyclic feature detection (Spooner & Mody, 2018). This Research is also applicable to National needs because Research is also being carried out to improve and produce Indigenously Manufactured Communication Equipments. AMC has an essential role in many military strategies. Electronic warfare (EW) consists of three components: electronic support (ES), electronic attack (EA) and electronic protect (EP). In ES, the information is gathered from radio frequency emissions. Automatic Modulation Classification is carried out after the signal detection is successfully completed. The resulting modulation information is useful for all the components in EW. It also has applications in Aircraft Communication using different Modulation Techniques for High Data Rate Communication for Aircraft to Aircraft Communication and Aircraft to Ground Communication and Drone Communication and Spectrum Sensing. Modulation Classification technique enables the transceiver to detect the Modulation type, the modulation type can be sent to Adaptive Modulator so that the signal can be modulated using the current modulation type, and the channel SNR can also be sensed and appropriate modulation can be utilised. BPSK is used for highly reliable communication, and higher order modulations like QAM can be used for high data rate communication.

Cyclostationary features and modulation classification present several challenges, including:

1.1 Complexity

Analyzing cyclostationary features requires sophisticated signal processing techniques and algorithms, making it computationally intensive.

1.2 Variability

Cyclostationary features can vary significantly due to factors such as environmental conditions, propagation effects, and interference, making it challenging to develop robust classification algorithms.

1.3 Dimensionality

The high dimensionality of cyclostationary feature spaces poses challenges for feature selection and classification, as it may lead to the curse of dimensionality and overfitting.

1.4 Dynamic Spectrum Allocation

Efficient spectrum allocation relies on accurate modulation classification to avoid interference and optimize spectrum utilization. However, the dynamic nature of spectrum allocation complicates the task, as modulation types and signal characteristics may change over time and across different frequency bands.

1.5 Interference and Noise

Cyclostationary features can be affected by interference and noise, which can degrade classification performance and make it difficult to distinguish between different modulation types.

1.6 Resource Constraints

In practical applications such as cognitive radio systems, there may be constraints on computational resources, power consumption, and processing time, which need to be considered when designing modulation classification algorithms based on cyclostationary features.

To address these challenges, researchers are continuously exploring new algorithms, techniques, and methodologies for efficient and accurate modulation classification using cyclostationary features. Additionally, collaboration between signal processing experts, communication engineers, and policymakers is essential to develop effective strategies for spectrum allocation and management considering modulation classification techniques.

2. LITERATURE REVIEW

2.1 Introduction

Opportunistic Spectrum Access is a modernized approach to tackle the spectrum congestion problem. Cognitive Radio (CR) is able to access spectrum hole, hence it can efficiently enhance electromagnetic spectrum usage. Cognitive radio (CR) is an innovation in communication technology, which is able to use different techniques for awareness of the surroundings. The main problem with the cognitive radio systems is to reduce interference. The Energy detection technique is a simple technique and it is efficient at higher SNR, but at lower SNR, its performance degrades. Adaptive modulator changes modulation type according to the SNR of the signal, and can also be used to maximize data rate. It is a part of modern systems. Dynamic Spectrum Access (DSA) is being utilized to overcome spectrum congestion. In a communication system, there is cooperation between the devices, so the receiver can demodulate a signal if the modulation format of the transmitted signal is known. The communication devices need knowledge of modulation type, nominal carrier frequency and modulation index in order to maintain stable communication systems. In digital communication systems (DCS), the receiver needs knowledge of modulation type, bits per symbol, deviation (for frequency modulated signals only), nominal carrier frequency, symbol constellation, frequency and the symbol rate.

In communication systems, the modulation classification was being done by using the banks of demodulator, where each of the banks was used for demodulating one modulation type. Demodulator output was the demodulated signal, corrupted by channel impairments. Implementation of modulation classification system is complicated, requiring storage. The modulated signals were being demodulated by demodulator's bank. With the advances in software-defined radios (SDRs), recently radio frequency (RF) communication devices can transmit in low power, alter the transmitting frequency, and change the modulation format during transmission. Adaptive modulation changes the data rate according to the channel conditions. If the modulation is being changed without sending the modulation information, then RF signals can be recognized by Automatic Modulation Classification. Automatic Modulation Classification (AMC) which is a widely used feature on receiver for adaptation. Link adaptation is also known as adaptive modulation and coding, develops an adaptive modulator scheme where a pool of different modulators have been installed in a system. It optimizes the transmission reliability and data rate by adaptively selecting modulation schemes according to the channel conditions. While transmitter is free to modulation format. The receiver needs to know the modulation format for demodulating the signal. For this purpose, modulation information is to be added in each frame, so that receiver can demodulate the modulated signal. This approach will affect the spectrum usage due to extra information about modulation in each frame. Automatic modulation classification is the solution.

Cyclic feature detection works well in noise, as well as carrier frequency offset (Federal Communications Commission, 2002). The techniques used for Modulation Classification include Energy Detection, Matched Filter Detection, Cyclostationary Feature Detection (Bhargavi & Murthy, 2010; Dobre et al., 2007). Out of these techniques Cyclostationary Feature Detection performs better (Bhargavi & Murthy, 2010; FCC, 2015; Yucek & Arslan, 2009). The automatic modulation classification (AMC) is performed between decoding of the signal carried by the RF wave and the demodulation process. AMC is a method for the classification of the modulation formats of the received signal on the receiver side. Automatic modulation classification has applications in the field of electronic warfare, military areas, electronic counter measures, public security, software defined radios and lately in cognitive radios. In military domains it is used for spectrum monitoring and communication intelligence, whereas in civil domain it includes data rate enhancement using Adaptive Modulation, spectrum management. Usage of Cognitive Radios in Communications is increasing because the Physical Layer research can be easily carried out by programming the physical layer using Software Defined Radio. With their ease of programmability and high bandwidth, they are ideal hardware platform choice for prototyping and research (Abid et al., 2022). In comparison to Application Specific Integrated Circuits (ASICs), Field Programmable Gate Array (FPGA) designs are cheaper and easier to develop due to their programmability, so they are also used in the development of low cost systems that have computing requirements beyond what a typical processor could offer. Historically, the FPGA in the Software Defined Radio was used for performing the detection of spectrum hole (Cabric et al., 2004), spectrum sensing based on energy (Mitola & Maguire, 1999), and a system performing the role of Cognitive Radio (Frasch, 2017). Energy Detection Technique performs better in high SNR (Khan & Ali, 2011; Memon et al., 2021).

Cyclic feature detection was being used to determine the type of modulation being performed (Digham et al., 2003), but recently the Cognitive Radios are utilizing this type of classification method for modulation classification (Bhargavi & Murthy, 2010). It has portrayed the capability of classifying BPSK, QPSK, MSK, and FSK with an accuracy of at least 95% if the SINR was greater than 1 dB. Furthermore, when 4th order cummulants were combined, QAM signals were able to be detected. Compressed Sensing is also being used in Cognitive Radio (Gardner, 1988). The energy detection technique and power-spectral density do not have accuracy of classifying Modulation technique that Cyclic feature detection provides for Spectrum Monitoring, even at low Signal Strength (Ramkumar, 2009). By the use of Compressed Sensing, Cognitive Radios do not need to reconstruct the original signal from the Primary User precisely, Compressed Sensing provides good way to reduce the computing power* required, especially for spectrum sensing. When applied to AMC, the accuracy is greatly decreased (Tian et al., 2012). The Fuzzy Logic Approach for Cooperative Sensing performs better in Centralized Cognitive Radio Network as compared to local individual decision (Lagunas & Najar, 2015).

2.2 Cognitive Radios

The improvement in the radios systems is that, the radios can now be programmed to learn and autonomously make "intelligent" decisions, like altering the transmit power to avoid wasting energy, which gives it the new name Cognitive Radio (CR). Another key component of Cognitive Radios is Dynamic Spectrum Access, which determines how they use and reuse spectrum. Most bands of the spectrum are allocated by the FCC (Xie & Wan, 2017), meaning that only licensed bands can be used for transmission on those frequency bands. As the need for the radio spectrum increases, the spectrum is becoming congested, in this situation limiting the bands to a single user becomes unrealistic. Cognitive Radios enable effective spectrum sharing without interfering with each other's communications. Normally, this function is performed by MAC Scheme, like Time Division Multiple Access or Frequency Division Multiple Access. The major goal of Cognitive Radio is to maintain interference free communication. Thus different access schemes for Cognitive Radio are needed, known as Dynamic Spectrum Access. There are three types of DSA: Underlay, Interweave, and Overlay (Zhang & Wang, 2011). Interweave involves finding spectrum holes, which are times when a channel not being used, and the Cognitive Radio can use the spectrum holes for communication. Overlay requires the Cognitive Radio to have knowledge about the Primary User's transmissions, and be able to decode them. Using this knowledge, the CR can transmit such that it does not interfere with the original message.

Finally, underlay DSA provides the CR opportunities to make use of extra power in the PU's transmission. While the Primary User's Signal Strength remains above a certain level, the Cognitive Radio is transmitting the data. Hence, Cognitive Radios are able to communicate for more time slots when comparing with interweave technique of DSA, and do not need to know as much about the PU's messages. However, in underlay technique, the Cognitive Radio needs a measure of the Signal to Interference plus Noise Ratio of the Primary User. These spectrum holes are known as black holes or white spaces (Federal Communications Commission, 2018; Goldsmith et al., 2009; Tahseen, 2011). Performance of the spectrum sensing decreases due to the fading, and shadowing (BTH, n.d.; Wang, Liu, Ma et al, 2023). Link adaptation (LA), also known as adaptive modulation and coding (AM&C), creates an adaptive modulation scheme where specified modulations are employed by the same system (Bilal & Khan, 2019; Puspitasari & Lee, 2023; Wang, You, Gao et al, 2023).

2.3 Modulation

For transmission of data over the air, radios modulate the information over a carrier wave. Modulation alters the spectrum of the data, thus improves the chance to recover the data at the receiver. The Modulation involves communication by using a High Frequency Carrier Wave, thereby reducing the size of antenna. The Examples include 2.4 GHz WiFi, or around 90 MHz for FM radio, TV Broadcast, Aircraft Communication, Drone Communication. Multiple Carrier Waves can be used for different uses as used in drones, the Control Frequency is 2.4 GHz, and Communication Frequency is 5.8 GHz. Modulation is the process in which a message signal is impressed on a reference electronic or optical carrier signal by varying any of the properties of the carrier signal according to message signal. Then the signal is modulated in-phase and quadrature components of a sinusoidal signal to represent the data which is being sent. In PSK, the amplitude or the frequency of the sinusoidal signal that is being transmitted does not change. The phase is changed according to the input bit stream.

2.4 Spectral Features

The spectrum provides the following parameters:

1. Amplitude.
2. Phase.
3. Frequency.

The amplitude, phase and frequency provide complete information about the modulated signal. These three parameters are obtained by obtaining Power Spectrum of the modulated signal. The spectrum is used for the classification of the modulation formats and demodulation and decoding of data.

2.4.1 Cyclo-Stationary Features

Modulated signals have some parameters which periodically change with time. Digital signal is modulated by periodic keying of amplitude, frequency and phase. The signals have the cyclo-stationary property that is used to classify modulation techniques. The Cyclic Domain Profile is extracted by using Cyclostationary Feature Detection Algorithm.

2.4.2 Time-Frequency Features

The Fourier transform (FT) displays the frequency components of the signal and cannot analyze the variations in time domain. For analyzing the frequency and time simultaneously, Short Time Fourier Transform can be utilized. Short Time Fourier Transform (STFT), is a time-frequency representation, in which a 2D representation is created by computing the Fourier Transform and using a sliding temporal window. By using the Short Time Fourier Transform (STFT) alteration of frequency with respect to time can be observed. The features that are extracted from the time-frequency analysis can help to classify the modulation formats. The Margenau-Hill distribution and Wigner- Vile distribution are also used in time-frequency analysis.

3. METHODOLOGY

3.1 1 Modulation Classification

Initially the Modulation Classification Algorithm was found to be capable of classifying the following Modulations:

1. BPSK.
2. QPSK
3. FSK
4. MSK

The Cyclic Domain Profiles were used to identify the modulation types.

The Parameters using which the Modulations have been simulated are given in Table 1.

Table 1. Modulation parameters

Modulations	**BPSK, QPSK, FSK, MSK**
Symbol Rates	1 kHz
Samples per Symbol	10
Intermediate Frequencies	2.5 kHz
TX Filter	Raised Cosine

3.1.1 1 Modulation Classification With Low Noise

In this process, initially the Message Signal is modulated into BPSK, QPSK, FSK, MSK Modulated Complex Baseband Waveform.

3.1.1.1 1 Modulation Classification With Noise

In this process, initially the Message Signal is modulated into BPSK, QPSK, FSK, MSK Modulated Complex Baseband Waveform.

The Parameters are given in Table 2.

Table 2. 1 Modulation classification

TX Filter	Root Raised Cosine
Eb / No	0dB – 20dB

3.2 2 Modulations Classification

The Modulation Classifier has been adjusted to classify 2 Modulations, the details are given upcoming sections. The Cyclic Domain Profiles were used to identify the modulation types.

The Parameters using which the Modulations have been simulated are given in Table 3.

Table 3. Parameters using modulations

Modulations		BPSK, QPSK, FSK, MSK	
Symbol Rates		1 kHz	10kHz
Samples per Symbol		8	
Filter Parameters	Stopband start	1.5 kHz	15 kHz
	Stopband Gain	-100 dB	-120 Db
Intermediate Frequencies		2.5 kHz	25 kHz
TX Filter		Root Raised Cosine	

3.2.1 2 Modulations Classification Without Noise

In this process, initially the Message Signal is modulated into BPSK, QPSK, FSK, MSK Modulated Complex Baseband Waveform. The 2 Modulated Complex Baseband Waveforms are then Upconverted. The Waveforms are then Downconverted to Baseband, then Resampled by using Sampling Rate 10 times of Received Symbol Rates and then Upconverted back to Intermediate Frequencies.

3.2.2 2 Modulation Classification With Noise

As soon as the modulations were classified by the Modulation Classifier, further work was done to channelize 2 different signals on Intermediate Frequencies by using Upconversion. By using Channelization technique, we are able to transmit 2 signals on the RF Channel. The RF Bandwidth can be efficiently utilized by sending data on multiple Intermediate Frequency Channels to enhance data rate. Different

Modulations can be applied to different channels as per requirement. Many Channels can be accessed simultaneously for more effective communication. Intermediate Frequency can be changed according to symbol rate. Then the Individual Signals at different Intermediate Frequencies can be separated by Downconverting to Baseband, then Resample the Signal, then again Upconvert to Intermediate Frequency.

The signals can be modulated independently on different symbol rates and modulations.

3.2.2.1 Adding Noise to Modulated Signal

A signal is being modulated with Digital Modulations, the Modulation can be selected during the simulation in the GUI at runtime. The AWGN Channel Impairments are added in the output complex waveform. The Eb/No can be changed to indicate change in channel condition as given in Table 4.

Table 4. Eb/No

Eb/No	0dB, 5dB, 10dB, 15dB, 20dB

3.2.2.2 Converting to Intermediate Frequencies

The Modulated Signals are upconverted to Intermediate Frequencies and then they are added. At Receiver they are then downconverted to their baseband signals and resampled and then again upconverted to the same Intermediate Frequency. This filters out the Required Signal. Then this filtered Signal on Intermediate Frequency is then sent Cyclostationary Feature Detector.

Table 5. Frequencies

Symbol Rates	**1 kHz**	**10 kHz**
Intermediate Frequencies	2.5 kHz	25 kHz
Resampling Rates	10 kHz	100 kHz

3.3 Modulation Classifiers

In Figure 1, a Modulation Classifier is shown. The Values of the Cyclic Domain Profile at specific points is used to classify different modulations. The values of the Array of CDP are being compared with fixed values for classification of Modulations.

Figure 1. Modulation classifiers

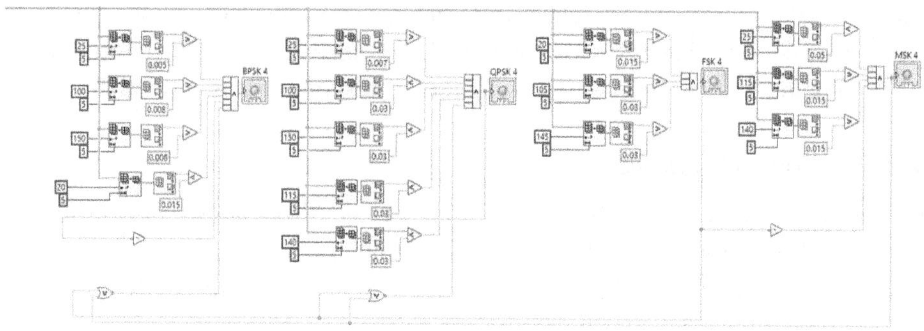

3.4 Adaptive Modulator

The Adaptive Modulation Scheme based on Eb/No. The functioning of Adaptive Modulator is shown in Figure 2. The Modulations change with respect to changing noise level.

Figure 2. Adaptive modulator

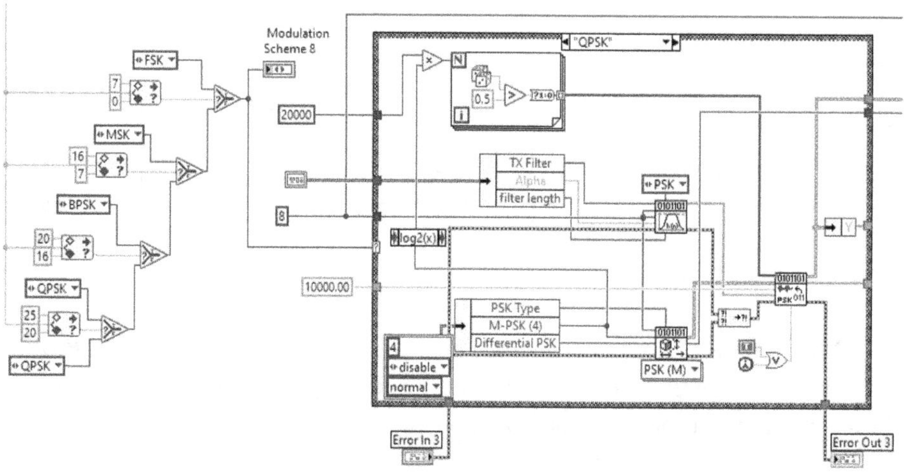

4. CELLULAR MODULATIONS CLASSIFICATION

The Cellular Modulations that can be classified are as follows:

1. GSM FSK

2. GSM MSK
3. EDGE 8PSK
4. WCDMA DL QPSK
5. WCDMA DL 16QAM
6. LTE DL 16QAM

In the LabVIEW Block Diagram given below, the TDMS waveform files of GSM FSK, GSM MSK, WCDMA DL QPSK, WCDMA DL 16QAM, and LTE DL 16QAM are in an Event Structure which will send the Cellular Modulated Waveform to Cyclostationary Feature Detector when the Cellular Modulation will be selected in the GUI during Simulation.

Figure 3. Cellular modulations input

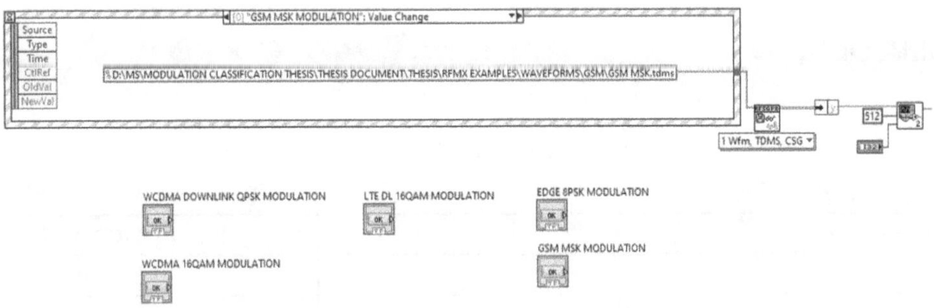

Table 6. Cellular standards

Cellular Standards	GSM, WCDMA, EDGE, LTE

4.1 GSM and EDGE

Table 7. GSM and EDGE specifications

Symbol Rate	**270.8 kHz**
Number of Symbols	1250

4.1.1 GSM MSK

4.1.1.1 Waveform Creation

Waveform of GSM MSK has been generated with the following Parameters

Table 8. GSM MSK waveform

Modulation	MSK

4.1.2 EDGE 8PSK

4.1.2.1 Waveform Creation

Waveform of EDGE 8PSK has been generated with the following Parameters

Table 9. EDGE 8PSK wavefzorm

Modulation	**8PSK**
Oversampling Factor	6

4.2 WCDMA DL

Table 10. WCDMA downlink

Cellular Standard	**WCDMA**
Chip Rate	3.84 MHz
Number of Frames	1
Link Direction	Downlink

4.2.1 WCDMA DL QPSK

4.2.1.1 Waveform Creation

Waveform of WCDMA DL QPSK has been generated with the following Parameters

Table 11. WCDMA downlink QPSK waveform

Modulation	QPSK

4.2.2 WCDMA DL 16QAM

4.2.2.1 Waveform Creation

Waveform of WCDMA DL 16QAM has been generated with the following Parameters

Table 12. WCDMA downlink 16QAM waveform

Modulation	16QAM

4.3 LTE DL 16QAM

4.3.1 Waveform Creation

Waveform of LTE DL 16QAM has been generated with the following Parameters

Table 13. LTE downlink 16QAM waveform

Cellular Standard	LTE
Modulation	16QAM
Channel Bandwidth	20 MHz
Link Direction	Downlink

5. RESULTS AND DISCUSSION

Initially only 1 Modulation could be classified by the classifier.

5.1 BPSK Modulation Classification With Low Noise

In Figure 4, 1 Modulation has been classified with low noise, which has been shown in the GUI in LabVIEW.

Figure 4. BPSK classification

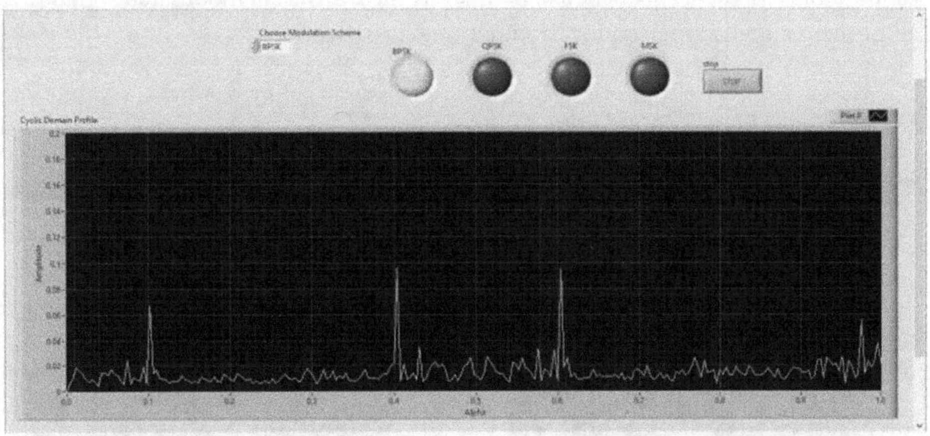

5.2 FSK Modulation Classification With Noise

In Figure 5, when AWGN Noise was added, 1 Modulation has been classified with noise, which has been shown in the GUI in LabVIEW.

Figure 5. FSK modulation classification with noise

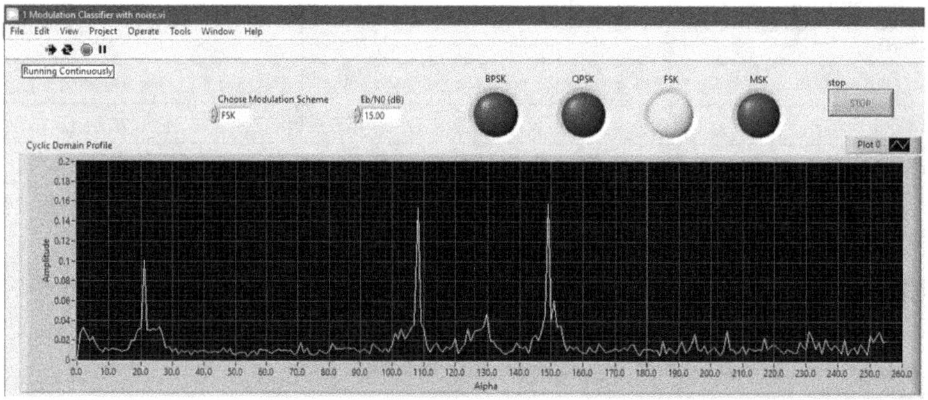

5.3 BPSK BPSK Modulations Classification Without Noise

The Modulation Classifiers have been tested without noise. There are 2 Modulations at different Intermediate Frequencies. The 2 Signals are first downconverted to their Baseband Frequency and then upconverted to their Intermediate Frequencies. In this way the 2 different modulated signals are separated and then CDP of these signals are used for Modulation Classification.

Figure 6. 2 Modulations classification without noise

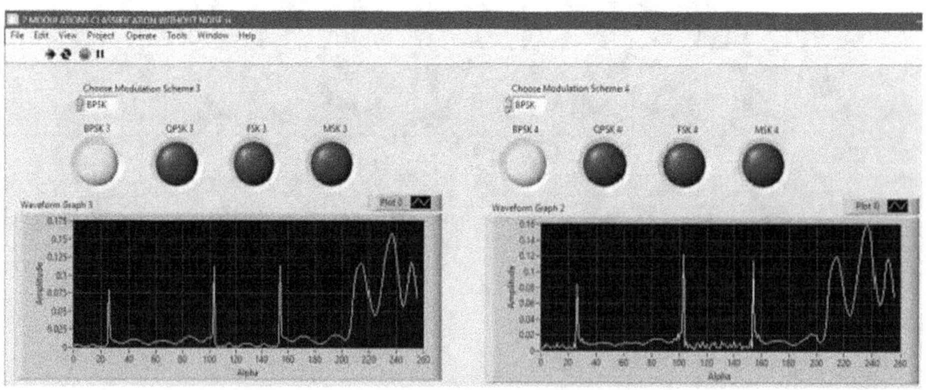

5.4 BPSK BPSK Modulations Classification in AWGN Channel

At 0Db in AWGN Channel, the modulation classification is shown in the figure below. The Modulations have been classified by the classifier based on the values of CDP. The CDP of BPSK at Symbol Rates of 10 kHz and 1 kHz have been displayed.

The Modulation Classifier has been tested with noise. There are 2 Modulations at different Intermediate Frequencies. The 2 Signals are first downconverted to their Baseband Frequency and then upconverted to their Intermediate Frequencies. In this way the 2 different modulated signals are separated and then CDP of these signals are used for Modulation Classification.

Figure 7. Two modulations classification without noise

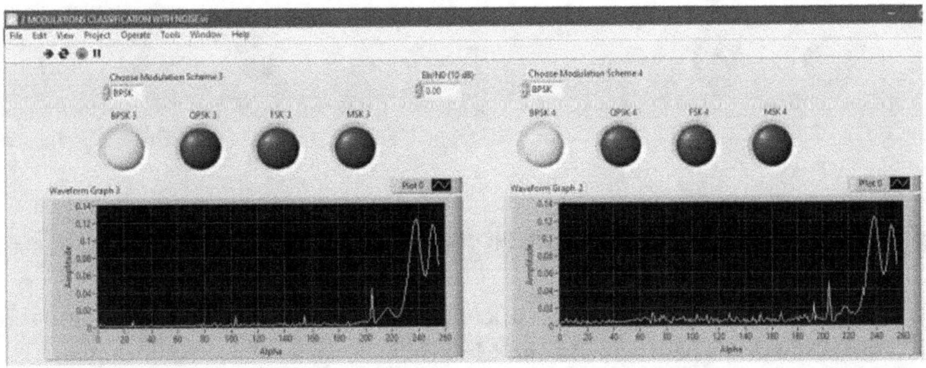

5.5 Adaptive Modulator

The Functioning of the Adaptive Modulator is shown in the figure below

Figure 8. Functioning of adaptive modulator

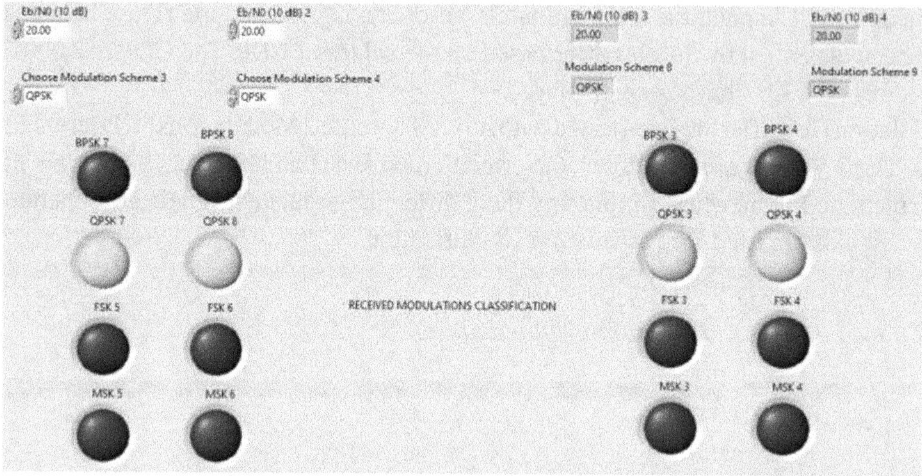

5.6 Comparison of the Classifiers in Noise

In Figure 9, a comparison of Average Classification Accuracy of different modulations at different noise levels from 0dB to 20dB between Old and Modified Classifiers have been shown.

Table 14. Classification comparison

Modulations	Old Classifier	Modified Classifier
BPSK	16%	80%
QPSK	26%	80%
FSK	32%	80%
MSK	31%	80%

Figure 9. Comparison of modulation classifiers

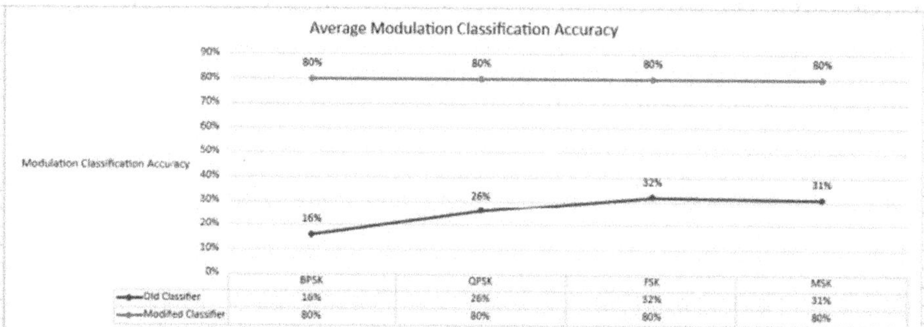

5.7 Cellular Modulation Classification

The output of the Cellular Modulations Classifier has been shown in the Figures, in which the Cellular Modulations have been classified using the classifier in the VI. The Indicators have been utilized to view which of the Modulation has been classified after Cyclostationary Processing. CDP Graph and IQ Vectors of the Cellular Modulations have also been shown in the Figures. In the Figure below, EDGE 8PSK has been classified, which has been displayed on LabVIEW GUI.

The Comparative Results of classification of Reference Cellular Modulations and Cellular Modulations in Noise have also been given in the table below and figures.

Table 15. Comparative results

	Reference Cellular Modulations Classification Accuracy	Cellular Modulations Classification Accuracy with Noise
GSM MSK	80%	70%
EDGE 8PSK	80%	70%
WCDMA QPSK DL	80%	70%
WCDMA 16QAM DL	80%	70%
LTE 16QAM DL	80%	70%

Figure 10. Cellular modulation classification accuracy

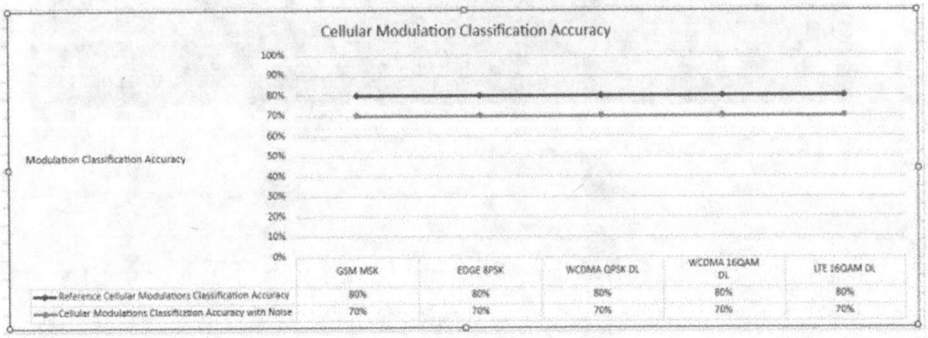

In Figure 11 and Figure 12, the Classifications of Reference EDGE 8PSK and EDGE 8PSK in Noise has been given, which has been simulated using LabVIEW.

Figure 11. Reference EDGE 8PSK classification

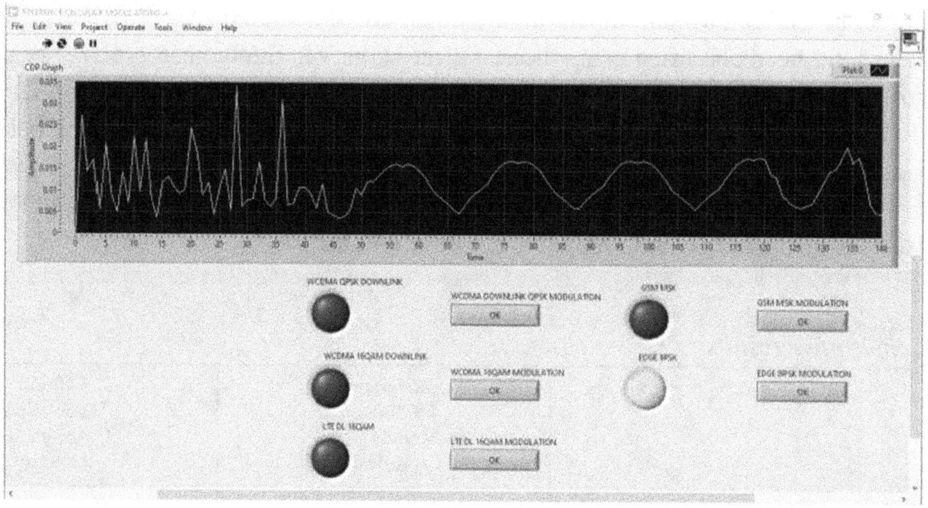

Figure 12. EDGE 8PSK classification in noise

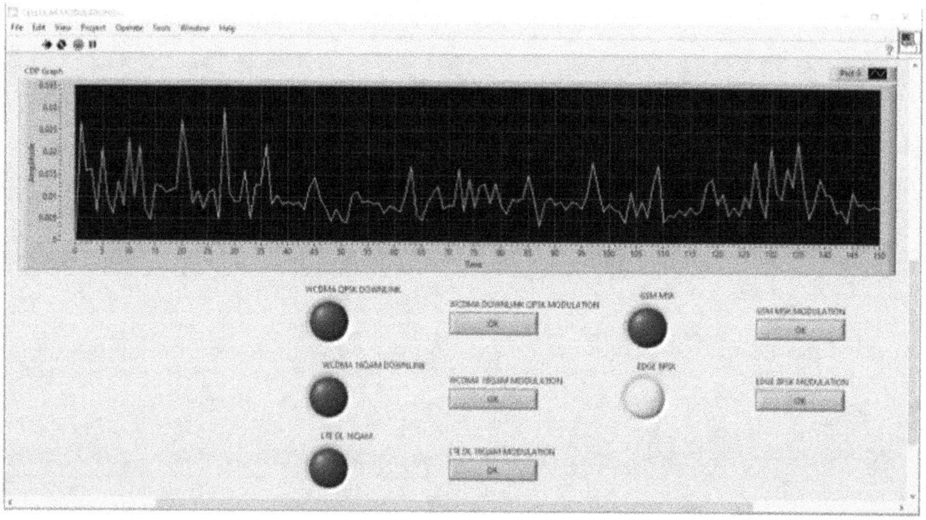

In Figure 13 and Figure 14, the Classifications of Reference GSM MSK and GSM MSK in Noise has been given, which has been simulated using LabVIEW.

Figure 13. Reference GSM MSK classification

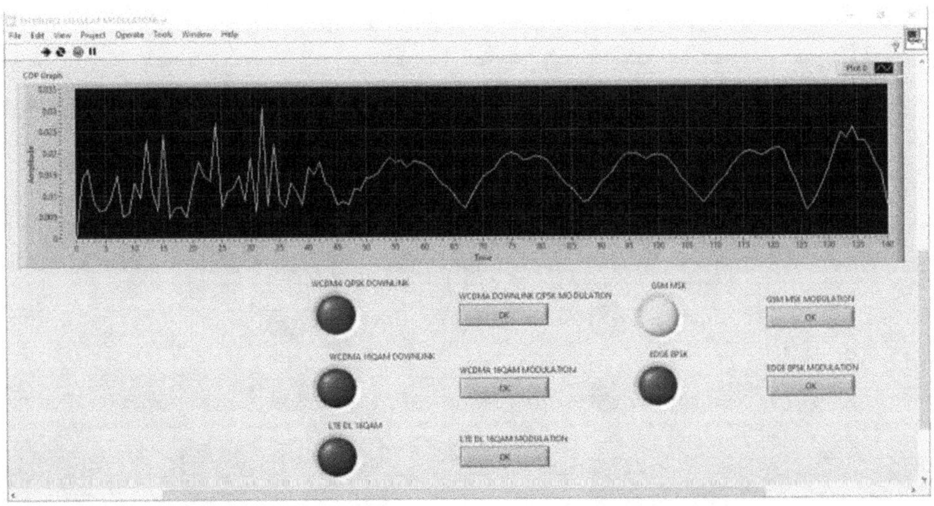

Figure 14. GSM MSK classification in noise

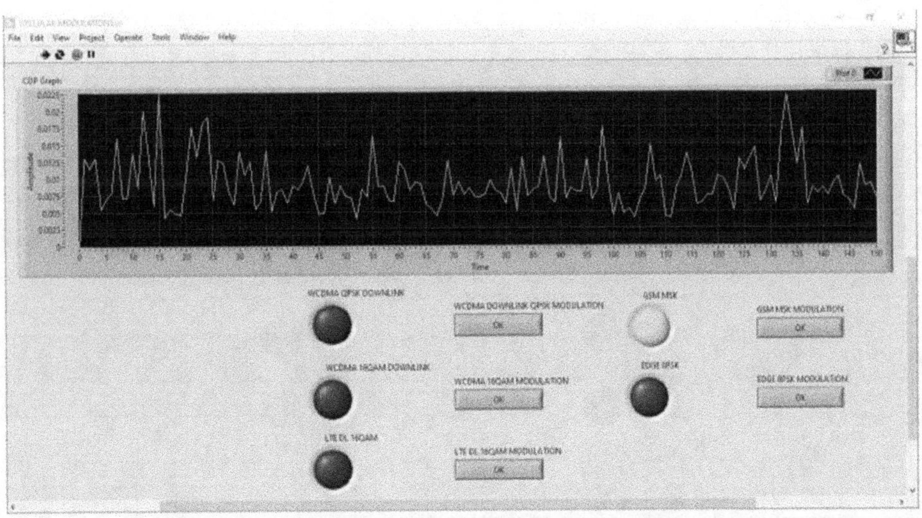

In Figure 15 and Figure 16, the Classifications of Reference LTE DL 16QAM and LTE DL 16QAM in Noise has been given, which has been simulated using LabVIEW.

Figure 15. Reference LTE DL 16QAM classification

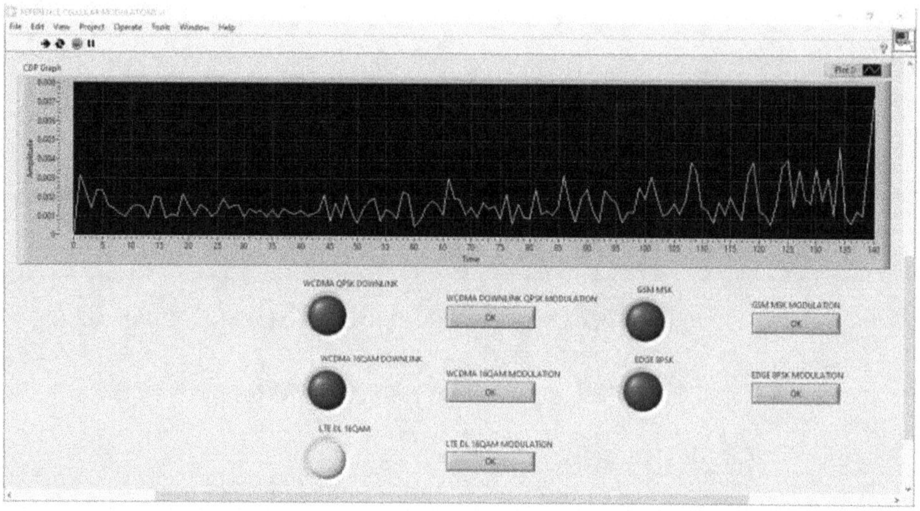

Figure 16. LTE 16QAM classification in noise

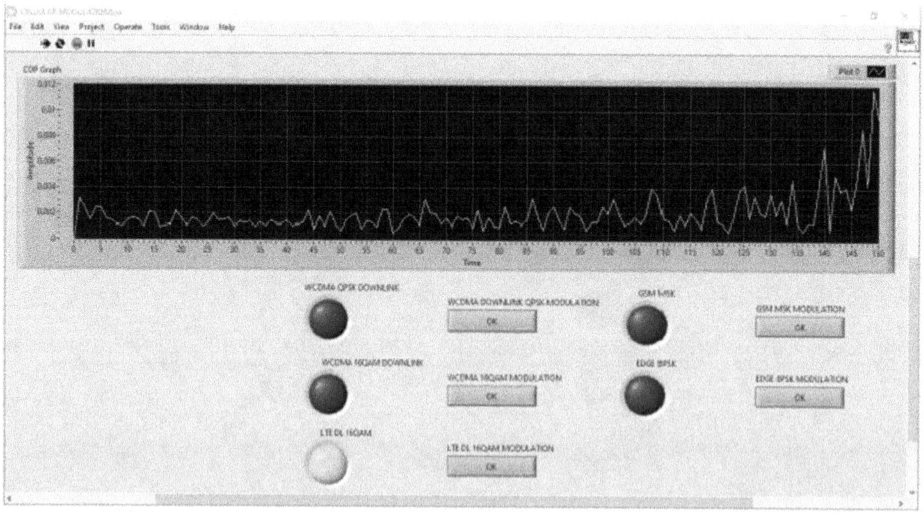

In Figure 17 and Figure 18, the Classifications of Reference WCDMA 16QAM DL and WCDMA 16QAM DL in Noise has been given, which has been simulated using LabVIEW.

Figure 17. Reference WCDMA 16QAM DOWNLINK classification

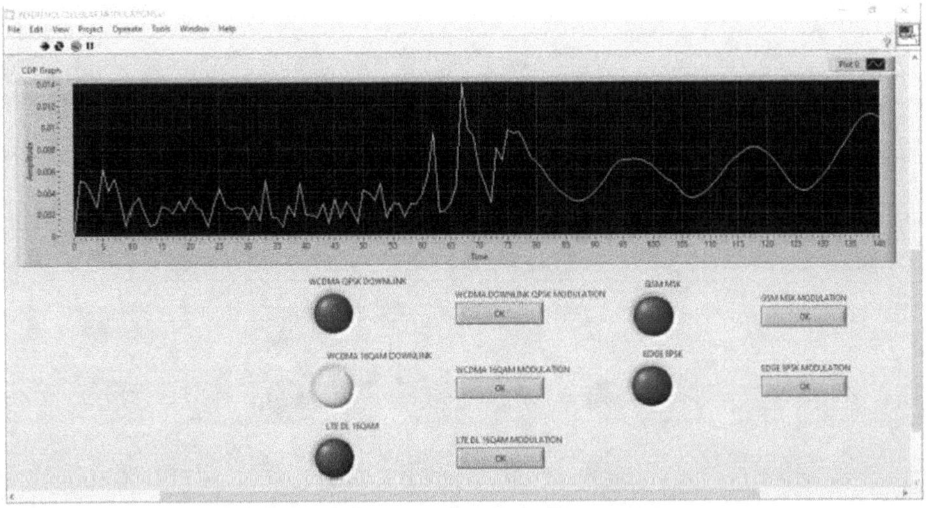

Figure 18. WCDMA 16QAM DOWNLINK classification in noise

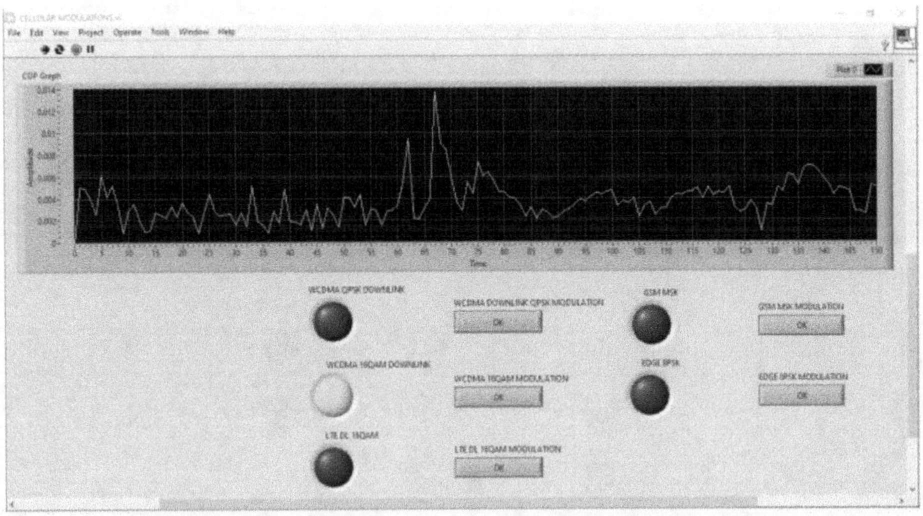

In the Figure 19 and Figure 20, the Classifications of Reference WCDMA QPSK DL and WCDMA QPSK DL in Noise has been given, which has been simulated using LabVIEW.

Figure 19. Reference WCDMA QPSK downlink classification

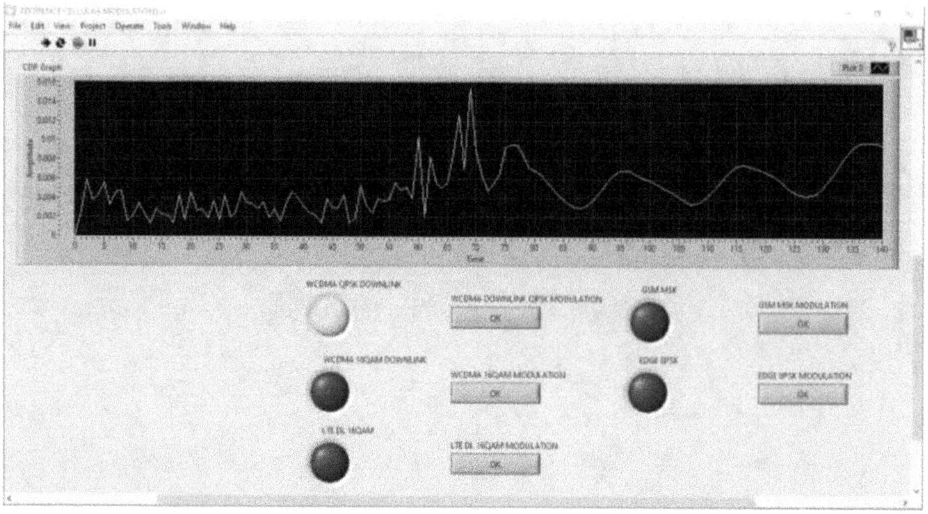

Figure 20. WCDMA QPSK DOWNLINK classification

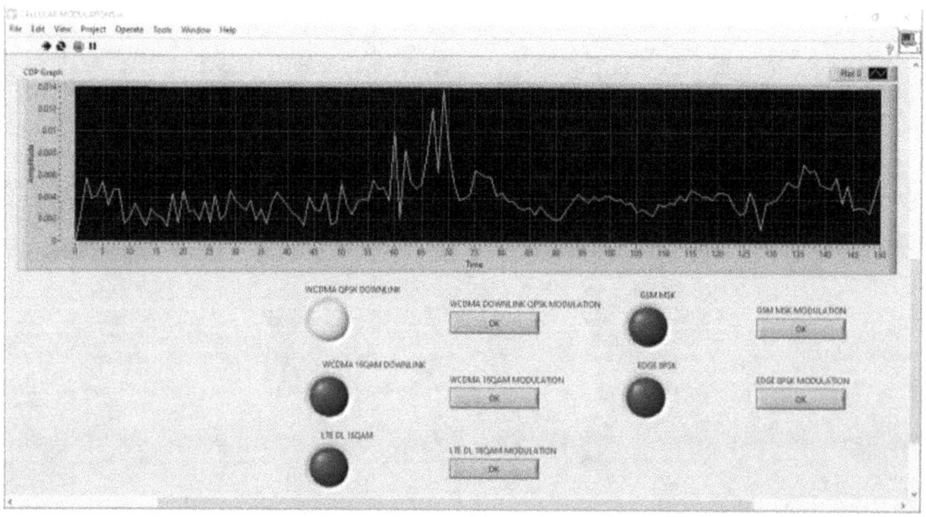

6. CONCLUSION

In this paper, Cyclostationary analysis of Digital Modulated Signals and its classification has been discussed with results and comparison, and based on these results an Adaptive Modulator has also been simulated. The Cyclostationary Features of the simulated signals are obtained using the Cyclostationary Feature Detection algorithm. The Adaptive Modulator for the simulated BPSK, QPSK, FSK and MSK modulated signals and is also capable of Adaptive Modulation based on Noise Level. The Modulation Classifier has also been modified for classifying Cellular Modulations including GSM FSK, GSM MSK, EDGE 8PSK, WCDMA Downlink QPSK, WCDMA Downlink 16QAM and LTE Downlink 16QAM. By utilizing the Modulation Classification and Adaptive Modulator we can achieve more reliable communication in different channel conditions and also adaptively modulate our signals as per requirements of channel to minimize the distortions caused by the noise.

REFERENCES

Abid, B., Khan, B. M., & Memon, R. A. (2022). Seismic Facies Segmentation Using Ensemble of Convolutional Neural Networks. *Wireless Communications and Mobile Computing*. Wiley. 10.1155/2022/7762543

An, T. T., & Lee, B. M. (2023). Robust automatic modulation classification in low signal to noise ratio. *IEEE Access : Practical Innovations, Open Solutions*, 11, 7860–7872. 10.1109/ACCESS.2023.3238995

Bhargavi, D., & Murthy, C. R. (2010). Performance comparison of energy, matched-filter and cyclostationarity-based spectrum sensing. *Signal Processing Advances in Wireless Communications (SPAWC), IEEE Eleventh International Workshop on*, (pp. 1 –5). IEEE. 10.1109/SPAWC.2010.5670882

Bilal, R., & Khan, B. M. (2019). Software-Defined Networks (SDN): A Survey. B. B. Gupta & Dharma P. Agrawal (eds.) *Handbook of Research on Cloud Computing and Big Data Applications in IoT.* IGI Global. 10.4018/978-1-5225-8407-0.ch023

BTH. (n.d.). *Applying OSTBC in Cooperative Cognitive Radio Networks - Arkiv EX.* Blekinge Tekniska Högskola. http://www.bth.se/fou/cuppsats.nsf/6753b78eb2944e0ac1256608004f0535/4830a668eb9bd48cc1257735004a7535?OpenDocument

Cabric, D., Mishra, S. M., & Brodersen, R. W. (2004). Implementation issues in spectrum sensing for cognitive radios. *Signals, Systems and Computers:Conference Record of the Thirty-Eighth Asilomar Conference*. IEEE. 10.1109/ACSSC.2004.1399240

Digham, F. F., Alouini, M.-S., & Simon, M. K. (2003). On the energy detection of unknown signals over fading channels. *Communications, ICC '03. IEEE International Conference*. IEEE. 10.1109/ICC.2003.1204119

Dobre, O., Abdi, A., Bar-Ness, Y., & Su, W. (2007, April). Survey of automatic modulation classification techniques: Classical approaches and new trends. *IET Communications*, 1(2), 137–156. 10.1049/iet-com:20050176

FCC. (2015). *Amendment of the commission's rules with regard to commercial operations in the 3550-3650 MHz band*. FCC.

Federal Communications Commission. (2002). *Revision of part 15 of the commission's rules regarding ultra-wideband transmission systems*. FCC.

Federal Communications Commission. (2018). FCC online table of frequency allocations. FCC. https://transition.fcc.gov/oet/spectrum/table/fcctable. pdf

Frasch, I. J. (2017). *Algorithmic framework and implementation of spectrum holes detection for cognitive radios.* [Master's thesis, Rochester Institute of Technology].

Gardner, W. (1988, August). Signal interception: A unifying theoretical framework for feature detection. *IEEE Transactions on Communications*, 36(8), 897–906. 10.1109/26.3769

Goldsmith, A., Jafar, S., Maric, I., & Srinivasa, S. (2009, May). Breaking spectrum gridlock with cognitive radios: An information theoretic perspective. *Proceedings of the IEEE*, 97(5), 894–914. 10.1109/JPROC.2009.2015717

Khan, B. M., & Ali, F. H. (2011). Mobility Adaptive Energy Efficient and Low Latency MAC for Wireless Sensor Networks. *2011 Fifth International Conference on Next Generation Mobile Applications, Services and Technologies*, Cardiff, UK. 10.1109/NGMAST.2011.46

Lagunas, E., & Najar, M. (2015, July). Spectral feature detection with sub-Nyquist sampling for wideband spectrum sensing. *IEEE Transactions on Wireless Communications*, 14(7), 3978–3990. 10.1109/TWC.2015.2415774

Memon, R. A., Qazi, S., & Khan, B. M. (2021). Design and Implementation of a Robust Convolutional Neural Network-Based Traffic Matrix Estimator for Cloud Networks. Wireless Communications and Mobile Computing. 10.1155/2021/1039613

Mitola, J., & Maguire, G. (1999, August). Cognitive radio: Making software radios more personal. *IEEE Personal Communications*, 6(4), 13–18. 10.1109/98.788210

Mohammadi, F. S., & Kwasinski, A. (2018). *Neural network cognitive engine for autonomous and distributed underlay dynamic spectrum access*. arXiv preprint arXiv:1806.11038.

Puspitasari, A. A., & Lee, B. M. (2023). A survey on reinforcement learning for reconfigurable intelligent surfaces in wireless communications. *Sensors (Basel)*, 23(5), 2554. 10.3390/s2305255436904758

Ramkumar, B. (2009, June). Automatic modulation classification for cognitive radios using cyclic feature detection. *IEEE Circuits and Systems Magazine*, 9(2), 27–45. 10.1109/MCAS.2008.931739

Spooner, C. M., & Mody, A. N. (2018, June). Wideband cyclostationary signal processing using sparse subsets of narrowband subchannels. *IEEE Transactions on Cognitive Communications and Networking*, 4(2), 162–176. 10.1109/TCCN.2018.2790971

Tahseen, M. (2011). *Analysis of OSTBC in Cooperative Cognitive Radio Networks using 2-hop DF Relaying Protocol - Arkiv EX*. Blekinge Tekniska Högskola.

Tian, Z., Tafesse, Y., & Sadler, B. M. (2012, February). Cyclic feature detection with sub-Nyquist sampling for wideband spectrum sensing. *IEEE Journal of Selected Topics in Signal Processing*, 6(1), 58–69. 10.1109/JSTSP.2011.2181940

Wang, C.-X., You, X., Gao, X., Zhu, X., Li, Z., Zhang, C., Wang, H., Huang, Y., Chen, Y., Haas, H., Thompson, J. S., Larsson, E. G., Renzo, M. D., Tong, W., Zhu, P., Shen, X., Poor, H. V., & Hanzo, L. (2023). On the road to 6g: Visions, requirements, key technologies and testbeds. *IEEE Communications Surveys and Tutorials*, 25(2), 905–974. 10.1109/COMST.2023.3249835

Wang, N., Liu, Y., Ma, L., Yang, Y., & Wang, H. (2023). Automatic modulation classification based on cnn and multiple kernel maximum mean discrepancy. *Electronics (Basel)*, 12(1).

Xie, L., & Wan, Q. (2017, June). Cyclic feature-based modulation recognition using compressive sensing. *IEEE Wireless Communications Letters*, 6(3), 402–405. 10.1109/LWC.2017.2697853

Yucek T. & Arslan, H. (2009). A survey of spectrum sensing algorithms for cognitive radio applications. *Communications Surveys Tutorials*. IEEE.

Zhang, H., & Wang, X. (2011). *A Fuzzy Decision Scheme for Cooperative Spectrum Sensing in Cognitive Radio.* Vehicular Technology Conference (VTC Spring). 10.1109/VETECS.2011.5956116

Chapter 13
QoS-Aware Smart Resource Allocation for Green Cognitive Radio Networks

Arifa Ahmed
https://orcid.org/0000-0002-6695-9973
Madanapalle Institute of Technology and Science, India

Deepak Mishra
University of New South Wales, Australia

Ganesh Prasad
National Institute of Technology, Silchar, India

Krishna Lal Baishnab
National Institute of Technology, Silchar, India

ABSTRACT

Cognitive radio (CR) technology has gained attention in wireless communications for its ability to enhance spectrum utilization efficiency through the implementation of dynamic spectrum access. This technology enables secondary users (SUs), or unlicensed users, to access the radio spectrum of primary users (PUs) in a limited manner that doesn't interfere with the PUs. The implementation of spectrum handoff can enhance SU's throughput performance however it may cause high energy consumption. This, in turn, can reduce device lifetime and affect system performance, particularly for battery-operated CR devices. This chapter focuses on optimizing the available resources in the green CR system to maximize energy efficiency (EE) and throughput performance while maintaining system reliability. It also explores using energy beamforming techniques to enhance the throughput performance of multi-antenna CR systems. Furthermore, this chapter addresses the challenges associated with SU's spectrum sensing by introducing a new transmission protocol designed for smart sensing implementation.

DOI: 10.4018/979-8-3693-2893-4.ch013

1. INTRODUCTION

This chapter investigates quality of service (QoS)-aware energy and throughput efficient cognitive radio (CR) networks, with the goal of optimizing resource allocation to attain maximum energy efficiency (EE) and throughput.

1.1 Background

The increasing demand for wireless radio spectrum is driven by the rapid expansion of wireless services. Static spectrum allocation policy leads to inadequate usage of the licensed spectrum bands (Ren et al, 2012). Recently, Federal Communications Commission has reported that 70% of the allocated spectrum in US is not utilized (FCC, 2002). In (McHenry, 2005), the author provides an estimate of spectrum utilization, examining the allocation and utilization of spectrum between 30 MHz and 3 GHz across six different locations. The findings highlight the low utilization of licensed spectrum, leading to spectrum scarcity. To address this spectrum underutilization problem, CR has been proposed. CR technology aims to enhance spectrum utilization by enabling dynamic spectrum access (DSA), allowing secondary users (SUs) or unlicensed users to opportunistically access unused spectrum without interfering with primary users (PUs) or licensed users (Shin et al, 2010).

1.2 Literature Review

The frame structure of a CR system typically comprises sensing and transmission slots. The duration of spectrum sensing and the transmit power of a CR user can significantly affect the EE performance of CR transmissions. There exist tradeoffs between the sensing duration and EE (Zhang et al, 2014; Pei et al, 2011; Zhang et al, 2016) as well as between the transmit power and EE (Zhang et al, 2016) of the CR networks. Zhang et al. (2016) and Awin et al. (2016) have jointly determined transmit power and sensing duration for maximizing EE for a CR system, however, the works do not consider spectrum handoff, and spectrum sensing is performed for a single channel. Though the work proposed by Zhang et al, 2014 has considered spectrum handoff, however, it primarily focuses on optimizing sensing duration to maximize the EE of the secondary network. Therefore, this chapter jointly optimizes the sensing duration and transmit power to maximize the EE of a CR system while considering spectrum handoff. The authors in (Arshid et al, 2022) introduces a cooperative spectrum sensing based second priority user transmission model for CR Networks, emphasizing EE through optimized sensing and handoff strategies. This approach effectively enhances throughput, reduces energy consumption, and minimizes handoff delay by addressing false alarms and miss-detection issues. However, it suggests further research to exploit unutilized spectrums efficiently and develop methods for reliable spectrum information sharing to mitigate energy wastage caused by malicious users.

The transmission performance of a multi-antenna CR system can be enhanced by employing a beamforming technique. In a MISO system, high-efficiency power transfer can be achieved by using a particular beamforming method called "energy beamforming" (Yang, Ho & Guan, 2014) at the transmitter side. This can significantly enhance the SNR performance at the receiver by boosting the overall received signal strength. The authors (Wang, Tao & Liu, 2012; Li, Zhang & Qin, 2016; Zheng, Krikidis & orn Ottersten, 2013; Noh & Oh, 2013) have implemented beamforming to optimize the SU's achievable rate or received SINR while ensuring the limitation on PU's rate. To reduce interference between secondary users in a multi-user OFDM-based secondary network, the base station utilizes ZF and regularized beamforming techniques (Agrahari, Varshney & Jagannatham, 2017). In it, a sum-rate maximization problem has been formulated to optimize the power allocation of the CR users under the interference constraint. Liu et al. (2016) have investigated the energy beamforming technique for a cooperative CR system to enhance power transfer efficiency. Here, the impact of beamforming on data rate is not observed rather it is used to maximize the harvested energy. From the state-of-the-art, energy beamforming has not been investigated to enhance the throughput performance of the CR system.

By optimizing the SU's throughput, the spectral efficiency of the CR system can be considerably enhanced (Zhang et al, 2015). Nonetheless, the throughput is affected by the following factors: sensing, channel estimation (CE), and frame lengths. The transmission time increases as the sensing duration decreases, but this also has an impact on the sensing quality, which lowers the CR system's throughput. In addition, beamforming can increase SU's throughput, but its effectiveness is reliant on CE's precision, which is defined by CE duration. Moreover, SU's throughput improves with frame duration for a fixed sensing period; however, interference also increases at the same time, resulting in a further reduction in throughput. To enhance the CR system's throughput performance, it is therefore essential to determine the joint optimal values of SU's spectrum sensing, CE, and frame durations. Salari & Chan (2023) present a framework for an energy harvesting cooperative cognitive radio network, wherein the primary network shares both energy and spectrum with the secondary network. The study aims to maximize the sum-rate of the secondary network by optimizing the distributed beamforming coefficients and energy harvesting time allocation factor, subject to power and interference constraints.

The efficacy of CR networks is significantly impacted by two critical channel sensing errors: mis-detection and false alarms. Mis-detection allows secondary users to transmit in occupied channels, interfering with primary network operations. Whereas, false alarms lead secondary users to avoid unoccupied channels, thereby decreasing secondary network throughput (Liang et al, 2008; Awin, Abdel-Raheem & Tepe, 2018). The unpredictable nature of PU activity in CR systems poses a serious challenge. PU activity can change unexpectedly during SU transmission, leading to interference beyond that caused by mis-detection (Ozcan, Gursoy & Tang, 2017). Precise knowledge of PU activity would significantly mitigate these challenges. This chapter explores a new protocol for the SU's transmission, in which the SU is notified of activity changes whenever the PU becomes idle. The SU can determine the PU's activity status without using spectrum sensing. But to receive this information, the SU has to pay the PU a pre-determined cost. With this strategy, the PU is guaranteed to be free from interference caused by the SU communication. Additionally, employing this method mitigates the risk of data transmissions being lost due to false alarms and saves time. Thus, this protocol plays a pivotal role in actualizing a smart sensing CR environment as it offers substantial throughput gain for SU and efficiently addresses problems associated with spectrum sensing.

1.3 Industrial Applications of CR Technology

Industrial applications of the research in CR include enhancing connectivity and communication stability in industrial IoT (IIoT) networks, enabling efficient and cost-effective power control for long-lasting IIoT devices, supporting smart city services such as advanced healthcare and automatic driving, and facilitating access to smart services for users. CR technologies, particularly in Low Power Wide Area Networks (LPWANs) like LoRa, Sigfox, LTE-M, and NBIoT, play a crucial role in enabling these applications by providing reliable and low-power communication solutions for industrial environments (Nurelmadina et al, 2021). Efficient spectrum management in IIoT networks optimizes wireless communication and enhances energy efficiency. Additionally, the research enables wireless power transfer using CR technology, reducing power consumption in IIoT devices. The use of dynamic spectrum access improves spectrum utilization and network performance in industrial wireless sensor networks. CR systems also enhance spectrum sharing and management in wireless industrial networks, leading to more efficient use of available resources. Furthermore, energy harvesting and CR strategies contribute to spectrum reutilization and lifetime extension in standard wireless networks. Overall, these applications highlight the potential of CR to enhance wireless communication in industrial environments, making processes more efficient and reliable (Fernando & Lăzăroiu, 2023).

Based on the motivation of the above literature survey, the following problem statements are formulated to design the throughput and energy-efficient CR systems. In Section 2, the sensing duration and transmit power of SU are jointly optimized to maximize the EE of the CR system while considering spectrum handoff. In Section 3, the SU's average throughput and interference energy received at the PU's receiver are analyzed in a multi-antenna CR system with a beamforming effect. In Section 4, a throughput maximization problem is investigated in a multi-antenna CR system as a joint optimization problem of spectrum sensing duration, CE duration, and frame duration under the interference constraint. Section 5 investigates the performance of a transmission protocol for a multi-antenna SU in terms of average throughput for implementing smart sensing. This chapter focuses on two distinct performance metrics: Section 2 examines the EE of the SU, while Sections 3, 4, and 5 analyze the SU's average throughput.

2. OPTIMIZATION OF ENERGY EFFICIENCY UNDER SPECTRUM HANDOFF SCHEME

This section presents a joint optimization problem for maximizing the energy efficiency of the CR systems under the spectrum handoff scheme. In this regard, first, the average throughput and energy consumption of secondary transmissions are determined and then the joint optimization problem is formulated. To find the optimal solution to this problem, the study explores the application of various swarm intelligence techniques.

2.1 Spectrum Sensing and Access Mechanism

A CR system is considered with multiple PUs and one pair of secondary transmitter (ST) and secondary receiver (SR). The primary network utilizes L contiguous orthogonal frequency channels (Bayhan & Alagoz, 2012), each possessing identical bandwidth W. The SU can opportunistically use the channels anytime they become idle, but the PUs have priority access to them. A central controller i.e. base station is considered in the primary network which assigns channels to the PUs when they request to transmit.

The ST conducts periodic sensing, with a sensing slot of time T_s incorporated into each frame of duration T. Using a wideband sensor, the ST concurrently detects the availability status of all the L channels. Assume that the ST is already using a channel. Based on the sensing results, the ST determines its next course of action. The available options are as follows: either continue transmission on the current channel i.e., operating channel (if it is available), remain idle and wait for the channel to become available (if the L channels are occupied), or switch to a different channel that is currently idle (if the current channel is occupied but another suitable channel is detected). The flowchart shown in Figure 1 illustrates this spectrum accessing procedure for data transmission in a frame in terms of three events: E1, E2, and E3. Figure 2 outlines the frame organization for these events, with T_{sw} denoting the channel switching latency.

Figure 1. ST's frame-based data transmission method for spectrum access

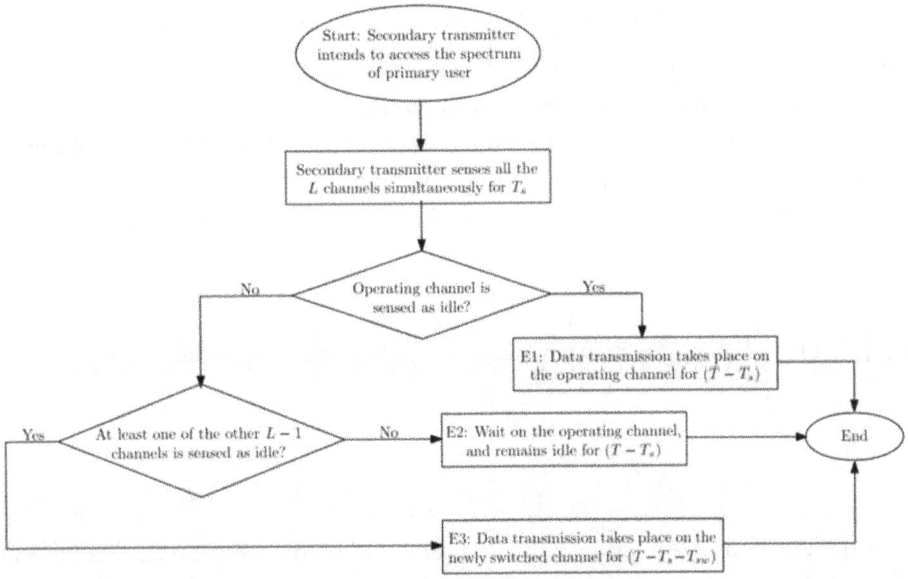

Figure 2. Frame structure for CR considering events E1, E2, and E3

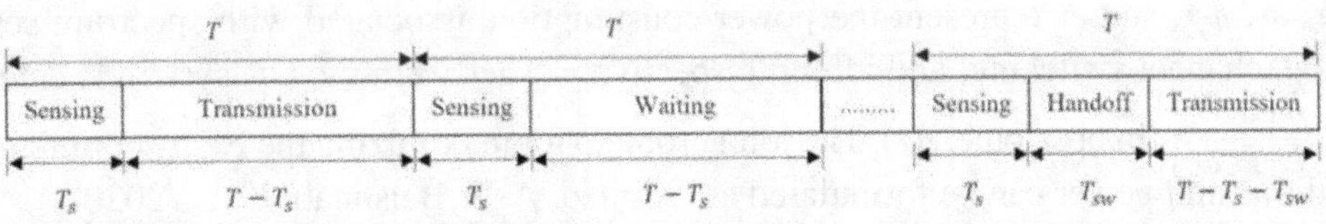

2.2 Energy Efficiency Maximization

The three events E1, E2, and E3 as seen in Figure 1 are assumed to occur with probabilities P_1, P_2, and P_3, respectively. These probabilities are determined as a function of T_s, as described in (Wang et al, 2011). The average throughput of secondary transmissions is expressed as (Ahmed, A. & Baishnab, K. L., 2020):

$$R[T_s, \phi_{ts}] = \frac{P_{S1}(T - T_s)(1 - q_l^s[T - T_s]) + P_3(1 - P_e)(T - T_s - T_{sw})(1 - q_l^s[T - T_s - T_{sw}])}{T} D_0 \tag{1}$$

Here, $P_{S1} = (1 - P_f)(1 - \rho)$, P_f is the false alarm probability, $\rho = \frac{\alpha_1}{T_{int}}$ is the traffic intensity of PU, $T_{int} = \alpha_0 + \alpha_1$ is the mean inter-arrival time of PU traffic, α_0 and α_1 are the mean durations of PU's idle state and busy state, respectively; $P_3 = (1 - P_1)\left(1 - (P_{S3} + P_{S4})^{L-1}\right)$, $P_1 = P_{S1} + P_{S2}$, $P_{S2} = (1 - P_d)\rho$, $P_{S3} = P_d\rho$, $P_{S4} = P_f(1 - \rho)$, P_d is the detection probability; $q_I^s[T - T_s]$ and $q_I^s[T - T_s - T_{sw}]$ are the reoccupation probability of the PU which is obtained by assuming a single PU state change during the secondary transmission for events E1 and E3, respectively; P_e represents the probability of ST switching to an occupied channel; $D_0 = \log_2\left[1 + \frac{\phi_{ts}}{\sigma_n^2}\right]$ bps/Hz is the ST's data rate, σ_n^2 is the variance of Gaussian noise, and ϕ_{ts} is the transmit power of the single-antenna ST.

Note that P_{S1}, P_{S2}, P_{S3}, and P_{S4} denote the probabilities of occurrence for four different scenarios which are obtained from the ST's spectrum sensing outcome and the actual state of PU. The detection probability and false alarm probability are represented as a function of sensing duration T_s as given by (Wang et al., 2011):

$$P_d[T_s] = Q\left[\frac{1}{\sqrt{2\gamma + 1}}\left(Q^{-1}[\bar{P}_f] - \sqrt{T_s f_s}\gamma\right)\right] \tag{2a}$$

$$P_f[T_s] = Q\left[\sqrt{2\gamma + 1}\, Q^{-1}[\bar{P}_d] + \sqrt{T_s f_s}\gamma\right] \tag{2b}$$

Here, $Q[\cdot]$ is the Gaussian Q-function; γ is the average PU's received SNR; f_s is the sampling frequency; and $\bar{P}_d$ and $\bar{P}_f$ are the target values for P_d and P_f, respectively. The overall average energy usage throughout the secondary transmission is calculated as follows (Ahmed, A. & Baishnab, K. L., 2020):

$$E[T_s, \phi_{ts}] = T_s\phi_s + P_1(\phi_{ts} + \phi_c)(T - T_s) + P_3\left((\phi_{ts} + \phi_c)(T - T_s - T_{sw}) + \phi_{sw}T_{sw}\right) + P_2\phi_i(T - T_s) \tag{3}$$

where ϕ_s, ϕ_c, ϕ_{sw}, and ϕ_i represent the power consumption associated with spectrum sensing, circuit operation, channel switching, and idling, respectively. The EE in the CR system is determined as $\eta[T_s, \phi_{ts}] = \frac{R[T_s, \phi_{ts}]}{E[T_s, \phi_{ts}]}$ (in bps/Joule/Hz). The joint problem for maximizing the EE by optimizing sensing duration and transmit power can be formulated as (Ahmed, A. & Baishnab, K. L., 2020):

(P0): $\underset{T_s, \phi_{ts}}{\text{maximize}}\, \eta[T_s, \phi_{ts}]$, subject to

C1: $P_d \geq \bar{P}_d$, C2: $0 < T_s \leq T$, C3: $\phi_{ts,\min} < \phi_{ts} \leq \phi_{ts,\max}$

Constraint $C1$ states that the detection probability P_d should meet or exceed its designated target value $\bar{P}_d$, the constraints $C2$ and $C3$ give a lower bound and an upper bound on T_s and ϕ_{ts}, respectively. Here, $\phi_{ts,\min}$ and $\phi_{ts,\max}$ are the minimum and maximum ST transmit powers, respectively. The lower bound on T_s is set such that $P_f \leq 0.5$. Thus, we have $T_s \geq (\alpha/\beta)^2$, where $\alpha = \sqrt{2\gamma + 1}\, Q^{-1}[\bar{P}_d]$ and $\beta = \gamma\sqrt{f_s}$. The problem (P0) can be reformulated as:

(P1): $\underset{T_s, \phi_{ts}}{\text{maximize}}\, \eta[T_s, \phi_{ts}]$, subject to

$\overline{C2}$: $(\alpha/\beta)^2 \leq T_s \leq T$, $C3$: $\phi_{ts,\min} < \phi_{ts} \leq \phi_{ts,\max}$

The complexity of the optimization problem (P1) makes it challenging to solve using classical optimization techniques. This study investigates the application of three distinct nature-inspired swarm-based optimization techniques: particle swarm optimization (PSO) (Shi & Eberhart, 1998; Shitharth et al, 2023), human behavior-based PSO (HPSO) (Liu et al, 2014), and the whale optimization algorithm (WOA) (Mirjalili & Lewis, 2016). These techniques are employed to find the optimal values of T_s and ϕ_{ts} that jointly maximize the EE function.

2.3 Numerical Results

In this section, we obtain various key insights on the performance metric, EE, which serves as an evaluation indicator. The results of the performance evaluation for PSO, HPSO, and WOA are detailed in Table 1 (Ahmed, A. & Baishnab, K. L., 2020). This table compares the performance of the algorithms based on two key metrics: fitness value (EE) and average convergence (conv.) time. To evaluate the performance of PSO, HPSO, and WOA, simulations are carried out four times for each algorithm. In each simulation, the following aspects are recorded: the optimal values of T_s, ϕ_{ts}, and EE (represented as $\eta_{\max}$), along with the convergence point and convergence time. The convergence point signifies the iteration number at which the convergence graph stabilizes to a value corresponding to the optimal solution. From Table 1, it is evident that the optimal T_s and ϕ_{ts} values, which yield the maximum EE, are consistent across all runs using the three techniques. Quantitatively, $T_s = 0.0121$ s, $\phi_{ts} = 0.7677$ W, and $\eta_{\max} = 6.7327$ bps/Joule/Hz. Using the exhaustive search method, Figure 3 displays a 3-D plot of EE versus both T_s and ϕ_{ts} to validate this optimal position (Ahmed, A. & Baishnab, K. L., 2020). The exhaustive search method (Nie et al, 2023) is selected due to its consistent delivery of superior results, despite its higher computational complexity. The EE is found to be highest at a specific point at (0.0121, 0.7677), which is also the best result obtained when employing all three swarm intelligence methodologies. Hence, it can be concluded that the mentioned three optimization methods are effective in identifying the optimal solution for the given joint optimization problem.

Table 1. Comparing the η_{max} and average convergence time performance of PSO, HPSO, and WOA.

Techniques	Run no.	Optimal T_s (s)	Optimal ϕ_{ts} (W)	$\eta_{\max}$ (bps /Joule/ Hz)	Conv. point (iter. no.)	Conv. time (s)	Avg. conv. time (s)
PSO	1	0.0121	0.7677	6.7327	23th	8.9966	10.2902
	2				27th	10.5055	
	3				30th	12.0732	
	4				25th	9.5853	

Techniques	Run no.	Optimal T_s (s)	Optimal ϕ_{ts} (W)	η_{max} (bps /Joule/ Hz)	Conv. point (iter. no.)	Conv. time (s)	Avg. conv. time (s)
HPSO	1	0.0121	0.7677	6.7327	39th	14.3662	14.1622
	2				35th	13.8986	
	3				41th	15.4832	
	4				34th	12.9007	
WOA	1	0.0121	0.7677	6.7327	11th	1.2618	1.4324
	2				18th	1.4971	
	3				13th	1.3995	
	4				19th	1.5712	

The convergence times of PSO, HPSO, and WOA are evaluated across all four simulation runs to compare their efficiency in reaching the convergence points. Table 1 shows that the average convergence time for PSO, HPSO, and WOA are 10.2902 s, 14.1622 s, and 1.4324 s, respectively. The research revealed that the WOA achieved a faster average convergence time compared to PSO and HPSO. Notably, all three techniques successfully converged to the same optimal solution for the proposed joint optimization problem. Next, we compare the performance results of the joint optimization scheme as obtained using the WOA for our proposed optimization problem with its two benchmark semi-adaptive schemes.

Figure 3. Energy efficiency η versus T_s and ϕ_{ts}

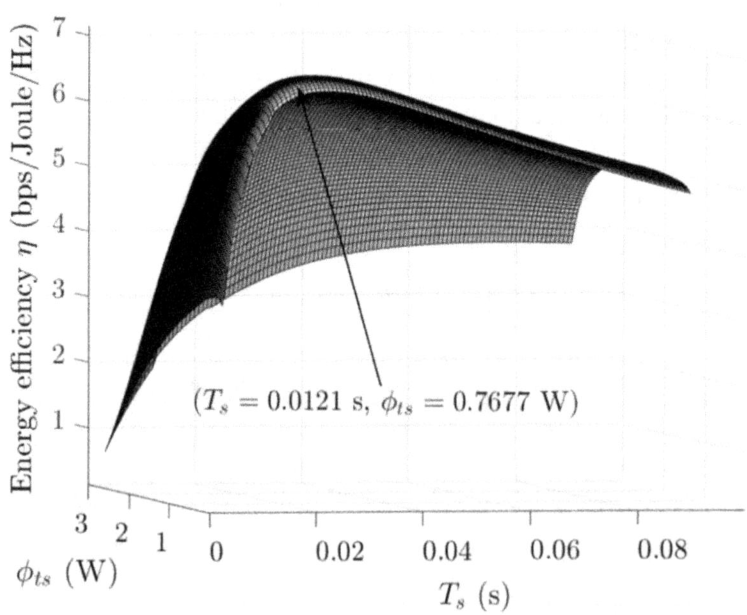

(a) Variation with traffic intensity ρ (b) Variation with number of channels L

Figure 4. Energy efficiency performance under the joint optimization scheme and the benchmark semi-adaptive schemes: OSD and OTP

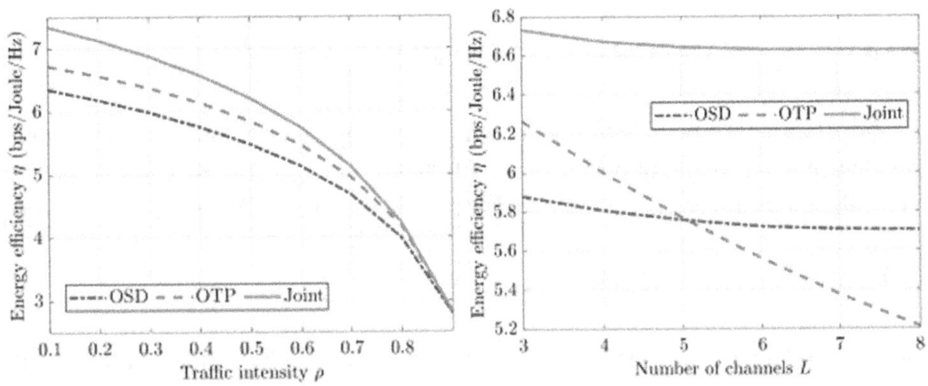

Using Figure 4, we compare the performance results between the joint optimization scheme as obtained using WOA for the problem (P0), and two benchmark semi-adaptive schemes, namely, optimal sensing duration (OSD) and optimal transmit power (OTP). Here, OSD performs optimization of T_s using WOA for a fixed transmit power $\phi_{ts} = 2$ W, whereas OTP performs optimization of ϕ_{ts} using WOA for a fixed sensing duration $T_s = 0.05$ s. It is seen from Figure 4(a) that OTP performs better than the OSD. While in Figure 4(b), OSD performs better than the OTP for $L > 5$. The reason is explained in detail as follows. In both OSD and OTP, the channel-switching probability P_3 increases with L which leads to an increment in the average throughput of secondary transmissions. However, in both benchmark semi-adaptive schemes, energy consumption increases with L due to two factors, namely, the increased P_3 and the increased ϕ_s. Thus, in both schemes, the increment in energy consumption is higher than that of the average throughput of secondary transmissions with increasing L. Therefore, in OTP, to maximize the EE, the ϕ_{ts} increases with L to improve the average throughput which leads to an increase in energy consumption. Nevertheless, in OSD, to maximize the EE, the T_s decreases with L to enhance the average throughput. Thus, the energy consumption in OSD does not increase as much as that of the OTP with increasing L. Therefore, for higher values of $L > 5$, OSD performs better than the OTP in providing EE. In both the figures, the joint optimization provides a significant performance improvement over OSD and OTP. We note that in Figure 4(a), the joint optimization provides an average improvement of about 11.3413% and 5.5347% over the OSD and OTP schemes, respectively. However, in Figure 4(b), this improvement becomes 15.4746% and 17.2441%, respectively.

In Section 2, an optimization problem is presented to maximize the EE of a CR system by jointly optimizing the sensing duration and transmit power while ensuring adequate protection to the PU and considering spectrum handoff. Swarm intelligence techniques, including PSO, HPSO, and WOA are employed to solve this problem. The study evaluates the effectiveness of these techniques in terms of maximum EE and average convergence time. The results indicate that all three techniques reach a unique optimal point where EE is maximized. However, WOA demonstrates superior performance over PSO and HPSO in terms of average convergence time. Besides, a comparison is performed between the joint optimization and the benchmark semi-adaptive schemes, and the numerical results show that the joint optimization outperforms the benchmark semi-adaptive schemes in providing EE. Next, this chapter investigates the performance of a multi-antenna CR system using the energy beamforming technique.

3. EVALUATION OF BEAMFORMING FOR MULTI-ANTENNA CR SYSTEMS

In this section, the energy beamforming technique is employed at the multi-antenna secondary transmitter to enhance the throughput performance of the secondary network. Closed-form expressions are developed for the interference energy received at the PU receiver and the average throughput of SU.

3.1 Three-Phase Transmission Model

A CR system is investigated that consists of a single PU, a single SU, and a shared receiver serving as the data destination, known as a data sink (DS) (Swathi & Begum, 2018), for both the PU and SU. The DS and the PU each employ a single antenna, while the SU utilizes a multiple-antenna system with M antennas (Ahmed, A., Mishra, D., Prasad, G. & Baishnab, K. L., 2021). The channel from the SU to the DS is designated as the downlink (DL). Conversely, the channel from the DS to the SU is referred to as the uplink (UL). Time-division duplexing (TDD) mode is used to communicate between the SU and DS. By utilizing the UL-DL channel reciprocity assumption, an estimate of CSI is derived (Kashyap, Björnson & Larsson, 2016; Mishra, D., & Johansson, H., 2019). This chapter explores a transmission strategy for a SU with multiple antennas to communicate with the DS in three steps. In the first phase of this transmission approach, the SU performs spectrum sensing. Energy detection method is considered by each of the antennas of the SU to detect the PU's status. The OR fusion strategy is used which integrates all of the findings from separate energy detectors to determine the ultimate conclusion on the presence of the PU. The overall detection probability and overall false alarm probability are given by $P_{od} = 1 - (1 - P_d)^M$ and $P_{of} = 1 - (1 - P_f)^M$, respectively.

It is assumed that each element of the channel vector $\mathbf{h}_d$ from M-antenna SU to the DS follows Rayleigh distribution. The channel vector is given by $\mathbf{h}_d = \sqrt{\beta_d}\,\mathbf{h}_{sd}$, where β_d and $\mathbf{h}_{sd}$ respectively denote the distance-dependent path loss and the scattering component of the channel $\mathbf{h}_d$. Each entry in $\mathbf{h}_{sd}$ is modeled as an i.i.d. zero mean unit variance circularly symmetric complex Gaussian random variable. In the second phase, the SU performs CE of $\mathbf{h}_d$ using pilots sent by the DS on the UL channel which is done with the least squares (Kashyap, Björnson & Larsson, 2016). The estimated channel $\hat{\mathbf{h}}_d$ is calculated as $\hat{\mathbf{h}}_d = \mathbf{h}_d + \frac{\mathbf{W}}{\sqrt{\phi_p T_e}}$, where ϕ_p is the pilot signal power, T_e is the CE duration, and $\mathbf{w}$ represents the additive white Gaussian noise (AWGN) vector at the SU with power σ_s^2. In the third phase, the SU employs beamforming to transmit data utilizing the estimated channel $\hat{\mathbf{h}}_d$. The average received SNR at the DS is given by $\frac{\phi_{tm}}{\sigma_d^2}\mathbb{E}\left\{\left|\frac{\hat{\mathbf{h}}_d^{\dagger}\mathbf{h}_d}{\hat{\mathbf{h}}_d}\right|^2\right\}$, where $\phi_{tm}\mathbb{E}\left\{\left|\frac{\hat{\mathbf{h}}_d^{\dagger}\mathbf{h}_d}{\hat{\mathbf{h}}_d}\right|^2\right\} = \phi_{tm}\left(\beta_d M - \frac{\beta_d \sigma_s^2 (M-1)}{\beta_d \phi_p T_e + \sigma_s^2}\right)$ is the average signal power received at the DS, ϕ_{tm} is the M-antenna SU's transmit power during downlink data transmission, and σ_s^2 and σ_d^2 are the noise power received at the SU and DS, respectively (Ahmed et al., 2021).

In this chapter, $(\cdot)^{\dagger}$ denotes the conjugate transpose, $|\cdot|$ represents the absolute value of a complex scalar, whereas $\|\cdot\|$ is used to indicate the Euclidean norm of a complex vector, and $\mathbb{E}\{\cdot\}$ denotes the expectation. Note that the accuracy of CE depends on CE duration. Increased CE duration translates to a more precise characterization of the CSI. Consequently, this enhances the beamforming technique's efficacy. Figure 5 (Ahmed et al., 2021) illustrates the frame structure employed by the SU. This frame has a total duration of T, where $T - T_s - T_e$ represents the portion allocated for data transmission.

Figure 5. SU's slotted frame structure for data transmission.

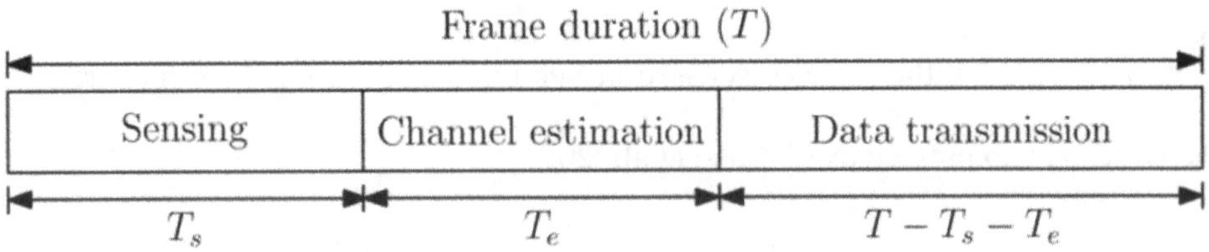

3.2 Performance Analysis

CR networks can experience interference from two main factors: mis-detection by the SU and the return of the PU during secondary transmission. To facilitate analysis and guarantee optimal sensing performance, a sensing duration T_s is chosen that makes sensing errors, like overall detection probability P_{od} and overall false alarm probability P_{of} very small. In this chapter, the sensing duration is set greater than or equal to T_s^{min}. This value represents the minimum sensing duration that ensures both the target overall detection probability $\bar{P}_{od}$ and the target overall false alarm probability $\bar{P}_{of}$. The expression for determining T_s^{min} is provided by (Ahmed et al., 2021) as given:

$$T_s^{min} \triangleq \frac{1}{\gamma^2 f_s}\left(Q^{-1}\left[1-(1-\bar{P}_{of})^{\frac{1}{M}}\right] - Q^{-1}\left[1-(1-\bar{P}_{od})^{\frac{1}{M}}\right]\sqrt{2\gamma+1}\right)^2 \quad (4)$$

Here, γ is the average PU's received SNR as discussed earlier which is assumed to be the same at each antenna of SU. Now, the average interference energy I_{avg} caused by the recurrence of PU is analyzed. Let $x[t]$ represent the interference duration which signifies the cumulative time duration when the PU remains active within the SU's transmission duration t, where $t \triangleq T - T_s - T_e$. It should be noted that this period t always commences when the PU is inactive. The average interference energy I_{avg} received at the DS is given by $I_{avg} = \mathbb{E}\left\{ x[t]\,\phi_{tm}\left|\frac{\hat{\mathbf{h}}_d^{\dagger}\mathbf{h}_d}{\hat{\mathbf{h}}_d}\right|^2 \right\}$, where $\phi_{tm}\left|\frac{\hat{\mathbf{h}}_d^{\dagger}\mathbf{h}_d}{\hat{\mathbf{h}}_d}\right|^2$ is the instantaneous power received at the DS. The closed-form expression of average interference duration is obtained as (Jiang et al., 2013):

$$\mathbb{E}\{x[t]\} = \rho(T - T_s - T_e) - \alpha_0\rho^2\left(1 - e^{-\frac{T-T_s-T_e}{\alpha_0\rho}}\right) \quad (5)$$

Since both $\phi_{tm}\left|\frac{\hat{\mathbf{h}}_d^{\dagger}\mathbf{h}_d}{\hat{\mathbf{h}}_d}\right|^2$ and $x[t]$ are independent, the average interference energy can be expressed as $I_{avg} = \mathbb{E}\{x[t]\}\,\phi_{tm}\mathbb{E}\left\{\left|\frac{\hat{\mathbf{h}}_d^{\dagger}\mathbf{h}_d}{\hat{\mathbf{h}}_d}\right|^2\right\}$. Now, using $\mathbb{E}\{x[t]\}$ in (5) and the expression of average signal power $\phi_{tm}\mathbb{E}\left\{\left|\frac{\hat{\mathbf{h}}_d^{\dagger}\mathbf{h}_d}{\hat{\mathbf{h}}_d}\right|^2\right\}$ received at the DS as discussed in Section 3.1, the closed-form expression of I_{avg} received at the DS can be expressed as (Ahmed et al., 2021):

$$I_{avg} = \left(\rho(T - T_s - T_e) - \alpha_0\rho^2\left(1 - e^{-\frac{T-T_s-T_e}{\alpha_0\rho}}\right)\right)\phi_{tm}\left(\beta_d M - \frac{\beta_d\sigma_s^2(M-1)}{\beta_d\phi_p T_e + \sigma_s^2}\right) \tag{6}$$

Now, the average throughput R_{avg} received at the DS can be calculated as the average of the ratio of the product of SU's effective transmission duration i.e., $T - T_s - T_e - x[t]$ and the corresponding received data rate $\log_2\left[1 + \frac{\phi_{tm}}{\sigma_d^2}\left|\frac{\hat{\mathbf{h}}_d^{\dagger}\mathbf{h}_d}{\hat{\mathbf{h}}_d}\right|^2\right]$ to the frame duration. Therefore, the R_{avg} in bps/Hz received at the DS is given as (Ahmed et al., 2021):

$$R_{avg} = \mathbb{E}\left\{\frac{T - T_s - T_e - x[t]}{T}\log_2\left[1 + \frac{\phi_{tm}}{\sigma_d^2}\left|\frac{\hat{\mathbf{h}}_d^{\dagger}\mathbf{h}_d}{\hat{\mathbf{h}}_d}\right|^2\right]\right\} \tag{7}$$

Because $\frac{T - T_s - T_e - x[t]}{T}$ and $\log_2\left[1 + \frac{\phi_{tm}}{\sigma_d^2}\left|\frac{\hat{\mathbf{h}}_d^{\dagger}\mathbf{h}_d}{\hat{\mathbf{h}}_d}\right|^2\right]$ are independent, the average throughput in (7) can be further expressed as $R_{avg} = \frac{T - T_s - T_e - \mathbb{E}\{x[t]\}}{T}\mathbb{E}\left\{\log_2\left[1 + \frac{\phi_{tm}}{\sigma_d^2}\left|\frac{\hat{\mathbf{h}}_d^{\dagger}\mathbf{h}_d}{\hat{\mathbf{h}}_d}\right|^2\right]\right\}$. However, since there is no closed-form expression for the R_{avg}. Therefore, Jensen's inequality is employed to derive a tight approximation of average throughput as given by $R_{avg} \approx \frac{T - T_s - T_e - \mathbb{E}\{x[t]\}}{T}\log_2\left[1 + \frac{\phi_{tm}}{\sigma_d^2}\mathbb{E}\left\{\left|\frac{\hat{\mathbf{h}}_d^{\dagger}\mathbf{h}_d}{\hat{\mathbf{h}}_d}\right|^2\right\}\right]$. Now, using the expression of $\phi_{tm}\mathbb{E}\left\{\left|\frac{\hat{\mathbf{h}}_d^{\dagger}\mathbf{h}_d}{\hat{\mathbf{h}}_d}\right|^2\right\}$ (cf. Section 3.1) and (5), the tight approximation of R_{avg} can be obtained as (Ahmed et al., 2021):

$$R_{avg} \approx \frac{(T - T_s - T_e)(1-\rho) + \alpha_0\rho^2\left(1 - e^{-\frac{T-T_s-T_e}{\alpha_0\rho}}\right)}{T}\log_2\left[1 + \frac{\phi_{tm}}{\sigma_d^2}\left(\beta_d M - \frac{\beta_d\sigma_s^2(M-1)}{\beta_d\phi_p T_e + \sigma_s^2}\right)\right] \tag{8}$$

3.3 Numerical Results

Here, we observe the performance metric, the SU's average throughput, which serves as an evaluation indicator, across various key system parameters. The validation of the expressions for interference energy I_{avg} in (6) and SU's average throughput R_{avg} in (8), across different values of mean inter-arrival time T_{int} and traffic intensity ρ is shown in Figure 6 (Ahmed et al., 2021). It is evident that there is a strong match between the simulation and analytical results. In the validation of I_{avg} and R_{avg}, the average root mean square error (RMSE) is 0.0873×10^{-6} and 0.1517, respectively.

Figure 6. Validation of average interference I_{avg} in (6) and average throughput R_{avg} in (8) for a T_s = 12 ms

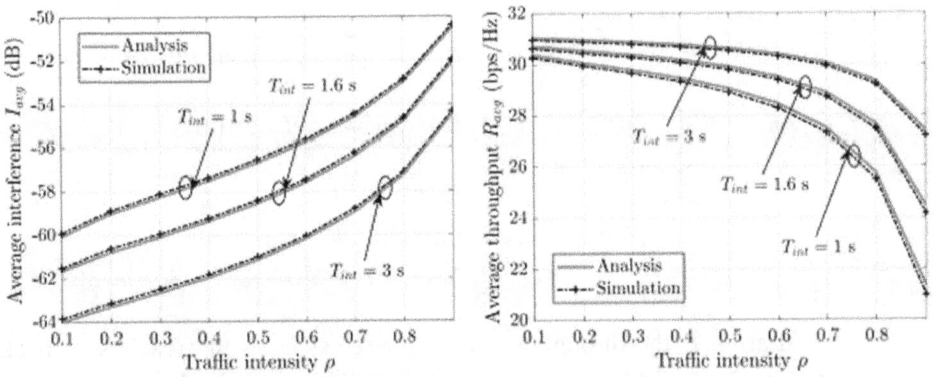

(a) Variation of I_{avg} with ρ and T_{int} (b) Variation of R_{avg} with ρ and T_{int}

For a given T_{int}, it is noticed that the mean ON duration α_1 increases while the mean OFF duration decreases with ρ. Therefore, as indicated by (5), the average interference duration increases with ρ, leading to an increase in the average interference energy I_{avg}, as depicted in Figure 6(a). Conversely, when the traffic intensity ρ is constant, both α_0 and α_1 increase with the mean inter-arrival time T_{int}. This leads to a lower likelihood of PU recurrence during SU data transmission, reducing the average interference energy I_{avg} with T_{int}. Since an increase in I_{avg} reduces the SU's average throughput R_{avg}, R_{avg} decreases with ρ for a fixed T_{int}. However, for a given ρ, R_{avg} increases with T_{int}, as illustrated in Figure 6(b).

Figure 7. Variation of R_{avg} with T_e for different values of T, where $T_s = T_s^{min}$

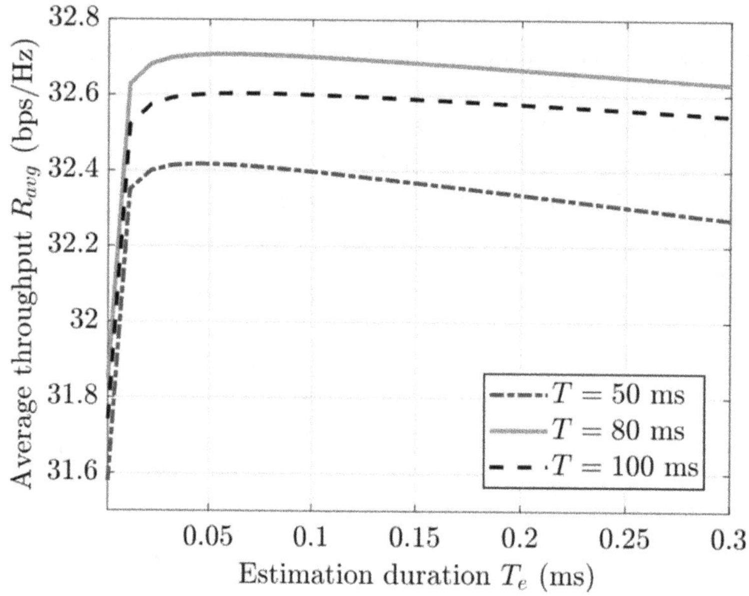

For a fixed frame duration *T*, the throughput initially rises sharply and reaches its peak with CE duration T_e, as depicted in Figure 7 (Ahmed et al., 2021). This increase in throughput is attributed to the average received signal power at the DS, which grows with T_e, thereby boosting the SU's average received throughput at the DS due to an increase in the received data rate. However, as T_e increases, the transmission duration decreases, impacting the throughput negatively. Consequently, the SU's throughput decreases after reaching a certain CE duration.

In this section, the interference energy received at the data sink and the SU's average throughput are studied considering the effects of CE and beamforming. Closed-form expressions for these metrics are derived. The proposed analysis is validated through simulation, and the numerical results offer insights into the impact of CE duration on the SU's throughput. It is observed that, for a given frame duration, there exists an optimal CE duration that achieves the best tradeoff. In the next section, the throughput maximization problem of the multi-antenna CR system is investigated.

4. OPTIMAL TIME ALLOCATION IN MULTI-ANTENNA CR SYSTEM

In this chapter, the available resources for the system outlined in the preceding section are optimized.

4.1 Cognitive Radio Timing Protocol for Throughput Maximization

This section explores a cognitive radio timing protocol (CRTP) designed to efficiently allocate time for various phases of CR operation with the goal of maximizing the SU's throughput. The optimization problem aims to maximize the average throughput R_{avg} as defined in (8), subject to the boundary constraints of the relevant variables and the limit constraint of average interference energy I_{avg} as described in (6). The joint optimization problem is formulated as given below (Ahmed, A., Mishra, D., Prasad, G. & Baishnab, K. L., 2022).

(P2): $\underset{T_s, T_e, T}{\text{maximize}} R_{avg}$, subject to

$$C1: \left(\rho(T - T_s - T_e) - \alpha_0\rho^2\left(1 - e^{-\frac{T-T_s-T_e}{\alpha_0\rho}}\right)\right)\phi_{tm}\left(\beta_d M - \frac{\beta_d\sigma_s^2(M-1)}{\beta_d\phi_p T_e + \sigma_s^2}\right) \leq Q_I,$$

$$C2: T_s \geq T_s^{min}, C3: T \leq T^{max}, C4: T \geq T_s, C5: T_e \leq T^{max} - T_s, C6: T_e \geq 0.$$

Ensuring adequate protection for the PU's communication against harmful interference from the SU requires satisfying constraint C1. This constraint ensures that the interference energy is kept below or equal to the tolerable threshold Q_I. Constraint C2 states that the sensing duration must exceed or equal its minimal value, denoted as T_s^{min}. Constraints C3 (or C5) and C4 (or C6) establish upper and lower bounds on T (or T_e), respectively. Here, T^{max} represents the maximum value for T.

As seen in (8), R_{avg} decreases continuously with T_s, regardless of the values of T_e and T. Thus, to maximize R_{avg}, the globally-optimal sensing duration T_s^* is determined from constraint C2 which is equal to the minimum sensing duration i.e. $T_s^* = T_s^{min}$.

After substituting $T_s = T_s^*$ into the problem (P2), it can be reformulated as a joint optimization problem (P3) involving two key variables T_e and T as shown below.

(P3): $\underset{T_e, T}{\text{maximize}} R_{avg}$, subject to

$$\overline{C1}: \left(\rho(T - T_s^* - T_e) - \alpha_0\rho^2\left(1 - e^{-\frac{T-T_s^*-T_e}{\alpha_0\rho}}\right)\right)\phi_{tm}\left(\beta_d M - \frac{\beta_d\sigma_s^2(M-1)}{\beta_d\phi_p T_e + \sigma_s^2}\right) \leq Q_I,$$

$$C3: T \leq T^{max}, \overline{C4}: T \geq T_s^*, \overline{C5}: T_e \leq T^{max} - T_s^*, C6: T_e \geq 0.$$

The optimization problem (P3) is nonconvex due to the jointly nonconvex nature of constraint $\overline{C1}$. To find the jointly optimal solution, a nontrivial technique is being explored.

4.2 Joint optimization of Frame Duration and CE Duration

Here, the individual optimization of frame duration and CE duration are first discussed. Later using those, a joint solution methodology is investigated to address the problem (P3).

4.2.1 Optimization of Frame Duration

Ahmed et al. (2022) outlined a method to determine the optimal frame duration T^* that maximizes the average throughput R_{avg} for a given CE duration T_e. This is achieved by evaluating problem (P3) as follows.

$$(\text{P4}): \underset{T}{\text{maximize}}\, R_{avg}, \text{ subject to } \overline{C1}, C3, \overline{C4}.$$

The problem (P4) is characterized as a generalized convex problem since the interference constraint $\overline{C1}$ is convex and the objective function R_{avg} is pseudoconcave in T. Therefore, solving the KKT conditions of the problem (P4) leads to achieving its globally optimal solution. To facilitate this, the Lagrangian function $\mathscr{L}(T,\lambda)$ of (P4) is derived as outlined below by Ahmed et al. (2022):

$$\mathscr{L}(T, \lambda) = R_{avg} - \lambda\left(\left(\rho(T - T_s^* - T_e) - \alpha_0\rho^2\left(1 - e^{-\frac{T-T_s^*-T_e}{\alpha_0\rho}}\right)\right)\phi_{tm}\left(\beta_d M - \frac{\beta_d\sigma_s^2(M-1)}{\beta_d\phi_p T_e + \sigma_s^2}\right) - Q_I\right) \tag{9}$$

where $\lambda \geq 0$ represents the Lagrange multiplier related to the constraint $\overline{C1}$, while constraints C3 and $\overline{C4}$ remain implicit. Ahmed et al. (2022) express the KKT conditions using (9) as depicted in (10).

$$\begin{aligned} \partial L/\partial T = {} & ln[1 + \phi_t m/(\sigma_d^2)(\beta_d M - (\beta_d\sigma_s^2(M-1))/(\beta_d\phi_P T_e + \sigma_s^2))] \\ & ((1-\rho)(T_s^* + T_e) - \alpha_0\rho^2(1 - e^{(-(}T - T_s^* - T_e)/(\alpha_0\rho))(1 + T/(\alpha_0\rho)))) \\ & 1/ln(2)T^2 - \lambda\phi_{tm}(\beta_d M - (\beta_d\sigma_s^2(M-1))/(\beta_d\phi_p T_e + \sigma_s^2))\rho(1 - e^{(-(T-T_s^*-T_e)/(\alpha_0\rho))})) = 0 \end{aligned} \tag{10a}$$

$$\lambda\left(\left(\rho(T - T_s^* - T_e) - \alpha_0\rho^2\left(1 - e^{-\frac{T-T_s^*-T_e}{\alpha_0\rho}}\right)\right)\phi_{tm}\left(\beta_d M - \frac{\beta_d\sigma_s^2(M-1)}{\beta_d\phi_p T_e + \sigma_s^2}\right) - Q_I\right) = 0 \tag{10b}$$

By substituting $\lambda = 0$ into (10a), the closed-form expression of T^* is obtained for the case $\lambda = 0$ as follows (Ahmed et al., 2022):

$$T^* = T_b \triangleq -\alpha_0\rho\left(1 + W_{-1}\left[\left(\frac{(1-\rho)(T_s^* + T_e) - \alpha_0\rho^2}{\alpha_0\rho^2}\right)e^{-\frac{T_s+T_e}{\alpha_0\rho}-1}\right]\right) \tag{11}$$

where $W_{-1}[\cdot]$ is the Lambert function. The point $(T_b, \lambda^* = 0)$ is considered a feasible KKT solution if the interference constraint $\overline{C1}$ is satisfied at $T = T_b$; however, it doesn't lead to the globally optimal solution otherwise. The closed-form expression of T^* for $\lambda>0$ is determined by solving (10b) as follows (Ahmed et al., 2022):

$$T^* = T_d \triangleq T_s^* + T_e + \frac{Q_I}{\rho J} + \alpha_0\rho\left(1 + W_0\left[-e^{-\frac{Q_I}{\alpha_0\rho^2 J}-1}\right]\right) \tag{12}$$

where $J = \phi_{tm}\left(\beta_d M - \frac{\beta_d\sigma_s^2(M-1)}{\beta_d\phi_p T_e + \sigma_s^2}\right)$ and $W_0[\cdot]$ denotes the Lambert function of the principal branch.

4.2.2 Optimization of CE Duration

To maximize the throughput R_{avg} for a specified frame duration T, the problem (P3) can be analyzed with respect to T_e as outlined by Ahmed et al. (2022).

$$(\text{P5}): \underset{T_e}{\text{maximize}}\, R_{avg}, \text{ subject to } \overline{C1}, \overline{C5}, C6.$$

Because of the concavity of R_{avg} and the non-convex nature of the constraint $\overline{C1}$ with respect to T_e, problem (P5) is classified as nonconvex. The optimal solution is determined using Algorithm 1 which employs the golden search method as presented by Ahmed et al. (2022).

Algorithm 1 iteratively reduces the search space Δ_{T_e} defined as the difference between the upper bound T_e^u and the lower bound T_e^l on T_e, by a factor of 0.618 based on R_{avg}. It ensures that the interference constraint $\overline{C1}$ is satisfied in every iteration. If the constraint is not met, R_{avg} is set to '0'. This algorithm concludes after N_g iterations, where the termination condition is satisfied with a tolerance of ξ when $\Delta_{T_e}(0.618)^{N_g} \leq \xi$.

4.2.3 Joint Optimization

As the joint optimization problem (P1) exhibits nonconvexity concerning both T_e and T, the jointly optimal solution is evaluated using an alternating optimization algorithm as given in Algorithm 2 (Ahmed et al., 2022). This algorithm begins with the starting points $T_e^{(0)}$ and $T^{(0)}$. Algorithm 2 calculates the optimal value of T at the jth iteration based on the previously determined optimal value of T_e at the $(j\text{-}1)^{th}$ iteration. With the obtained optimal value of T, the optimal T_e is computed at the jth iteration. This algorithm continues iterating until the difference between the objective function values at the current iteration and the previous iteration exceeds an acceptable tolerance threshold ζ.

4.3 Numerical Results

Figure 8. Performance comparison of joint optimization algorithm against the exhaustive search

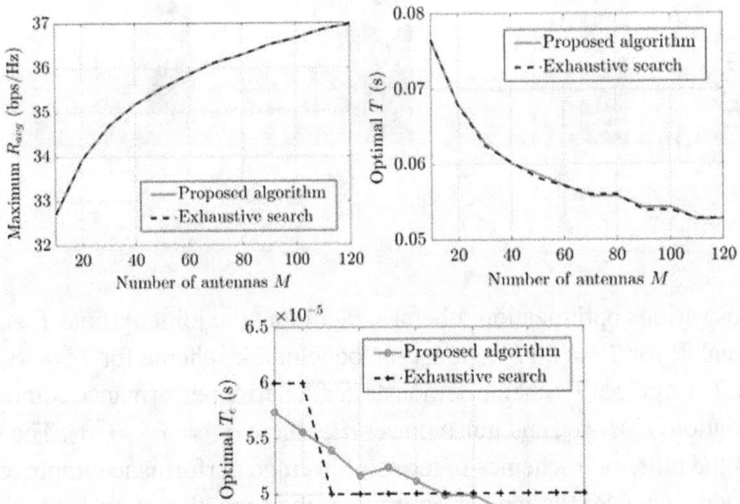

(a) Variation of maximum throughput with M; (b) Variation of optimal frame duration with M; (c) Variation of optimal CE duration with M

In this section, we analyze the average throughput of SU which serves as an evaluation indicator across various key system parameters. Figure 8 illustrates the optimal values of average throughput, frame duration, and CE duration for various values of M obtained using the exhaustive search method and Algorithm 2, as described by Ahmed et al. (2022). The results of Algorithm 2 closely match the exhaustive search at a given M, as shown in Figure 8(a) and Figure 8(b). On the other hand, in Figure 8(c), a minor gap is noticeable between the results obtained using the two algorithms for each given M. This difference arises from the smaller optimal value of T_e and the step size used in the exhaustive search to determine T_e. Nonetheless, as depicted in Figure 8(a), this minor deviation does not impact the SU's maximum average throughput. The above discussion highlights that the proposed joint optimization algorithm effectively determines the jointly optimal values of T and T_e.

Figure 9. Performance comparison of different optimization schemes, where $\phi_p = -30$ dB

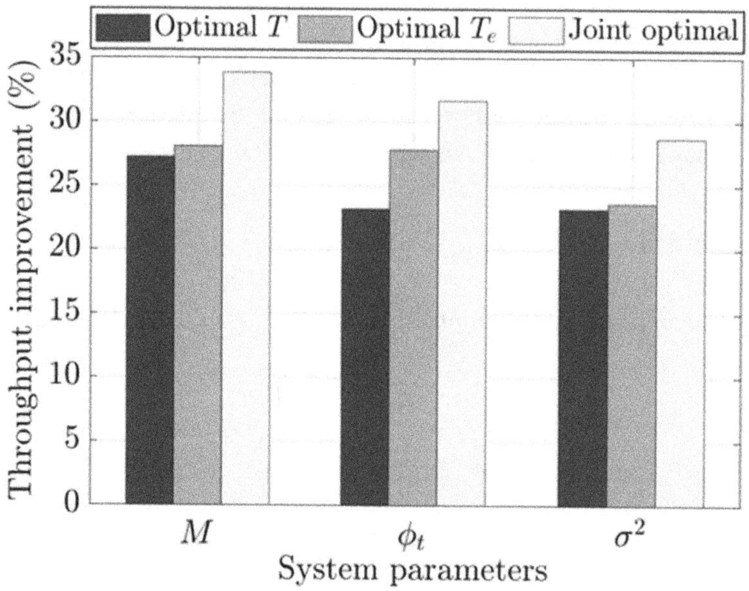

Figure 9 compares various optimization schemes, including the joint optimal T_e and T, optimal T for $T_e = 2T_s^*$, and optimal T_e for $T = 10T_s^*$ against the benchmark scheme for $T = 10T_s^*$ and $T_e = 2T_s^*$ (Ahmed et al., 2022). The optimal T_e scheme demonstrates superior performance compared to the optimal T scheme for the variation of M, ϕ_{tm}, and noise power σ^2, where $\sigma^2 = \sigma_s^2 = \sigma_d^2$. The joint optimization scheme outperforms the other two schemes in terms of average performance improvement.

This section introduces a CRTP aimed at optimizing the time allocation for various phases of CR operation discussed in Section 3, with the aim of maximizing the average throughput of the SU. Specifically, the closed-form expressions of the interference energy received at the DS and the SU's average throughput as obtained in Section 3 are considered. Subsequently, a joint optimization problem to maximize the SU's throughput is established across key variables: sensing duration, CE duration, and frame duration while maintaining the interference limits. This problem is then reformulated as a joint optimization problem in frame duration and CE duration with the optimal sensing duration already determined. Due to the nonconvex nature of this problem, an alternating optimization algorithm is proposed to determine its joint optimal solution. The presented numerical results offer insights into the impact of various key system parameters on the optimal values of average throughput, CE duration and frame duration. The comparison of the performance improvement achieved by various optimization schemes against a benchmark scheme reveals that the proposed joint optimization scheme significantly outperforms the individual optimization schemes. Additionally, the correctness of the proposed joint optimization algorithm has been numerically verified. The results indicate that the optimal values obtained using the proposed algorithm closely match those obtained through the exhaustive search method. The following section introduces a transmission protocol for the multi-antenna SU designed to enable smart sensing. Its performance is then analyzed in terms of average throughput.

5. TRANSMISSION IN SMART SENSING CR

5.1 Transmission Protocol for the Smart Sensing CR System

In this section, for performance analysis, the same multi-antenna CR system discussed in Section 3 is considered. In a smart sensing CR system, spectrum sensing is not conducted by the SU (Ahmed, A., Mishra, D., Prasad, G., Baishnab, K. L. & Tourki, K., 2021). It is considered that whenever the PU switches from a busy to an idle state it informs the SU regarding the length of its idle state. As a result, the SU can transmit data directly when the PU is idle. Furthermore, since CE is incorporated after each coherence duration T_c, each idle state in this case is split up into many coherence blocks for SU transmission. The SU first estimates the channel between itself and the DS for the duration T_e in each coherence block of an idle state. For the remainder of the coherence block, data is then transmitted using that estimated channel. To estimate the channel $\mathbf{h}_d$, the system relies on the assumption of UL-DL channel reciprocity as illustrated previously. The SU employs a maximum ratio transmission (MRT) based precoder to improve beamforming gain by using the estimated channel $\hat{\mathbf{h}}_d$ while transmitting data to the DS on the DL.

5.2 Transmission Protocol for the Conventional CR System

Here, to determine the activity status of the PU, the multi-antenna SU executes spectrum sensing after each frame duration T. An energy detection mechanism is used by each SU antenna to identify the activity status of the PU. The OR fusion rule is applied which aggregates the individual decisions of energy detectors to get the final decision on spectrum sensing in terms of P_{od} and P_{of}. To initiate communication with the sink node, the SU first conducts spectrum sensing for a duration T_s. If the PU is not detected, the SU then spends another period i.e., T_e to estimate the channel $\mathbf{h}_d$ as explained earlier. Finally, the SU utilizes beamforming based on the estimated channel $\hat{\mathbf{h}}_d$ to transmit data for the remaining duration $T - T_s - T_e$. Here, $T = T_c$, indicating that the channel $\mathbf{h}_d$ necessitates estimation after each frame duration T (Ahmed, A. et al., 2021).

5.3 Average Throughput in Smart Sensing CR System

Let the length of an idle state be denoted as T_I, where $\mathbb{E}\{T_I\} = \alpha_0$. The SU estimates the channel in each coherence block of an idle state, leading to the expression of SU's average transmission duration in an idle state as (Ahmed, A. et al., 2021):

$$\mathbb{E}\{T_{ti}\} = \mathbb{E}\left\{T_I - \frac{T_I T_e}{T_c}\right\} = \alpha_0\left(1 - \frac{T_e}{T_c}\right) \tag{13}$$

The average throughput of SU received at the data sink can be calculated as (Ahmed, A. et al., 2021):

$$R_p = \frac{\mathbb{E}\{T_{ti}\}}{\alpha_0}\mathbb{E}\left\{\log_2\left[1 + \frac{\phi_{tm}}{\sigma_d^2}|\boldsymbol{\Upsilon}_d|^2\right]\right\} \tag{14}$$

where $\boldsymbol{\Upsilon}_d \triangleq \frac{\hat{\mathbf{h}}_d^{\dagger}\mathbf{h}_d}{\hat{\mathbf{h}}}$. Using Taylor series approximation, the mean of the logarithmic term present in (14) can be approximated as follows (Ahmed, A. et al., 2021):

$$\mathbb{E}\left\{\log_2\left[1 + \frac{\phi_{tm}}{\sigma_d^2}|\boldsymbol{\Upsilon}_d|^2\right]\right\} \approx \log_2\left[1 + \frac{\phi_{tm}}{\sigma_d^2}\left(\beta_d M - \frac{\beta_d\sigma_s^2(M-1)}{\beta_d\phi_p T_e + \sigma_s^2}\right)\right] - \frac{\phi_{tm}^2\beta_d^2\left(\beta_d\phi_p T_e M\left(\beta_d\phi_p T_e + 2\sigma_s^2\right) + \sigma_s^4\right)}{2\ln(2)\sigma_d^4\left(\beta_d\phi_p T_e\left(1 + \frac{\beta_d\phi_{tm}M}{\sigma_d^2}\right) + \sigma_s^2\left(1 + \frac{\beta_d\phi_{tm}}{\sigma_d^2}\right)\right)^2} \tag{15}$$

Let the approximation obtained in (15) be denoted by C_0. Using this notation C_0 and (13), the tight closed-form approximation of R_p in (14) can be obtained as (Ahmed, A. et al., 2021):

$$R_p \approx \left(1 - \frac{T_e}{T_c}\right)C_0 \tag{16}$$

5.4 Average Throughput in Conventional System

In a conventional CR system, the SU transmits data to the sink node under two scenarios. The first case occurs when the PU is absent and the SU correctly detects the PU's activity. The second case is considered when the PU is present and the SU wrongly detects PU's activity. The collision probabilities in the first and second scenarios, respectively, caused by multiple changes in the PU's activity status within the SU's transmission duration are expressed as (Ahmed, A. et al., 2021):

$$P_c^{S1} = \rho - \frac{\alpha_0\rho^2}{T - T_s - T_e}\left(1 - e^{-\frac{T-T_s-T_e}{\alpha_0\rho}}\right) \tag{17a}$$

$$P_c^{S2} = \rho + \frac{\alpha_1(1-\rho)^2}{T - T_s - T_e}\left(1 - e^{-\frac{T-T_s-T_e}{\alpha_0\rho}}\right) \tag{17b}$$

Using (15) and (17), the tight closed-form approximation of SU's average throughput in the conventional CR system is computed as (Ahmed, A. et al., 2021):

$$R_c \approx \frac{T - T_s - T_e}{T}\left((1-\rho)(1-P_{of})\left(C_0(1-P_c^{S1}) + C_1 P_c^{S1}\right) + \rho(1-P_{od})\left(C_0(1-P_c^{S2}) + C_1 P_c^{S2}\right)\right) \quad (18)$$

Here, the approximation for $\mathbb{E}\left\{\log_2\left[1 + \frac{\phi_{tm}}{\sigma_d^2 + P_p}|\Upsilon_d|^2\right]\right\}$ is denoted by C_1, which can be expressed when σ_d^2 in (15) is replaced by $\sigma_d^2 + P_p$, and P_p denotes the PU signal power. Next, simulation is performed to determine the average throughput R_p.

5.5 Numerical Results

Here, we examine the SU's average throughput as an evaluation indicator across different system parameters. To validate the expression of R_p, simulation results are first generated to determine R_p, and subsequently, the expression of R_p obtained from (16) is validated.

Simulation results are generated using Algorithm 3 (Ahmed, A. et al., 2021) to determine the average throughput R_p. In this context, two arrays for the PU's OFF (idle) state $\mathbf{y}$ and ON (busy) state $\mathbf{z}$ are first generated, as illustrated in Step 1. In each OFF state $\mathbf{y}(i)$, where $i \in \{1, 2,...., N\}$ and N is the number of elements of $\mathbf{y}$, the number of bits transmitted R_i is determined by the main loop. Here, the estimated channel $\hat{\mathbf{h}}_d$ is calculated using the expression defined in Section 3.1. The algorithm ends with an evaluation of R_p. Figure 10 is used to evaluate the average throughput R_p as derived in (16) for varying values of T_c and M (Ahmed, A. et al., 2021). A close match between the analytical and simulation results is noted with a RMSE of 0.0301.

Figure 10. Validation of average throughput R_p in (16)

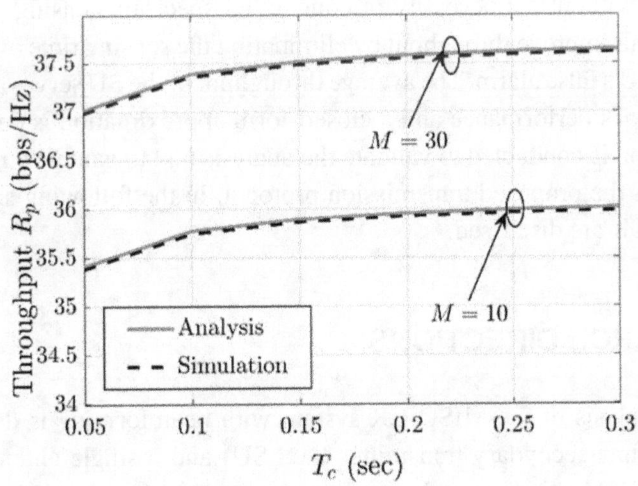

In Figure 10, the variation of R_p with T_c is investigated for various values of M. It is observed that the average throughput R_p increases with T_c as the transmission duration in a coherence block rises with T_c for a given M. Additionally, for a given T_c, it is noticed that the R_p increases with M. This happens because the average received power at the DS rises with M.

Figure 11. Comparing percentage gain in average throughput, where noise power $\sigma^2 = \sigma_s^2 = \sigma_d^2$

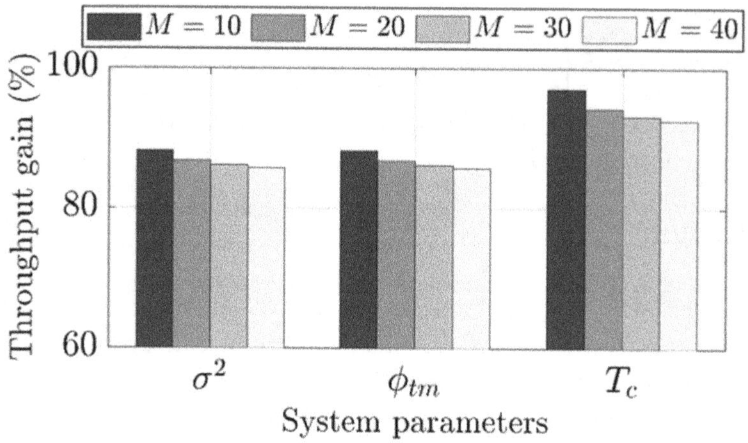

Figure 11 illustrates the performance gain that the proposed transmission protocol achieves over the conventional CR system (Ahmed, A. et al., 2021). In it, the throughput performance of the proposed transmission protocol is compared to that of the conventional one. This comparison is shown for specific values of M with varying system parameters. It is observed that the average performance gain falls with M for a given system parameter. This is because, as the sensing duration decreases with M, the SU's transmission duration increases which leads to an increase in the R_c.

In section 5, a transmission protocol is introduced for SU to implement smart sensing. In this protocol, the PU assists the SU in determining its activity status, enabling the SU to obtain precise information about the PU's activity without the necessity of conducting spectrum sensing. It is expected that this transmission protocol will improve throughput by eliminating the sensing time overhead and throughput loss resulting from the SU's false alarm. The average throughput of the SU serves as the evaluation metric for assessing the protocol's performance and a closed-form approximation is derived for this purpose. A numerical investigation is conducted to validate the analysis and to gain new insights into the performance improvements of the proposed transmission protocol. In the following section, potential future research directions for CR are discussed.

6. FUTURE RESEARCH DIRECTIONS

The performance analysis of the MISO CR system with beamforming is done for a network that consists of a multi-antenna secondary transmitter (i.e., SU) and a single antenna secondary receiver (i.e., DS). This work can be carried out with a multi-antenna secondary transmitter and a multi-antenna secondary receiver i.e., MIMO (Begum & Sharmila, 2020) CR systems. Extending the analysis from a MISO CR system to a MIMO CR system can provide valuable insights into the performance of CR systems with multiple antennas at both the transmitter and receiver sides. MIMO systems offer increased diversity and spatial multiplexing gains, potentially improving the overall performance of the CR system.

It would be interesting to explore the further optimization and adaptation of beamforming techniques for MIMO CR systems, aiming to enhance both spectrum efficiency and reliability.

Extending the performance analysis of the MISO CR system to include scenarios with a single-antenna multiple secondary receivers and a single-antenna multiple PU system would provide valuable insights into the system's behavior in more complex environments. Incorporating the effects of random arrivals and departures of multiple PUs during the SU's frame duration would further enhance the realism of the analysis, allowing for a more comprehensive understanding of the system's performance under dynamic PU activity.

We intend to carry out the EE performance of the MISO CR system with beamforming and to formulate an EE maximization problem to jointly optimize the SU's transmit power and underlying time resources while satisfying the minimum sensing quality, interference, and transmit power constraints. By optimizing the SU's transmit power and time resources, EE maximization ensures efficient use of limited energy resources. This is crucial for prolonging the battery life of CR devices, especially in energy-constrained environments like wireless cognitive sensor networks (Kshirsagar et al, 2020).

Incorporating the energy harvesting technique for further improvement in the EE performance of the CR system under the spectrum handoff scheme is a strategic decision. This approach can enhance the sustainability and reliability of CR systems by reducing their dependence on traditional power sources. It can also lead to cost savings, making it a valuable addition to CR technology, especially in scenarios where energy efficiency is crucial, such as in IoT (Al-Ani et al, 2023) deployments.

The transmission protocol developed for the smart sensing CR system presents a significant opportunity to influence future energy trading protocol designs. It can act as a benchmark for designing protocols that facilitate agreements between PU and SU, allowing them to determine an optimal mutual cost. This cost should be mutually acceptable to the SU, while also meeting the PU's requirements for sharing necessary information, thereby fostering efficient spectrum utilization and collaboration in CR networks.

7. CONCLUSION

The main objective of this chapter is to design energy and throughput-efficient CR systems. The throughput performance of the secondary network can be improved by performing spectrum handoff, nonetheless, this approach may lead to increased energy consumption. Therefore, it is crucial to develop an energy-efficient CR system while executing spectrum handoff. In this chapter, the available resources in different CR systems: multi-channel CR systems and single-channel MISO CR systems are optimized to maximize the EE and throughput performances while maintaining system reliability. This chapter includes the beamforming technique to investigate the interference energy and average throughput of the multi-antenna SU. Additionally, a transmission protocol is discussed to address the spectrum sensing-related issues of the SU, such as sensing time overhead and false alarms, which have the potential to degrade the throughput of the CR systems. The performance of this transmission protocol in terms of average throughput is analyzed and tight closed-form approximation is derived. Thereafter, the average throughput is compared to that of the conventional CR system to obtain performance gain.

However, it is crucial to recognize the limitations of our study. Our system model assumes a simplified CR network with only one PU and one SU. While this simplification has facilitated a clear and tractable analysis, it may not comprehensively represent the intricacies of real-world CR networks, where multiple PUs and SUs coexist. Future research should therefore aim to extend our work to more realistic scenarios, considering a higher number of users and dynamic spectrum availability.

REFERENCES

Agrahari, A., Varshney, P., & Jagannatham, A. K. (2017). Precoding and downlink beamforming in multiuser MIMO-OFDM cognitive radio systems with spatial interference constraints. *IEEE Transactions on Vehicular Technology*, 67(3), 2289–2300. 10.1109/TVT.2017.2768823

Ahmed, A., & Baishnab, K. L. (2020). Joint optimal design of sensing time and transmission power for maximizing energy efficiency in cognitive radio system. *Wireless Personal Communications*, 110(4), 1839–1857. 10.1007/s11277-019-06814-z

Ahmed, A., Mishra, D., Prasad, G., & Baishnab, K. L. (2021, January). Performance analysis of multi-antenna CR system with beamforming under various traffic scenarios. In *2021 IEEE 18th Annual Consumer Communications & Networking Conference (CCNC)* (pp. 1-6). IEEE. 10.1109/CCNC49032.2021.9369528

Ahmed, A., Mishra, D., Prasad, G., & Baishnab, K. L. (2022). Cognitive Radio Timing Protocol for Interference-Constrained Throughput Maximization. *IEEE Transactions on Cognitive Communications and Networking*, 8(2), 989–1004. 10.1109/TCCN.2021.3137487

Ahmed, A., Mishra, D., Prasad, G., Baishnab, K. L., & Tourki, K. (2021, December). Primary Activity Awareness Based Secondary Transmission Gains in Smart Sensing CR. In *2021 IEEE Globecom Workshops (GC Wkshps)* (pp. 1-6). IEEE. 10.1109/GCWkshps52748.2021.9681938

Al-Ani, A. K., Ul Arfeen Laghari, S., Manoharan, H., Selvarajan, S., & Uddin, M. (2023). Improved Transportation Model with Internet of Things Using Artificial Intelligence Algorithm. *Computers, Materials & Continua*, 76(2).

Arshid, K., Jianbiao, Z., Hussain, I., Pathan, M. S., Yaqub, M., Jawad, A., Munir, R., & Ahmad, F. (2022). Energy efficiency in cognitive radio network using cooperative spectrum sensing based on hybrid spectrum handoff. *Egyptian Informatics Journal*, 23(4), 77–88. 10.1016/j.eij.2022.06.008

Awin, F., Abdel-Raheem, E., & Ahmadi, M. (2016). Joint optimal transmission power and sensing time for energy efficient spectrum sensing in cognitive radio system. *IEEE Sensors Journal*, 17(2), 369–376. 10.1109/JSEN.2016.2627884

Awin, F., Abdel-Raheem, E., & Tepe, K. (2018). Blind spectrum sensing approaches for interweaved cognitive radio system: A tutorial and short course. *IEEE Communications Surveys and Tutorials*, 21(1), 238–259. 10.1109/COMST.2018.2863681

Bayhan, S., & Alagoz, F. (2012). Scheduling in centralized cognitive radio networks for energy efficiency. *IEEE Transactions on Vehicular Technology*, 62(2), 582–595. 10.1109/TVT.2012.2225650

Begum, M. B., & Sharmila, K. (2020). Improving performance of Full Duplex Device to Device Communication Underlying Cellular Networks. *International Journal of Innovative Research in Technology*, 7(2), 436–445.

F. C. Commission (2002). *Spectrum policy task force report*. (FCC 02-155). FCC.

Fernando, X., & Lăzăroiu, G. (2023). Spectrum sensing, clustering algorithms, and energy-harvesting technology for cognitive-radio-based internet-of-things networks. *Sensors (Basel)*, 23(18), 7792. 10.3390/s2318779237765848

Jiang, C., Chen, Y., Liu, K. R., & Ren, Y. (2013). Renewal-theoretical dynamic spectrum access in cognitive radio network with unknown primary behavior. *IEEE Journal on Selected Areas in Communications*, 31(3), 406–416. 10.1109/JSAC.2013.130307

Kashyap, S., Björnson, E., & Larsson, E. G. (2016). On the feasibility of wireless energy transfer using massive antenna arrays. *IEEE Transactions on Wireless Communications*, 15(5), 3466–3480. 10.1109/TWC.2016.2521739

Kshirsagar, P. R., Manoharan, H., Al-Turjman, F., & Maheshwari, K. K. (2020). Design and testing of automated smoke monitoring sensors in vehicles. *IEEE Sensors Journal*, 22(18), 17497–17504. 10.1109/JSEN.2020.3044604

Li, Q., Zhang, Q., & Qin, J. (2016). Beamforming for information and energy cooperation in cognitive non-regenerative two-way relay networks. *IEEE Transactions on Wireless Communications*, 15(8), 5302–5313. 10.1109/TWC.2016.2555913

Liang, Y. C., Zeng, Y., Peh, E. C., & Hoang, A. T. (2008). Sensing-throughput tradeoff for cognitive radio networks. *IEEE Transactions on Wireless Communications*, 7(4), 1326–1337. 10.1109/TWC.2008.060869

Liu, H., Chen, J., Ding, G., Tsiftsis, T. A., & Rowell, C. (2016, March). Antenna beamforming for energy harvesting in cognitive radio networks. In *2016 IEEE MTT-S International Wireless Symposium (IWS)* (pp. 1-4). IEEE. 10.1109/IEEE-IWS.2016.7585472

Liu, H., Xu, G., Ding, G. Y., & Sun, Y. B. (2014). Human behavior-based particle swarm optimization. *TheScientificWorldJournal*.24883357

McHenry, M. A. (2005). *NSF spectrum occupancy measurements project summary*. (Shared Spectrum Company Report).

Mirjalili, S., & Lewis, A. (2016). The whale optimization algorithm. *Advances in Engineering Software*, 95, 51–67. 10.1016/j.advengsoft.2016.01.008

Mishra, D., & Johansson, H. (2019). Optimal channel estimation for hybrid energy beamforming under phase shifter impairments. *IEEE Transactions on Communications*, 67(6), 4309–4325. 10.1109/TCOMM.2019.2901790

Nie, D., Yu, W., Ni, Q., Pervaiz, H., & Min, G. (2023). Cluster control and energy consumption minimization for cooperative prediction based spectrum sensing in cognitive radio networks. *IEEE Transactions on Communications*, 71(9), 5580–5594. 10.1109/TCOMM.2023.3287531

Noh, J. H., & Oh, S. J. (2013). Cognitive radio channel with cooperative multi-antenna secondary systems. *IEEE Journal on Selected Areas in Communications*, 32(3), 539–549. 10.1109/JSAC.2014.140314

Nurelmadina, N., Hasan, M. K., Memon, I., Saeed, R. A., Zainol Ariffin, K. A., Ali, E. S., Mokhtar, R. A., Islam, S., Hossain, E., & Hassan, M. A. (2021). A systematic review on cognitive radio in low power wide area network for industrial IoT applications. *Sustainability (Basel)*, 13(1), 338. 10.3390/su13010338

Ozcan, G., Gursoy, M. C., & Tang, J. (2017). Spectral and energy efficiency in cognitive radio systems with unslotted primary users and sensing uncertainty. *IEEE Transactions on Communications*, 65(10), 4138–4151. 10.1109/TCOMM.2017.2691342

Pei, Y., Liang, Y. C., Teh, K. C., & Li, K. H. (2011). Energy-efficient design of sequential channel sensing in cognitive radio networks: Optimal sensing strategy, power allocation, and sensing order. *IEEE Journal on Selected Areas in Communications*, 29(8), 1648–1659. 10.1109/JSAC.2011.110914

Ren, P., Wang, Y., Du, Q., & Xu, J. (2012). A survey on dynamic spectrum access protocols for distributed cognitive wireless networks. *EURASIP Journal on Wireless Communications and Networking*, 2012(1), 1–21. 10.1186/1687-1499-2012-60

Salari, S., & Chan, F. (2023). Maximizing the sum-rate of secondary cognitive radio networks by jointly optimizing beamforming and energy harvesting time. *IEEE Transactions on Vehicular Technology*, 72(6), 8128–8133. 10.1109/TVT.2023.3238347

Shi, Y., & Eberhart, R. (1998, May). A modified particle swarm optimizer. In *1998 IEEE international conference on evolutionary computation proceedings. IEEE world congress on computational intelligence (Cat. No. 98TH8360)* (pp. 69-73). IEEE. 10.1109/ICEC.1998.699146

Shin, K. G., Kim, H., Min, A. W., & Kumar, A. (2010). Cognitive radios for dynamic spectrum access: From concept to reality. *IEEE Wireless Communications*, 17(6), 64–74. 10.1109/MWC.2010.5675780

Shitharth, S., Yonbawi, S., Manoharan, H., Alahmari, S., Yafoz, A., & Mujlid, H. (2023). Physical stint virtual representation of biomedical signals with wireless sensors using swarm intelligence optimization algorithm. *IEEE Sensors Journal*, 23(4), 3870–3877. 10.1109/JSEN.2022.3233407

Swathi, T., & Begum, M. B. (2018). Maximum Connected Load Balancing Cover Tree Algorithm for Wireless Sensor Network. *Journal of Switching Hub*, 3(3), 1–5.

Wang, R., Tao, M., & Liu, Y. (2012). Optimal linear transceiver designs for cognitive two-way relay networks. *IEEE Transactions on Signal Processing*, 61(4), 992–1005. 10.1109/TSP.2012.2226452

Wang, S., Wang, Y., Coon, J. P., & Doufexi, A. (2011). Energy-efficient spectrum sensing and access for cognitive radio networks. *IEEE Transactions on Vehicular Technology*, 61(2), 906–912. 10.1109/TVT.2011.2180549

Yang, G., Ho, C. K., & Guan, Y. L. (2014). Dynamic resource allocation for multiple-antenna wireless power transfer. *IEEE Transactions on Signal Processing*, 62(14), 3565–3577. 10.1109/TSP.2014.2328980

Zhang, J., Zheng, F. C., Gao, X. Q., & Zhu, H. B. (2014). Sensing-energy efficiency tradeoff for cognitive radio networks. *IET Communications*, 8(18), 3414–3423. 10.1049/iet-com.2014.0400

Zhang, L., Xiao, M., Wu, G., Li, S., & Liang, Y. C. (2016). Energy-efficient cognitive transmission with imperfect spectrum sensing. *IEEE Journal on Selected Areas in Communications*, 34(5), 1320–1335. 10.1109/JSAC.2016.2520166

Zhang, W., Wang, C. X., Chen, D., & Xiong, H. (2015). Energy-spectral efficiency tradeoff in cognitive radio networks. *IEEE Transactions on Vehicular Technology*, 65(4), 2208–2218. 10.1109/TVT.2015.2432073

Zheng, G., Krikidis, I., & Ottersten, B. (2013). Full-duplex cooperative cognitive radio with transmit imperfections. *IEEE Transactions on Wireless Communications*, 12(5), 2498–2511. 10.1109/TWC.2013.032513.121464

Chapter 14
Revolutionizing Waste Management Through the Integration of IoT and Deep Learning Technology

V. Sathana
K. Ramakrishnan College of Engineering, India

Sherwin Prince Daniel
K. Ramakrishnan College of Engineering, India

C. B. Srikanth
K. Ramakrishnan College of Engineering, India

K. Senthilvel
K. Ramakrishnan College of Engineering, India

D. Santhoshraj
K. Ramakrishnan College of Engineering, India

ABSTRACT

The constantly growing volume of waste produced daily by people and businesses has made the management of waste in urban contexts a challenging issue. These days, it's common practice to combine deep learning and the internet of things model with efficient waste management tactics to offer flexible real-time data monitoring and categorization solutions. The proposed methodology involved IoT-enabled garbage management which is used to detect the level of thrash and to gather images of the garbage objects from the onboard camera. In this model, the bounding-box regressor and the object classifier are removed. Instead, the garbage detection and segregation process uses the backbone model architecture of TensorFlow on ResNet. It is trained to detect and recognize different types of waste items in images. The trained deep learning model is integrated into the trash bin to categorize waste products into relevant groups (such as biodegradable and non-biodegradable). The method outperforms other existing methods in terms of garbage segmentation and categorization.

DOI: 10.4018/979-8-3693-2893-4.ch014

1. INTRODUCTION

In our constantly expanding country, the matter of waste management has grown significantly to be a major issue. Numerous tons of mixed wasteful rubbish, all over the place including organic, recyclable, and hazardous materials, are produced in India. Reports indicated that the population growth will result in more utility demand and use, as well as higher garbage generation at a proportionate pace. The incorrect handling, collecting, and disposal of these wastes harms all living things, contaminating the air, water, and soil and possibly causing fatal illnesses in humans. We can turn these wastes into resources if they are properly and efficiently collected. The implementation of cutting-edge technology like the Internet of Things (IoT) and deep learning has revolutionized waste management in various communities. This addresses the appropriate methods for gathering and breaking down waste to optimize its beneficial effects while effectively reducing its real amount. The development of Deep learning has led to the rapid growth of several techniques for various visual recognition tasks, including the automatic identification of images. This paper proposes a hybrid approach such as an IOT and Deep Learning-based waste management system. The remainder of the document was arranged as follows: Section 2 presents the problem definition and its implications. Section 3 illustrates the objective of the proposed work. Section 4 describes the overall methodology and implementation process in detail. Section 5 represents the experimental results of the proposed system, showcasing their impact. Section 6 illustrates the conclusion with significant findings.

2. RELATED WORK

To better understand the current systems and determine what kind of algorithm may be used to get the best outcomes, several papers were examined (Thamarai & Vankamamidi, 2023). Using CNN models that have already been trained, like InceptionV3 and Inception ResNet, identify garbage items in comments. Using a convolutional neural network (CNN), this system gathers home garbage and separates it into organic and inorganic wastes. The organic waste is used to generate electricity, while the inorganic garbage is disposed of in a recycling bin. The model reduced costs and saved time. Faisal (2022) comments on the identification of microscopic rubbish items using the enhanced Refinement approach with the central network being VGG16. The results of classification are obtained by applying non-maximal suppression (NMS). The object detection module (ODM), transfer connection block (TCB), and anchor refinement module

(ARM) are the three categories into which the improved Refinement method falls. This approach did not detect larger item sizes. A dataset (the TrashNet dataset) comprising over 500 photographs of recycled items from six classifications (paper, glass, plastic, metal, cardboard, and rubbish) was gathered by Wang et al. (2021). They examined two models with architectures similar to AlexNet but smaller: SVM and TrashNet. Sheng et al. (2020) clever waste management approach combined TensorFlow-based deep learning with LoRa communication. Tensorflow is used to implement real-time object classification and detection while the presented technique communicates the sensor data. The waste canister has multiple chambers designed to separate waste materials, including paper, metal, plastic, and common waste sections controlled by a servo motor. Through the use of a DL technique, Bobulski and Kubanek (2019) investigate a waste segregation strategy for effective disposal and recycling. The Darknet neural network's YOLOv3 algorithm was utilized to train custom datasets.

3. PROBLEM DEFINITION

In modern urban environments, waste management presents a critical challenge, leading to environmental degradation, health risks, and inefficient resource utilization. Traditional waste management systems often suffer from a lack of real-time monitoring, resulting in ineffective collection schedules, overflowing bins, and heightened operational costs. To address these pressing issues, a smart bin setup utilizing Internet of Things (IoT) technology can offer innovative solutions to revolutionize waste management practices. It is very necessary to sort waste materials according to their properties (such as biodegradable, and non-biodegradable) to make recycling and disposal easier. To maintain a clean and hygienic environment in the area around us, efficient recycling initiatives and treatment strategies are to be used to optimize resource recovery and reduce environmental contamination. Therefore, we are integrating technology to automatically sort garbage into different bins to optimize waste management. Compared to the human method, this system can sort rubbish waste more rapidly and precisely.

4. OBJECTIVE OF THE PROPOSED WORK

The primary objective of this project is to develop and implement a smart waste management system tailored for pin-based waste segregation, leveraging IoT sensors and deep learning algorithms. Specific objectives include:

- **Automate Waste Segregation:** Develop a system capable of automatically segregating biodegradable and non-biodegradable waste materials at the source using pin-based mechanisms.
- **Utilize IoT Sensors:** Deploy IoT sensors within waste collection bins to monitor and collect real-time data on waste composition, enabling accurate classification and analysis.
- **Implement Deep Learning Algorithms:** Develop and train advanced deep learning algorithms to analyze sensor data and differentiate between biodegradable and non-biodegradable waste materials with high accuracy.
- **Enable Automated Waste Disposal:** Integrate IoT sensors and deep learning algorithms to automate the waste disposal process, minimizing the need for manual intervention and streamlining waste management operations.
- **Promote Efficient Resource Utilization:** Through precise waste segregation and automated disposal, promote efficient resource utilization by minimizing waste and maximizing recycling and reuse opportunities.
- **Evaluate System Performance**: Assess the effectiveness and efficiency of the smart waste management system in improving waste management practices, resource utilization, and environmental sustainability.
- **Optimize System Operation:** Continuously optimize the operation of the smart waste management system based on real-world feedback, data analysis, and technological advancements to enhance performance and reliability.

4. PROPOSED WORK

The existing waste management solutions used are complex to deploy and manage. Many of the infrastructure and software systems may pose compatibility and interoperability challenges. The goal of this work is to develop a deep learning model for an intelligent waste management system using TensorFlow on the Resnet model and the Internet of Things. This proposal introduces an innovative solution to optimize waste classification efficiency through the integration of on-board artificial intelligence (AI) within waste bins. The overarching objective is to overcome challenges related to limited database connectivity in scattered urban environments while simultaneously reducing latency in waste classification processes. By leveraging onboard AI, this approach aims to enhance the effectiveness and sustainability of waste management systems.

4.1 Design of System Model

Fig. 1 shows the IoT-enabled system model. The Arduino is integrated with Python-written detection and control algorithms serving as the main processing hub for the garbage sorting and categorization system. To identify waste, the trained model is combined with the camera. In addition, the non-detectable garbage in the waste placement area is found using an ultrasonic sensor as a result, the waste is transferred to the garbage Compartment. Servo motors are utilized in the system's classification phase to transfer waste into the waste compartment. The key elements of the proposed methodology include Image pre-processing, feature extraction, Segmentation, and Classification. To create trash classification and categorization, the model also incorporated electronic parts. The classified waste information is transmitted to the Arduino Nano, which coordinates the operation of Misra et al. (2018) IoT devices such as servo motors and buzzers. Based on the classification results, the Arduino Nano controls the servo motor to position collection bins or compartments accordingly for sorting different waste types. Additionally, an ultrasonic sensor (AlFoudery et al., 2018) in the trash disposal area locates the non-detectable junk, and as a result, the waste is sent to the garbage compartment. During the system's classification phase, servo motors are used to move waste into the waste compartment by following these steps:

- Place trash on the movable platform.
- Waste is detected by a sensor.
- The camera takes a picture of the trash and sends it to a microcontroller.
- Micro Controller performs waste image classification
- If waste = Biodegradable then Open the first compartment's retractable lid and the retractable platform.
- Otherwise, If the garbage isn't biodegradable, then open the other retractable lids on the metal compartment and open the retractable platform.
- End with an open retractable platform.
- Close the retractable platform and lid.

Figure 1. Overview of IoT-enabled waste management system

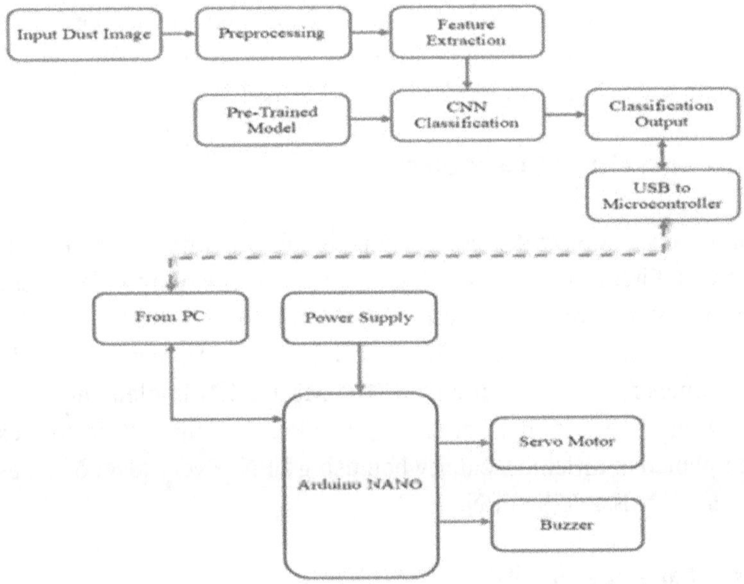

The key elements of the proposed methodology are shown in Figure 2 including Image Augmentation, Garbage Segmentation, the Resnet model, and the Internet of Things (IoT) enabled trash bins to identify waste. To create trash classification and categorization, hardware integration is required for the model. The model is also incorporated with a list of the electronic parts.

4.2 Data Collection

The database, which is organized into categories and contains a collection of images, was gathered from the Garbage Classification AI Challenge\Dockship.io website as shown in Figure 2. There are over 2500 images in this dataset, all of which have been reduced in size to 512 x 512.

Figure 2. (a) Bio-degradable; (b) non-biodegradable

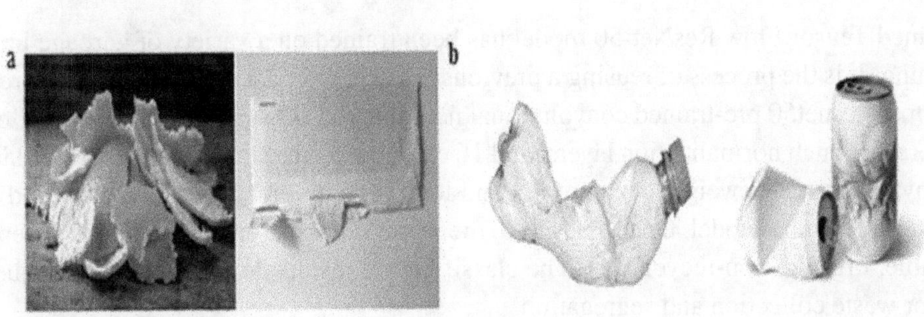

This approach used an innovative waste detection and segregation system that utilizes an independent algorithm for waste detection, separate from the base network responsible for feature extraction. The system employs an object detector alongside the ResNet50 architecture as the feature extractor, culminating in a final fully connected classification layer. By analyzing waste images, the system is trained to classify and segregate waste based on its visual characteristics.

4.3 Segmentation and Object Detector

The system incorporates an object detector responsible for identifying waste objects within images. This detector locates waste items and provides localization information to the subsequent stages of the system. Since Mask RCNN requires dimensions of $1024 \times 1024 \times 3$, we have resized the images to fit the network's required input size. Every input image is resized for the feature map. The input for subsequent network layers comes from this feature map. The Mask RCN implementation uses the backbone as the ResNet model with the feature pyramid network (FPN) to help with feature extraction. Garbage can be accurately segmented at different scales when using FPN. Every level of access to low-level and high-level features is made possible by FPN.

4.4 Feature Extractor (ResNet50)

ResNet50 serves as the feature extractor, extracting relevant features from the waste images. These features are crucial for accurately classifying and segregating waste items.

Fully Connected Classification Layer: The final layer of the system consists of a fully connected classification layer. Here, the extracted features are utilized to classify waste items and determine their appropriate segregation category.

4.5 Training Process

The system undergoes a training process wherein it learns to associate specific visual characteristics with different types of waste. This training enables the system to accurately classify, and segregate waste based on input images.

4.6 Segregation

A pre-trained Tensor Flow ResNet-50 model has been trained on a variety of garbage images using transfer learning. It is the process of reusing a previously learned model for a new task that requires less processing time. Resnet50 pre-trained convolutional neural network comprises 53 convolutional layers, 49 ReLu layers, 53 batch normalization layers, and 16 extra layers making up a total of 171 layers in this model. The hyperparameter weight decay of 0.4 and learning rate of 0.0001 was employed during the training of a deep learning model. Once classified, the waste items are categorized into different classes (e.g., recyclable, organic, non-recyclable). The classification results determine the subsequent actions to be taken for waste collection and segregation.

4.7 Proposed Algorithm

Input: Garbage Image

Output: Segregation of Garbage Image

The TensorFlow framework helps in building Neural Network models to classify images. These networks are made up of Convolutional Neural Networks (CNN). The TensorFlow Object Detection procedure makes use of a series of steps as follows:

1. Image Enrichment (Augmentation of Data)
2. Reshape the input to fit the target convolutional layer.
3. The tf.nn.conv1d() method is used to generate the convolutional layer.
4. The pooling process is conducted after the convolution procedure. To generate the pooling layer, use tf.nn.maxpool().
5. Steps 3 and 4 are repeated for more convolution and pooling layers.
6. The outputs from the pooling and convolution layers are applied as input to the flattening layer and the output is then sent to the fully connected layer.
7. The tf.matmul() function creates the fully connected layer (FC), the tf.nn.relu() function adds an activation function, such as ReLu, and the tf.nn.dropout() techniqueapplies a dropout operation.
8. The final class prediction is made using the tf.matmul() technique.
9. Finally, the TensorFlow variables contain weights and biases.
10. Dense class is used for fully connected layers.
11. The output layer makes use of the sigmoid function.
12. Stop

6. RESULTS AND DISCUSSION

The proposed method tested with 25 garbage images previously not seen by the model. The performance of the model was observed by examining how precisely the model detects and locates the correct class of objects in a particular test image. Python-based Computer Vision and Deep Learning libraries will be exploited for the development and experimentation of the work. Tools such as Anaconda Python and libraries such as TensorFlow keras with CNN will be utilized for this process. Training will be conducted on NVIDIA GPUs. The dataset consists of garbage waste-trained images that have been considered. The module is trained repetitively to attain maximum accuracy. If a new image is given to the module its features get compared with the features that are already trained in the datasets and provide the appropriate result. Fig. 3 illustrates the visual representation of loss value for the training and testing of the dataset. The output expected from the proposed system is that it captures the image of the garbage waste one after the other, identifies the image, and classifies it into biodegradable and non-biodegradable. Fig.4 represents the segregation of waste using the trained waste detection model. The results reported better classification accuracy for the model. In particular, the model achieved the highest overall classification accuracy of 96.64%. 80% of the data was passed to the model for training with batch size 4 and epochs 10 and it resulted in an accuracy value of 0.958 and loss value of 0.1964.

20% of the data was passed to the model for testing with batch size 4 and epochs 10 and it resulted in an accuracy value 0.954 and value 0.4004.

Figure 3. Visual representation of training and testing loss behavior vs epoch

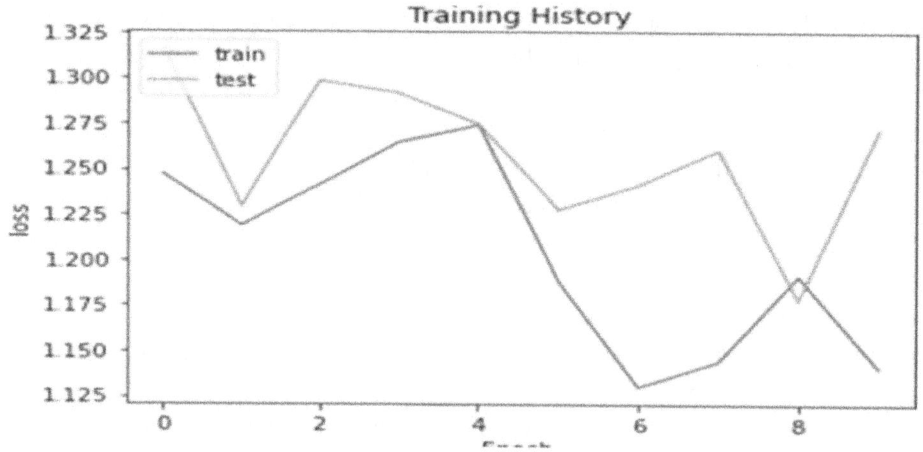

Figure 4. (a) bio waste and (b) non-bio waste

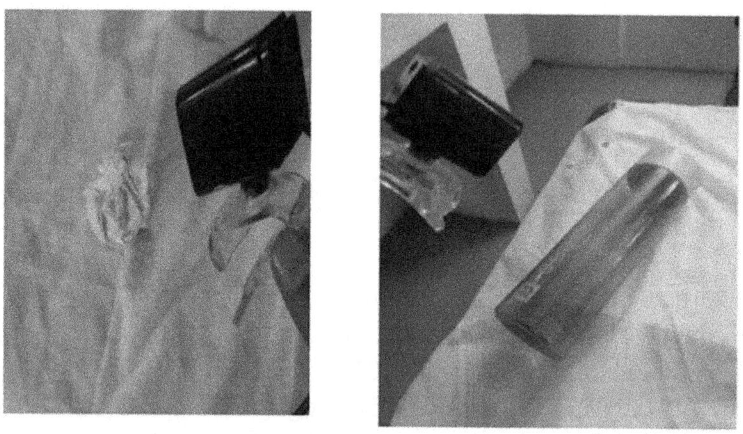

7. CONCLUSION

The integration of IoT and deep learning technologies in smart waste management systems offers a sophisticated solution for addressing the challenges of waste segregation and disposal. By enabling real-time detection and classification of waste types, automated waste deposition mechanisms, and centralized control and monitoring, the proposed system enhances operational efficiency and promotes sustainable waste management practices. Embracing such innovative solutions is crucial for mitigating the environmental impact of waste disposal and fostering a cleaner, healthier environment for future generations. The proposed system utilizes IoT sensors and deep learning algorithms to detect and classify waste types in real time. This enables prompt and accurate segregation of biodegradable and non-biodegradable waste, optimizing resource utilization and promoting sustainability. A centralized control unit orchestrates an automated pin-opening mechanism, facilitating the seamless deposition of waste into appropriate bins. This automated process streamlines waste disposal and minimizes the need for manual intervention, enhancing operational efficiency. The centralized control unit serves as the system's brain, coordinating the operation of IoT sensors, deep learning algorithms, and automated mechanisms. It ensures smooth communication and integration of system components, facilitating effective waste management. Garbage may be automatically classified with the use of this technology, reducing the need for human involvement and preventing contamination and illness. When we compared the output to the trash dataset, we got a good accuracy rate. The experimental results show how well the proposed method works to detect and categorize garbage. After the model is fully trained, it produces good accuracy values above 95% and losses below 50%. Therefore, it can accurately determine if garbage is biodegradable and non-biodegradable. Furthermore, the integration of cloud-based analytics and data visualization tools represents a transformative step towards optimizing waste management operations. By harnessing the power of cloud computing and advanced analytics, the system can gain valuable insights into waste generation patterns. These insights enable dynamic optimization of waste collection routes and schedules, leading to increased operational efficiency and reduced environmental footprint.

REFERENCES

AlFoudery, A., Alkandari, A. A., & Almutairi, N. M. (2018). Trash Basket Sensor Notification Using Arduino with Android Application. *Indonesian Journal of Electrical Engineering and Computer Science*, 10(1), 120–128. 10.11591/ijeecs.v10.i1.pp120-128

Bobulski, J., & Kubanek, M. (2019). Waste Classification System Using Image Processing and Convolutional Neural Networks. In *International Work Conference on Artificial Neural Networks* (pp. 350-361). Springer. 10.1007/978-3-030-20518-8_30

Faisal, S. (2022). Deep Learning based small object detection and classification model for Garbage Waste Management in Smart Cities and IoT Environment. *Applied Science*.

Misra, D., Das, G., Chakrabortty, T., & Das, D. (2018). An IoT-based waste management system monitored by cloud. *Journal of Material Cycles and Waste Management*, 20(3), 1574–1582. 10.1007/s10163-018-0720-y

Sheng, T.J., Islam, M.S., Misran, N., Baharuddin, M.H., Arshad, H., Islam, M.R., Chowdhury, M.E., Rmili, H., & Islam, M.T. (2020). An internet of things based smart waste management system using LoRa and tensorflow deep learning model. *IEEE Access*. IEEE.

Thamarai, M. & Vankamamidi, S. (2023). Smart self-power generating Garbage Management System using Deep Learning for Smart Cities. *Microprocessors and Microsystems*. Elsevier.

Wang, J. Q., Qu, C., Ran, X., Liu, C., & Chen, B. (2021). *Smart Municipal Waste Management System, Based on Deep Learning and IoT, Waste Management. Elsevier.*

Chapter 15
Spectrum Sensing and Sharing for Internet of Things

Poornima Pandian
https://orcid.org/0000-0002-5751-0859
Sri Sai Ram Engineering College, India

Chithra Selvaraj
MIT Campus, Anna University, India

S. Priyanga
https://orcid.org/0000-0002-2395-7492
SRM Institute of Science and Technology, India

ABSTRACT

The limited availability of spectrum and the reduced use of allotted spectral resources by the licensed users (primary users) have led to the development of technology termed as cognitive radio, which aims to provide unlicensed users with a wider range of spectral awareness through the use of a variety of spectrum sensing and sharing options. Cognitive radio technology can help the unlicensed users to discover and make use of the available spectrum through the process of spectrum sensing, thereby improving spectrum utilization and reducing spectrum scarcity across wireless networks. One of the primary growth determinants for utilizing the large number of internet of things (IoT) networks and quality mobile broadband services is the availability of enough spectrum resources, but in practice, spectrum is usually scarce, due to the adoption of static spectrum allocation policy. Cognitive radio technology can enable efficient and strategic frequency band usage by utilizing available licensed channels, creating a vast spectrum that can support widespread adoption for IoT networks.

1. INTRODUCTION

An exemplary growth is being seen in the number of mobile and Internet of Things (IoT) devices. A growing number of sophisticated apps and services are also constantly being developed as a result of improving these systems' performance. A significant rise in wireless data traffic and, as a consequence, spectrum occupancy are the results of these growth trends, which presents new and significant

DOI: 10.4018/979-8-3693-2893-4.ch015

limitations to be dealt with. Ensuring highly efficient and dependable communications in locales that are dynamic, heterogeneous, and increasingly complex, with numerous radio systems coexisting, is one of the problems at stake. In order to do this, a number of cognitive radio (CR) strategies have been introduced. These techniques would enable the devices to sense and evaluate the spectrum, make judgments about its use, and modify their communication settings in accordance with that knowledge and their prior experience (Wunsch et al., 2013). It is not suitable to carry out this analysis in the cloud provided the spectrum scarcity issue, as this would result in poor latency and excessive network traffic from the data uploading. Because of this, and the advancements in the functionality and performance of end devices, edge computing is becoming more and more significant in domains such as the Internet of Things (Premsankar et al., 2018).

The multiplicity of IoT connectivity methods has led to an increase in the quantity of hardware that is connected. Numerous types of electronic equipment fall under this category; some examples are smart home appliances, smartphones, and other types of smart equipment. These high-tech gadgets are capable of communicating with other devices, collect data, interpret it, and take necessary action. This places the internet of things at the core of economics, both now and in the future. Furthermore, competition for a finite spectrum has intensified due to the internet-of-things devices' exponential expansion. The use of cognitive radio appears to be a viable option in relation to spectrum availability, one that might potentially accommodate both current and future internet-of-things devices. This represents a thorough comprehension of the potential interactions between cognitive radio and the internet of things. When cognitive radio technology is integrated with the Internet of Things (IoT), a system known as the cognitive radio-based Internet of Things (CRIoT) is created. Numerous applications—including sensitive to time applications like modern technology and intelligent transportation—will make use of CRIoT. Numerous network requirements, such as channel assignment delay, reliability, energy efficiency, end-to-end latency and high throughput, must be addressed by the combined technologies. A number of communication protocols and technologies are needed to adjust to growing connectivity. Among the various Internet of Things apps and services are CR applications for Bluetooth, ZigBee, and Wi-Fi inside smart environments. IoT networks need to do a lot of things. Interference-free channels will be gathered while the Primary User is present, and any modifications to the IoT operational channel will be monitored by the PU continuously.

2. SPECTRUM REQUIREMENTS OF 5G COMMUNICATIONS

One issue impeding the widespread adoption of 5G and beyond-5G based IoT devices is the scarce availability of the transmission spectrum as shown in figure 1. The massive demand for the spectrum makes it more difficult to assign the frequency band that is available. At the same time, a wide range of frequency bands continue to be underutilized for a variety of reasons. It is projected that 5G and beyond-5G communication networks would take use of the underutilized frequency range and boost bandwidth by addressing the problem of spectrum scarcity for the billions of anticipated Internet-connected devices.

A massive number of devices running in millimeter wave and THz frequency, which are currently unregulated with the exception of some proprietary cellular networks, will be supported by ultra-dense radio access networks, which are expected to be supported by the 5G and beyond-5G Internet of Things (Yu & Zikriya, 2020). An approach to help with these problems is spectrum sharing. Basically, the idea behind spectrum sharing is the sharing of spectrum among several networks while adhering to a specific temporal and spatial arrangement. Through utilizing the available bands, or underutilized spectrum, CRNs successfully implement such spectrum sharing technique. Context-aware sharing can be done intelligently by the cognitive radios (CRs), which are based on the software-defined radios (SDR) concept. Reducing the spectrum underusage and spectrum crunch are the two main objectives of the CRs in 5G and beyond-5G.

Figure 1. CR based IoT system model

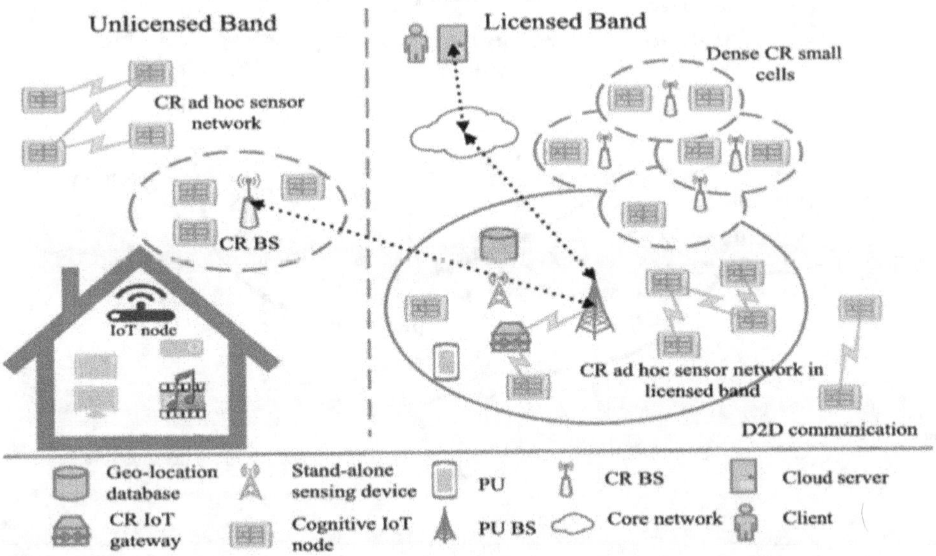

3. AN OVERVIEW OF COGNTIVE RADIO

The cognition cycle comprises of four important stages namely: spectrum sensing, spectrum sharing, spectrum decision and spectrum mobility. The figure.2 illustrates the process involved in the cognition cycle.

The CR is capable of the following two things:

- *Cognitive ability:* The capacity to perceive the available spectrum environments and then determining the available spectral bands, especially the underused spectral bands. The CR has the ability to identify whether the desired spectral band is busy or idle at the given moment.
- *Re-configurability:* This feature enables the radio to dynamically configure the system in order to connect with a different set of frequencies and utilize technologies that are specific to the underlying.

Figure 2. Cognition cycle

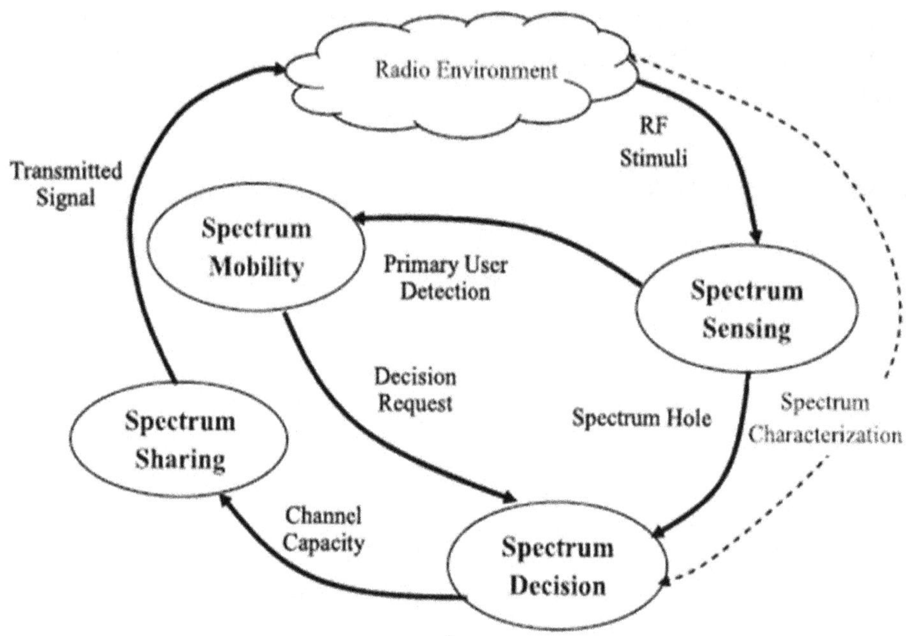

(Akyildiz et al., 2009)

4. RECENT ADVANCEMENTS IN SPECTRUM SENSING TECHNIQUES:

Deep Learning (DL) and Machine Learning (ML) are two of the most effective methods for handling challenging classification issues is given in figure 3. More precisely, they have been used in wireless communication to guarantee high-quality services for mobile consumers and to effectively manage the spectrum and power resources. Improving spectrum sensing performance is one of the goals of applying ML and DL in the CR domain. Two stages are typically used in learning techniques: learning and prediction. The information supplied during the learning phase of spectrum sensing applications relates to the Primary User (PU) features and the Secondary User (SU) sensing parameters (such as the Test Statistics, SNR, etc.), whereas the information provided during the prediction phase may pertain to the sensing outcome, power efficiency, the operational model to be used, and other challenges (Nasser *et al.,* 2021).

In traditional spectrum sensing, SU must decide the test statistic threshold before deciding whether the PU is present or not. The target false alarm and detection probabilities may be used to calculate this threshold. Therefore, a number of statistical characteristics pertaining to the channel, the noise, and the PU signal needs to be determined initially. The requirement for a priori statistical knowledge can be fulfilled by means of ML and DL. Most of the work in the literature focuses on fine-tuning ML or DL systems using numerical statistics of two hypotheses: H0, which assumes that PU is absent, and H1, which assumes that PU is active is represented in figure 4.

Figure 3. Machine learning based spectrum sensing

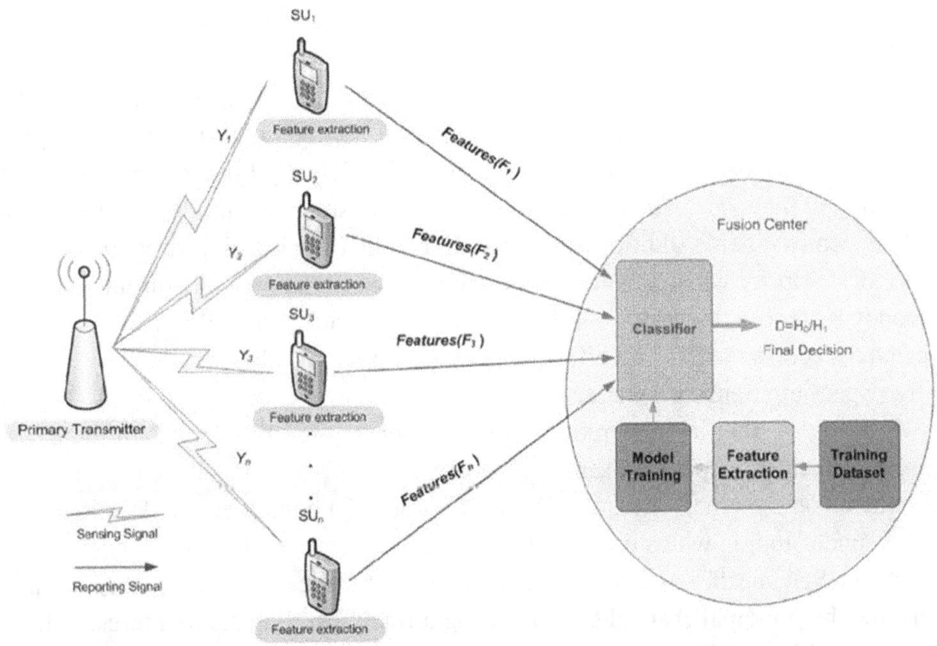

When a single node detects the spectrum and decides for itself that is referred to as local sensing. Under low SNR conditions, by training with the retrieved cyclic features of the PU signal is utilized to separate the H0 and H1 hypotheses. This ensemble classifier uses the AdaBoost algorithm and is based on decision trees. Three methods are provided for addressing wideband SS: k-means, expectation maximization, and neural networks. The methods are applied to a wideband spectrum to identify the existence of one or more PUs. Hybrid Spectrum Sensing (HSS) has been suggested as a way to improve the ML system's accuracy while determining the PU status. HSS uses many detectors to simultaneously detect the presence of PU. HSS has the ability to offset a detector's shortcomings with the benefits of another. For example, Auto Correlation based Detector (ACD) compensates for the noise uncertainty that Energy Detector (ED) experiences at low SNR. Conversely, the low oversampling rate of the PU signal negatively affects ACD but not ED.

Artificial neural networks have been used to carry out HSS. Training is carried out utilizing the Test Statistics of two detectors associated with H0 and H1, whereby uses ED and CSD statistics and uses ED and likelihood ratio statistics. The existing works extends the HSS-based DL for six detectors, demonstrating how well these methods operate for PU detection at very low SNRs. The idea of one-class-based learning is implemented through the use of HSS. Data from many detectors are gathered under H0 to understand the detecting system of the H0 class. No PU-related data is required during the training phase; only noise-related data is used. PU is considered to be active during the prediction phase when the predictor finds an outlier of the H0 class.

Moreover, spectrum perceptions, or the forecasting of the PU channel occupancy, are accomplished by the use of ML and DL. Using learning strategies, one may create a geo-frequency-temporal map of the PU actions, for instance. The spectrum sensing performance is improved and this map guides spectrum access. In order to lower the handover rate and prevent interference, SU can choose the communication channels carefully by using the forecast of the spectrum vacancy. On the other hand, in order to make spectrum sensing before accessing, SU may choose to target a channel during a handover where PU is anticipated to be absent. As a result, the handover is quicker and safer.

Furthermore, in situations where PU is expected to remain absent on the channel for an extended amount of time, SU may choose to extend the time interval between two successive spectrum sensing operations, which will conserve energy and improve spectral efficiency. From one geolocation to another, the channel prediction may vary. Therefore, based on its assessment of the PU's conduct, a mobile SU would move between channels. Additionally, it was shown that the learning strategies worked well for controlling the SUs' priority when it came to using the available channels. Regarding the shared use of the few frequency resources, the SUs could cooperate or compete. To equitably allocate the available frequency resources across the SUs, an effective accessing policy is also necessary.

Cooperative Spectrum Sensing scheme:

Several SUs work together in cooperative Spectrum Sensing (SS) to decide on the PU state in the end. There are two distinct cooperative sensing schemes:

(1) Hard conclusion Scheme, where the Fusion Center (FC) combines the decisions of all SUs to arrive at a final conclusion, in which each SU takes a decision on the PU state independently.

(2) Soft Decision Scheme, in which the FC compiles Test Statistics computed at the SUs and combines them to determine the principal channel by comparing a final Test Statistic to a threshold.

In cooperative SS, ML and DL are introduced to address a number of concerns, including linked results between collaborating SUs, malicious findings presented by some SUs, and elevating the credibility of SUs near PU over those farther away. To differentiate among the H0 and H1 hypotheses in a cooperative SS, few machine learning (ML) techniques such as the K-Means and Support-Vector Machine (SVM) are also incorporated. The system is trained using two low-dimension probability vectors associated with both H0 and H1 of ED. The threshold curve between the H0 and H1 clusters is set using SVM. The K-nearest-based machine learning is used for cooperative SS. The suggested process is split into two stages: training and classification. The dependability of every CR user while reporting to the fusion centre throughout the training phase is taken into account in the overall decision on the PU existence or absence made at the conclusion of the classification phase.

Figure 4. Comparison between traditional and machine learning spectrum sensing

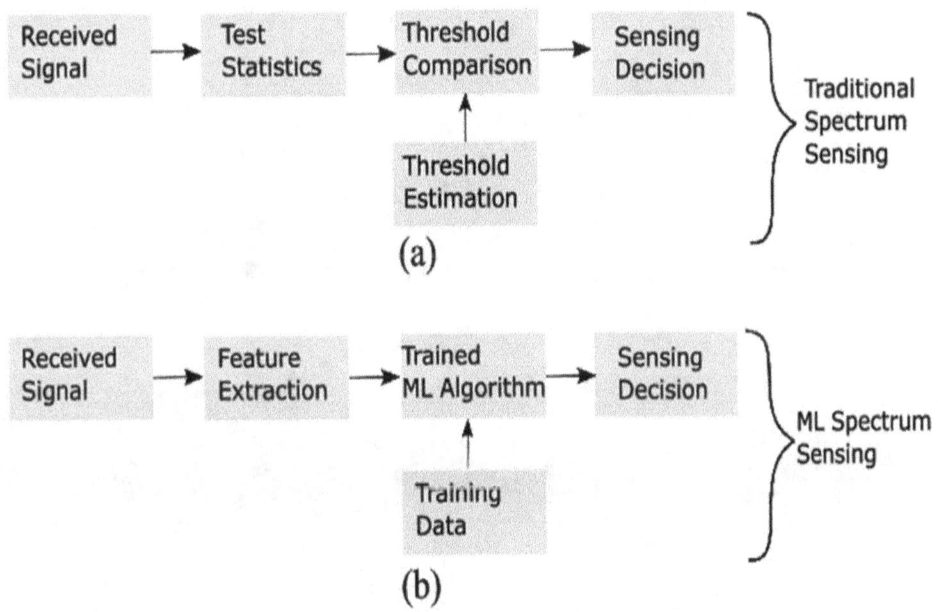

In this work (Lee *et al.*, 2019), a cooperative SS based on convolutional neural networks is suggested, wherein the outputs of the SUs are merged using Hard and Soft Combining Schemes. To increase the system's resilience, the channels' spectral and spatial correlations are taken into consideration. The Fusion Center may get reports of connected choices from the nearby SUs. When these SUs show extreme fading, this has a detrimental effect on the sensing performance (Khalid *et al.,* 2011). In this work (Shi *et al.,* 2019), the authors suggest utilizing cooperative learning techniques for SS in a non-orthogonal multiple access setting to get over the physical layer complexity of that kind of access strategy. The cooperative SS is carried out in (Ning *et al.,* 2020) using reinforcement learning. The dynamic scanning preference list is created by scanning the PU channels using the reinforcement learning. This list aids in lowering access latency and scanning overhead.

5. SPECTRUM SENSING FOR INTERNET OF THINGS

IoT applications are expected to be used in nearly every aspect of life, including smart homes, e-health applications, traffic, agricultural, and monitoring purposes. IoT and CR can greatly benefit from one another is shown in figure 5. On the one hand, the CR can monitor the PU channel by taking use of the widespread WSN/IoT deployment. One new business model that appears is Spectrum Sensing as a Service (SSaaS) (Al-Kofahi *et al.,* 2020). However, there is a great demand for spectrum resources due to the vast number of IoT devices. Because of its dynamic ability to facilitate spectrum sharing, CR technology can be viewed as a potential solution in this situation (Moon, 2017).

Figure 5. Illustration of communication between CR and IoT devices

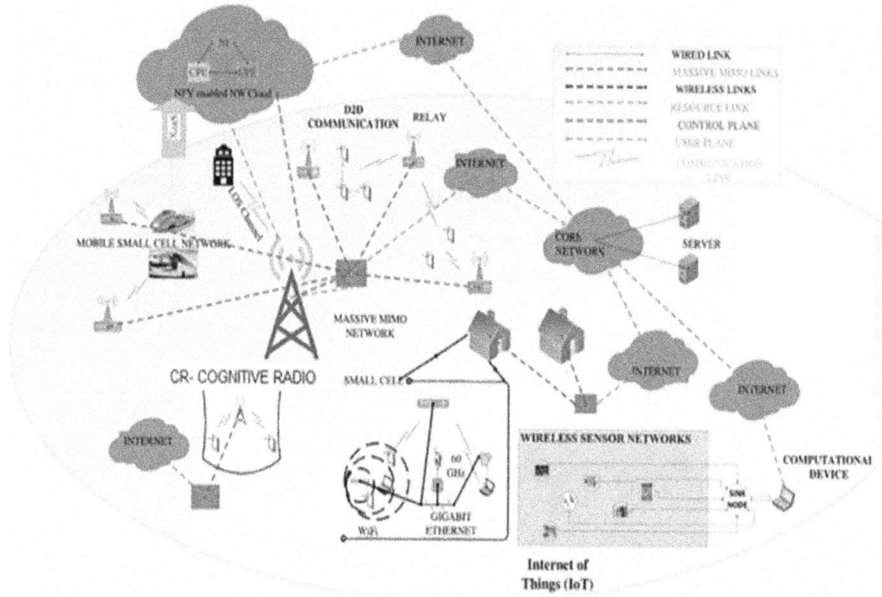

When SSaaS is taken into account, the WSN node(s) associated with the SSaaS provider perform the SS process, not the secondary network's SU. The SSaaS provider of the PU channel state subsequently notifies the secondary network. As a result, the secondary network decides whether to transmit on that channel. Since SU may function across a variety of frequency bands, WSN ought to be equipped with the electrical machinery required to reconfigure in response to the secondary network's requests for sensing jobs. Furthermore, as the SU should always be up to date and the SS information on the PU channel should be pertinent, the WSN network's on-demand responsiveness is crucial. Because LoRaWAN and WSN do not provide on-demand data communication, this difficulty is hampered by IoT network protocols and WSN. A few changes to the LoRaWAN protocol were proposed in order to accommodate the on-demand SS. The enormous range of battery-powered nodes places a severe energy constraint on the selection of WSN nodes that carry out the SS. To choose which nodes will carry out the SS, a number of methodologies can be used, including random selection, SNR-based selection, battery energy level-based selection, and hybrid criteria-based selection. Every strategy has an effect on the network's longevity as well as the performance of the SS. An SNR-based method, for example, guarantees good SS performance, but it may negatively affect the nodes' lifetime. Poor SS performance could result from the battery's energy level-based selection, which ignores the primary signal's SNR. Alternatively, to guarantee the satisfaction of both parties, it is crucial to review the contract between the SSaaS provider and the secondary network. For example, in (Bayhan *et al.*, 2019) when nodes of the WSN are compensated only if they successfully execute sensing, blockchain technology is utilized to determine the necessary SS parameters.

6. DYNAMIC SPECTRUM SHARING

Improving the use of the restricted frequency resources is necessary due to the growing popularity of WSN/IoT devices. It has been suggested to use CR in dynamic spectrum sharing to get around the restrictions on the frequency resources that are accessible in the context of WSN. The concept of white holes which plays a significant role in providing dynamic spectrum access is given in Figure 6.

Figure 6. Concept of the spectrum holes

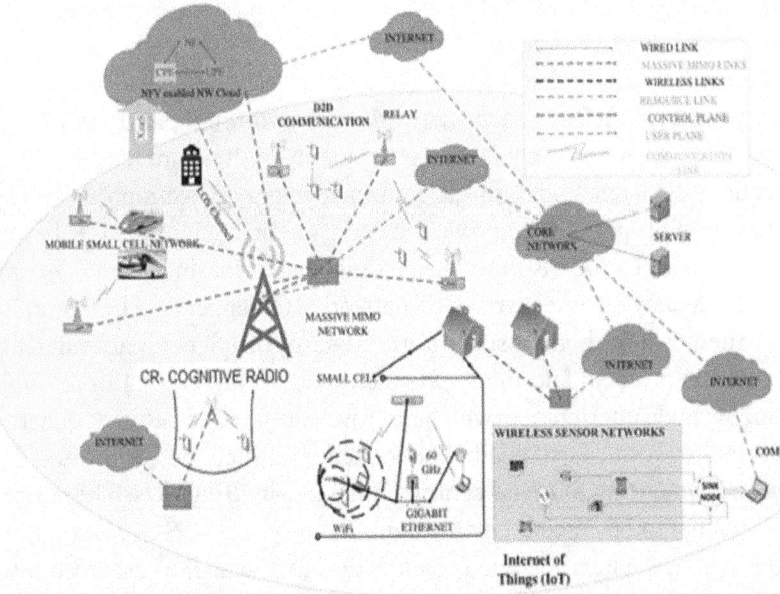

CR views WSN nodes as SUs while using Dynamic Spectrum Sharing for WSN communication. Since more energy is required to complete the SS duty, the WSN nodes take on additional energetic responsibilities by monitoring the PU channel. Because the nodes in IoT/WSN networks are working under protocols that increase their lifespan to several years, energy consumption is actually very important. The lifespan of a sensor, for example, may be longer than five years in Low Power Wide Area Network (LPWAN) Internet of Things networks because of the transmission standards, particularly the minimal activity period that is 1% or fewer, as in LoRa devices. It is difficult to adapt the CR mechanism to IoT networks because of a number of problems, including giving the nodes access to suitable frequency bands, implementing the SS capability, choosing the nodes that will be in charge of spectrum sensing, etc. The application of CR in an IoT network and the standardization of CR-based IoT are the subjects of several studies. They describe the extra characteristics, such hardware implementation and protocol implementation, needed to carry out the CR process. This involves varying the frequency spectrum that is used based on the process of sensing. In fact, the proliferation of IoT devices necessitates multiband access, and having many frequency channels accessible facilitates easier and more effective channel switching.

The optimization of energy-throughput for the CR-IoT is examined in (Ejaz *et al.,* 2015), where cooperative SS is utilized. Multiple SUs/nodes in the SS must participate in cooperative sensing. Even though sensing has improved, cooperative sensing may present the participating SUs with more energy consumption difficulties because of the continual SS and reporting procedures. In (Li *et al.,* 2018), the adoption of cooperative or local sensing is examined in various sensing scenarios and used to slotted-ALOHA protocol-based CR-based NB-IoT technology. According to the study of (Li *et al.,* 2018), using cooperative SS is only appropriate in low SNR scenarios, while using local SS is more convenient when SNR is reasonably large. Once the CR feature is put into practice, other difficulties arise, like boosting throughput and spectrum efficiency as well as lowering wireless node energy consumption. A dynamic technique for CR-based nodes to access both licensed and unlicensed bands is proposed in (Moon, 2017) under the name "dynamic licensed-unlicensed access". Through sharing available legal bands, such a system may reduce the demand on the unlicensed bands (where WSN/IoT nodes often operate). In (Ejaz *et al.,* 2015), a wireless sensor network energy-throughput trade-off is put forth in relation to cooperative SS and low-power, specialized equipment. With regard to secondary throughput and primary signal detection, the latter study aims to minimize energy consumption.

Energy Harvesting has been proposed to improve the energy efficiency of the CR-based IoT system by offsetting the extra energy used by the CR functions, like spectrum sharing and SS. In order to handle the enormous demand on frequency resources from IoT networks, CR appears to be a promising method. However, CR-IoT needs to demonstrate both its spectrum and energy efficiency, particularly for industrial applications. The transmission protocol and the deployment pattern are crucial in minimizing energy consumption and minimizing node interference, which improves the CR-IoT network's energy efficiency.

But there are still a number of issues that need to be looked into, such the best protocol for exchanging data between central entity-sensing SUs and central entity-transmitting SUs. The sensing SUs must first be informed of the channel to be detected by the central entity. These SUs should then reply with their choices. Time and frequency synchronization, such as control channel, is enforced by this method. Second, once the available channel has been determined, the SUs or nodes that require transmission should be notified of sufficient available channels so they can adjust their circuits appropriately. The right electronic circuitry that can accommodate the dynamic reconfiguration is required for this setup. Furthermore, to meet the node requirements for interference reduction and data transmission, the central entity must manage resources efficiently.

Figure 7. Working of CR-IoT nodes

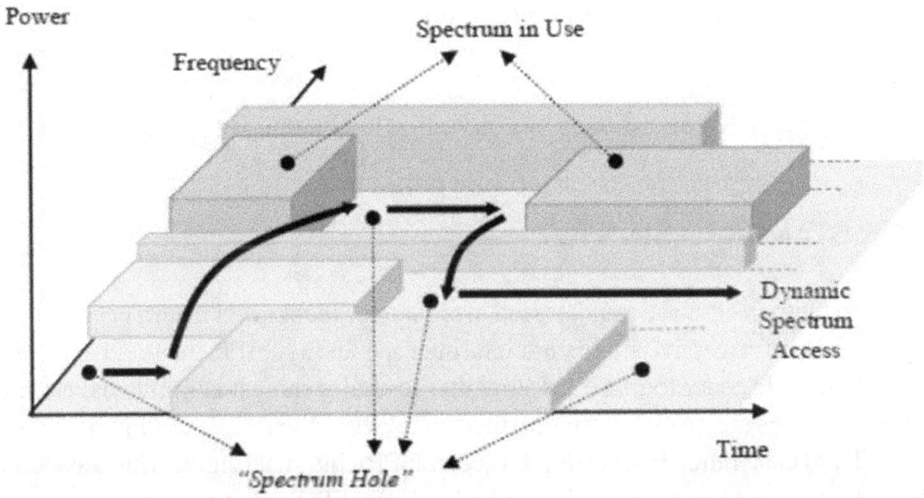

The primary and secondary users (IoTs) collaborate when it comes to spectrum sharing is shown in Figure 7. In another word, the primary network or networks lease its unused slots to the Internet of Things (IoTs) while maintaining the PUs' quality of service. A few of these protocols are covered in this section. In order to increase the number of utilities for both the PUs and the SUs, a cooperative transmission model between the PUs and the SUs by leasing the spectrum from the primary network to the Sus is suggested. The resource sharing issue is resolved by the application of game theory. To obtain the necessary optimal leasing strategy, selective collaboration and the Nash Bargaining Solution (NBS) are employed.

A cooperative two-player negotiating game between the PU and the SUs is used to model the context. A Request-to-Send (RTS) broadcast by a PU notifies its Primary Receiver (PR) of its minimum transmission rate at the start of the process. The PR then determines the transmission rate by estimating the channel's state. When the appropriate transmission rate is reached over the channel, the PR responds with a Clear-to-Send (CTS). Every SU receives both RTS and CTS. When the PU's transmission rate falls below the minimum transmission rate, it signifies that it will collaborate with an SU. The optimum optimization that can maximize the total of utility for both the PU and the SU is chosen by each SU after calculating the parameters of the bargaining game and the optimal strategy. The utility of the PU is the total benefit from both power savings and transmission rate changes, whereas the utility of the SU is the result of both transmission rate and energy consumption. Following the setting of a timer by each SU, the first to transmit a Partner-to-transmit (PTS) with the optimal cooperation strategy between the two will be the winner. Eventually, the SU will be selected to share the channel with that PU after the PU and the other SUs receive the PTS message.

A proposed distributed cognitive random-access algorithm was constructed. Since mobile IoTs cannot effectively sense the spectrum, the authors conjectured that different mobile IoTs can connect with the CRIoT network's wired spectrum sensors. There exist observation errors in spectrum occupancy due to variations in the distances between IoTs and sensors. According to the authors, IoTs ought to modify their likelihood of transmission in accordance with the interference detected by the sensor. Ensuring that the needs of the primary users are met while simultaneously increasing the area's total IoT data rates and spectrum consumption is the goal. Using the sensor's reported interference level, stochastic geometry was employed to create an optimum interference distribution at each Internet of Things. After determining the ideal transmission probabilities for each IoT using the determined conditional interference distribution, a straightforward algorithm was developed to give a probability to each IoT. The new algorithm, called area spectral efficiency, was assessed by contrasting it with the traditional ALOHA architecture.

7. NEED FOR STANDARDISING IOT

The Internet of Things needs to be standardised in order to be recognised from a personal and business aspect. The prevention of excessive fragmentation, interoperability, efficiency and scalability are the main recommendations. In order to make IoT a reality, several technical committees, working groups, and standardisation bodies are involved. Among them are the European Telecommunications Standards Institute (ETSI), Third Generation Partnership Project (3GPP), Internet Engineering Task Force (IETF), IEEE, and the National Institute of Standards and Technology (NIST). A few are focusing on the Internet of Things as a whole. Portions of these initiatives focused on communication technologies, including near field communications (NFC) and radio frequency identification (RFID). The process of developing communication protocols is ongoing, and several standardisation organisations are working to make sensors suitable for the Internet of Things.

In the current era of underutilised spectrum, cognitive radio is considered to facilitate increased spectrum utilisation. CRNs are used for a variety of industrial, commercial, and personal uses. The FCC in the US is thinking of allowing it in unlicensed VHF and UHF TV frequencies. Globally, several standardisation organisations are developing CRN standards for DSA and coexistence. Because of research, the first worldwide standard for TV white space (TVWS) operation is IEEE 802.22.

There is a lot of work involved in making CR-based IoT a reality. It is challenging to classify research directions because CR-based IoT frameworks are still in their early stages. It is recommended that they take a step forward and globalise their activities, as there are organisations like IEEE, the International Telecommunication Union (ITU), and 3GPP that are working on both CRNs and IoT independently.

It will be preferable to incorporate IoT into this framework while efforts are underway to expand the bands available for CRNs. The coexistence of CR-based IoT in existing regulatory assignments could be another way to solve the problem. Although RFID and NFC are short-range technologies, they are relevant to research in IoT standards for communication techniques. Short- to medium-range coexistence has been taken into account in CRN standards, and specific long-range standards, such TVWS, have been developed. We propose that the long-range capacity of CRNs be considered in IoT standards. In a similar vein, CRN protocols are at a progressing stage, while little progress has been made with IoT protocols. We believe that this aspect of CRN standardisation can also help CR-based IoT.

8. APPLICATIONS OF CR-IOT

We already have real-world IoT applications in the healthcare sector. A patient's surroundings and body are outfitted with smart sensors to track vital indicators like blood pressure, temperature, glucose levels, and more. The medical personnel regularly monitors the parameters through remote monitoring.

Although there are currently wireless solutions available, it is crucial to provide seamless monitoring. For this reason, medical personnel should receive healthcare information without the necessity for spectrum assignment. This can be accomplished over great distances using CR-based IoT frameworks withou worrying about spectrum availability shown in figure 8.

Figure 8. IoT applications

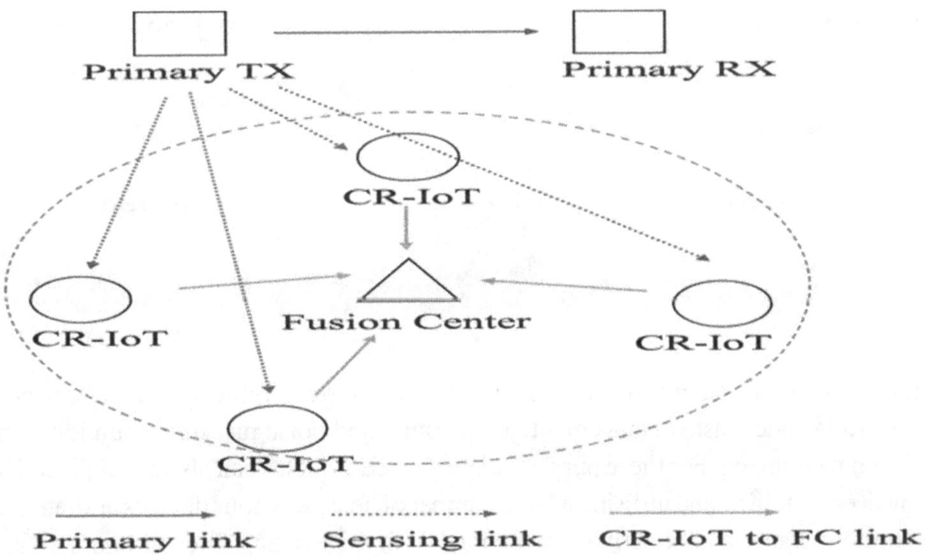

CR based IoT has gained interest in recent years. Intelligent transport systems (ITS) are one example, which combine sensors based on the road and vehicles for many uses like traffic monitoring and electronic ringing. The 30 MHz and 75 MHz bands, as well as the 5.8–5.9 GHz spectrum, are designated for Dedicated Short Range Communication (DSRC), which has a channel width of 10 MHz in the US and Europe, respectively. Only limited data may be delivered by this short-range connection, which is utilised for alerts and emergencies. Long coverage necessitates a large number of sensors and roadside units (RSUs), which will ultimately be quite expensive. Furthermore, assigning communication strategies is not given much thought for this reason. We believe that CRNs can address these issues in a more appropriate way.

One other example from the social domain is a traffic management system as shown in Figure 9 such as smart traffic lights. Using CR-based Internet of Things devices for business, like providing wireless broadband to customers, is another example of a social activity.

Figure 9. Different sensors involved in IoT applications

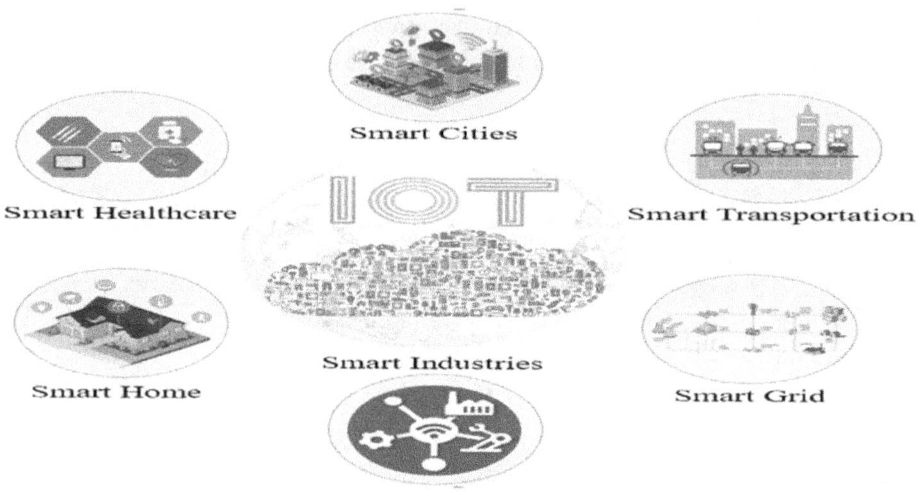

Applications of IoT in the environmental field are already being developed and are very popular. Key areas in this regard include waste management, temperature and noise measures, humidity estimations, and CO2 emission monitoring. For these purposes, specialised vehicles can also be employed in addition to sensors. However, installing and utilising a large number of heterogeneous devices at strategic locations is necessary for dependable and effective monitoring. With the existing fixed assignment approach, it is challenging to assign spectrum to this many devices. IoT features with cognitive capabilities may be able to alleviate network congestion and shortages.

IoT-based solutions are currently in use within homes and buildings, and it is anticipated that as technology advances, the IoT era will become essential in the future. Sensors and other devices will be installed in homes to help with daily tasks and enhance quality of life. There are already examples of home automation and home energy management, such as smart fridges and smart lights, in place. For these purposes, WiFi access points are typically deployed; yet, this might result in significant interference in the industrial, scientific, and medical (ISM) band. Adding cognitive capabilities to the sensors to ensure non-interference in ISM bands is a superior choice.

In modern times, the smart grid is a developing paradigm. Customers seek information about their energy usage at all times and locations. This suggests that IoT will be required in smart grids in the future. The ability to carry massive amounts of data over long distances without interference from many metres or devices in a constrained radio bandwidth is a significant disadvantage. There are problems with current wireless approaches in this area. These issues can be solved with wired methods like DSL and optical fibre or wireless methods like cellular, but installing cable or fibre or buying spectrum will cost a lot of money. Thus, the only workable option available to us are CRNs.

An integrated approach to urban development known as "smart city" integrates Internet of Things (IoT) with information and communication technology (ICT) systems. Offering consumers e-services for an enhanced, environmentally responsible lifestyle is the driving force behind smart cities. An information and communication system will be the foundation for this since it will be necessary to have constant connectivity. User engagement and data collection will be crucial components as well. CRNs have the ability to facilitate connection.

Less reliance on humans has been the trend recently, which has brought us to the Internet of Vehicles (IoV) paradigm, where vehicle control is accomplished through the integration of embedded systems, controls, and communications. It is anticipated that IoV would make its own travel decisions. In the future, safe navigation might be made possible by users' intents, sensors mounted on cars, and information sent between vehicles. Due to their extended range and interference-free spectrum sensing, CRNs can be a good solution to the problem of mobile vehicle spectrum availability in the Internet of Vehicles.

REFERENCES

Akyildiz, I. F., Lee, W. Y., & Chowdhury, K. R. (2009). CRAHNs: Cognitive radio ad hoc networks. *Ad Hoc Networks*, 7(5), 810–836. 10.1016/j.adhoc.2009.01.001

Al-Kofahi, O. M., Almasaeid, H. M., & Al-Mefleh, H. (2020). Efficient on-demand spectrum sensing in sensor-aided cognitive radio networks. *Computer Communications*, 156, 11–24. 10.1016/j.comcom.2020.03.032

Bayhan, S., Zubow, A., Gawlowicz, P., & Wolisz, A. (2019). Smart contracts for spectrum sensing as a service. *IEEE Transactions on Cognitive Communications and Networking*, 5(3), 648–660. 10.1109/TCCN.2019.2936190

Ejaz, W., & Ibnkahla, M. (2017). Multiband spectrum sensing and resource allocation for IoT in cognitive 5G networks. *IEEE Internet of Things Journal*, 5(1), 150–163. 10.1109/JIOT.2017.2775959

Ejaz, W., Shah, G. A., Hasan, N. U., & Kim, H. S. (2015). Energy and throughput efficient cooperative spectrum sensing in cognitive radio sensor networks. *Transactions on Emerging Telecommunications Technologies*, 26(7), 1019–1030. 10.1002/ett.2803

Janu, D., Singh, K., & Kumar, S. (2022). Machine learning for cooperative spectrum sensing and sharing: A survey. *Transactions on Emerging Telecommunications Technologies*, 33(1), e4352. 10.1002/ett.4352

Jothiraj, S., & Balu, S. (2019). A novel linear SVM-based compressive collaborative spectrum sensing (CCSS) scheme for IoT cognitive 5G network. *Soft Computing*, 23(18), 8515–8523. 10.1007/s00500-019-04097-x

Kelaidonis, D., Somov, A., Foteinos, V., Poulios, G., Stavroulaki, V., Vlacheas, P., & Giaffreda, R. (2012, November). Virtualization and cognitive management of real world objects in the internet of things. In *2012 IEEE International Conference on Green Computing and Communications* (pp. 187-194). IEEE. 10.1109/GreenCom.2012.37

Khalid, L., & Anpalagan, A. (2011). Cooperative sensing with correlated local decisions in cognitive radio networks. *IEEE Transactions on Vehicular Technology*, 61(2), 843–849. 10.1109/TVT.2011.2177871

Khan, A. A., Rehmani, M. H., & Rachedi, A. (2017). Cognitive-radio-based internet of things: Applications, architectures, spectrum related functionalities, and future research directions. *IEEE Wireless Communications*, 24(3), 17–25. 10.1109/MWC.2017.1600404

Lee, W., Kim, M., & Cho, D. (2019). Deep Cooperative Sensing: Cooperative Spectrum Sensing Based on Convolutional Neural Networks. *IEEE Transactions on Vehicular Technology*, 68(3), 3005–3009. 10.1109/TVT.2019.2891291

Li, T., Yuan, J., & Torlak, M. (2018). Network throughput optimization for random access narrowband cognitive radio Internet of Things (NB-CR-IoT). *IEEE Internet of Things Journal*, 5(3), 1436–1448. 10.1109/JIOT.2017.2789217

Miz, V., & Hahanov, V. (2014, September). Smart traffic light in terms of the cognitive road traffic management system (CTMS) based on the Internet of Things. In *Proceedings of IEEE east-west design & test symposium (EWDTS 2014)* (pp. 1-5). IEEE.

Moon, B. (2017). Dynamic spectrum access for internet of things service in cognitive radio-enabled LPWANs. *Sensors (Basel)*, 17(12), 2818. 10.3390/s1712281829206215

Nasser, A., Al Haj Hassan, H., Abou Chaaya, J., Mansour, A., & Yao, K. C. (2021). Spectrum sensing for cognitive radio: Recent advances and future challenge. *Sensors (Basel)*, 21(7), 2408. 10.3390/s2107240833807359

Ning, W., Huang, X., Yang, K., Wu, F., & Leng, S. (2020). Reinforcement learning enabled cooperative spectrum sensing in cognitive radio networks. *Journal of Communications and Networks (Seoul)*, 22(1), 12–22. 10.1109/JCN.2019.000052

Pati, B. M., Kaneko, M., & Taparugssanagorn, A. (2020). A deep convolutional neural network based transfer learning method for non-cooperative spectrum sensing. *IEEE Access : Practical Innovations, Open Solutions*, 8, 164529–164545. 10.1109/ACCESS.2020.3022513

Premsankar, G., Di Francesco, M., & Taleb, T. (2018). Edge computing for the Internet of Things: A case study. *IEEE Internet of Things Journal*, 5(2), 1275–1284. 10.1109/JIOT.2018.2805263

Shah, M. A., Zhang, S., & Maple, C. (2013, September). Cognitive radio networks for Internet of Things: Applications, challenges and future. In *2013 19th International Conference on Automation and Computing* (pp. 1-6). IEEE.

Shi, Z., Gao, W., Zhang, S., Liu, J., & Kato, N. (2019). AI-enhanced cooperative spectrum sensing for non-orthogonal multiple access. *IEEE Wireless Communications*, 27(2), 173–179. 10.1109/MNET.001.1900305

Somov, A., Dupont, C., & Giaffreda, R. (2013, July). Supporting smart-city mobility with cognitive Internet of Things. In *2013 Future Network & Mobile Summit* (pp. 1-10). IEEE.

Wunsch, F., Paisana, F., Rajendran, S., Selim, A., Alvarez, P., Muller, S., Koslowski, S., Van den Bergh, B., & Pollin, S. (2017). DySPAN Spectrum Challenge: Situational Awareness and Opportunistic Spectrum Access Benchmarked. *IEEE Transactions on Cognitive Communications and Networking*, 3(3), 550–562. 10.1109/TCCN.2017.2745682

Yu, H., & Zikria, Y. B. (2020). Cognitive radio networks for internet of things and wireless sensor networks. *Sensors (Basel)*, 20(18), 5288. 10.3390/s2018528832947832

Zikria, Y. B., Ali, R., Afzal, M. K., & Kim, S. W. (2021). Next-generation internet of things (iot): Opportunities, challenges, and solutions. *Sensors (Basel)*, 21(4), 1174. 10.3390/s2104117433562343

Compilation of References

Abbas, N., Nasser, Y., & Ahmad, R. (2015). Recent advances on artificial intelligence and learning techniques in cognitive radio networks. *EURASIP Journal on Wireless Communications and Networking*, 2015(1), 2015. 10.1186/s13638-015-0381-7

Abeer, F. (2018). Hierarchical Cluster-Based Cooperative Spectrum Sensing in Cognitive Radio Networks Using Adaptive Threshold. *IEEE 61st International Midwest Symposium on Circuits and Systems (MWSCAS)*. IEEE.

Abid, B., Khan, B. M., & Memon, R. A. (2022). Seismic Facies Segmentation Using Ensemble of Convolutional Neural Networks. *Wireless Communications and Mobile Computing*. Wiley. 10.1155/2022/7762543

Agiza, H. N., Bischi, G. I., & Kopel, M. (1999). Multistability in a dynamic Cournot game with three oligopolists. *Mathematics and Computers in Simulation*, 51(1-2), 63–90. 10.1016/S0378-4754(99)00106-8

Agrahari, A., Varshney, P., & Jagannatham, A. K. (2017). Precoding and downlink beamforming in multiuser MIMO-OFDM cognitive radio systems with spatial interference constraints. *IEEE Transactions on Vehicular Technology*, 67(3), 2289–2300. 10.1109/TVT.2017.2768823

Ahmed, A., Mishra, D., Prasad, G., & Baishnab, K. L. (2021, January). Performance analysis of multi-antenna CR system with beamforming under various traffic scenarios. In *2021 IEEE 18th Annual Consumer Communications & Networking Conference (CCNC)* (pp. 1-6). IEEE. 10.1109/CCNC49032.2021.9369528

Ahmed, A., Mishra, D., Prasad, G., Baishnab, K. L., & Tourki, K. (2021, December). Primary Activity Awareness Based Secondary Transmission Gains in Smart Sensing CR. In *2021 IEEE Globecom Workshops (GC Wkshps)* (pp. 1-6). IEEE. 10.1109/GCWkshps52748.2021.9681938

Ahmed, A., & Baishnab, K. L. (2020). Joint optimal design of sensing time and transmission power for maximizing energy efficiency in cognitive radio system. *Wireless Personal Communications*, 110(4), 1839–1857. 10.1007/s11277-019-06814-z

Ahmed, A., Mishra, D., Prasad, G., & Baishnab, K. L. (2022). Cognitive Radio Timing Protocol for Interference-Constrained Throughput Maximization. *IEEE Transactions on Cognitive Communications and Networking*, 8(2), 989–1004. 10.1109/TCCN.2021.3137487

Ait Omar, D., Outanoute, M., Baslam, M., Fakir, M., & Bouikhalne, B. (2017). Joint Price and QoS Competition with Bounded Rational Customers. In El Abbadi, A., & Garbinato, B. (Eds.), *Networked Systems* (pp. 457–471). Springer International Publishing. 10.1007/978-3-319-59647-1_33

AIT. (2023). *About.* Artificial Intelligence Topics. https://aitopics.org/search

Akbari, K., & Abouei, J. (2018). Signal Classification for Detecting Primary User Emulation Attack in Centralized Cognitive Radio Networks. Electrical Engineering (ICEE), Iranian Conference On. 10.1109/ICEE.2018.8472515

Akeela, R., & Dezfouli, B. (2018). Software-defined Radios: Architecture, state-of-the-art, and challenges. *Computer Communications*, 128(March), 106–125. 10.1016/j.comcom.2018.07.012

Akyildiz, I. F., Lee, W. Y., & Chowdhury, K. R. (2009). CRAHNs: Cognitive radio ad hoc networks. *Ad Hoc Networks*, 7(5), 810–836. 10.1016/j.adhoc.2009.01.001

Alam, I., Sharif, K., Li, F., Latif, Z., Karim, M. M., Biswas, S., Nour, B., & Wang, Y. (2020). A Survey of Network Virtualization Techniques for Internet of Things Using SDN and NFV. *ACM Computing Surveys*, 53(2), 1–40. 10.1145/3379444

Al-ani, A., Ul, S., Laghari, A., Manoharan, H., Selvarajan, S., & M. Uddin, M. (2023). *Improved Transportation Model with Internet of Things Using Artificial Intelligence Algorithm*. Tech Science Press. .10.32604/cmc.2023.038534

Al-Ani, A. K., Ul Arfeen Laghari, S., Manoharan, H., Selvarajan, S., & Uddin, M. (2023). Improved Transportation Model with Internet of Things Using Artificial Intelligence Algorithm. *Computers, Materials & Continua*, 76(2).

Alawad, H., Kaewunruen, S., & An, M. (2020). Learning From Accidents: Machine Learning for Safety at Railway Stations. *IEEE Access : Practical Innovations, Open Solutions*, 8, 633–648. 10.1109/ACCESS.2019.2962072

Aldino, A. A., & Pratiwi, E. D., Setiawansyah, Sintaro, S., & Dwi Putra, A.(2021). Comparison of Market Basket Analysis to Determine Consumer Purchasing Patterns using FP-Growth and Apriori Algorithm. In *proceedings of 2021 International Conference on Computer Science, Information Technology, and Electrical Engineering (ICOMITEE)*. IEEE. 10.1109/ICOMITEE53461.2021.9650317

ALFoudery, A., Alkandari, A. A., & Almutairi, N. M. (2018). Trash Basket Sensor Notification Using Arduino with Android Application. *Indonesian Journal of Electrical Engineering and Computer Science*, 10(1), 120–128. 10.11591/ijeecs.v10.i1.pp120-128

Alhasan, A., Hussein, M. H., Audah, L., Al-Sharaa, A., Ibrahim, I., & Mahmoud, M. A. (2023). A case study to examine undergraduate students' intention to use internet of things (IoT) services in the smart classroom. *Education and Information Technologies*, 28(8), 1–24. 10.1007/s10639-022-11537-z

Ali, A., Yaqoob, I., Ahmed, E., Imran, M., Kwak, K. S., Ahmad, A., Hussain, S. A., & Ali, Z. (2018). Channel Clustering and QoS Level Identification Scheme for Multi-Channel Cognitive Radio Networks. *IEEE Communications Magazine*, 56(4), 164–171. 10.1109/MCOM.2018.1700752

Ali-Ahmad, H., Cicconetti, C., De La Oliva, A., Mancuso, V., Sama, M. R., Seite, P., & Shanmugalingam, S. (2013). An SDN-based network architecture for extremely dense wireless networks. *SDN4FNS 2013 - 2013 Workshop on Software Defined Networks for Future Networks and Services*. IEEE. 10.1109/SDN4FNS.2013.6702534

Alissa, K., Alyas, T., Zafar, K., Abbas, Q., Tabassum, N., & Sakib, S. (2022). Botnet Attack Detection in IoT Using Machine Learning. *Computational Intelligence and Neuroscience*.

Al-Kofahi, O. M., Almasaeid, H. M., & Al-Mefleh, H. (2020). Efficient on-demand spectrum sensing in sensor-aided cognitive radio networks. *Computer Communications*, 156, 11–24. 10.1016/j.comcom.2020.03.032

Almufarreh, A., & Arshad, M. (2023). Promising emerging technologies for teaching and learning: Recent developments and future challenges. *Sustainability (Basel)*, 15(8), 6917. 10.3390/su15086917

Almuqren, L., Maray, M., Alotaibi, F. A., Alzahrani, A., Mahmud, A., & Rizwanullah, M. (2024). Optimal Deep Learning Empowered Malicious User Detection for Spectrum Sensing in Cognitive Radio Networks. *IEEE Access : Practical Innovations, Open Solutions*, 12, 35300–35308. 10.1109/ACCESS.2024.3367993

Al-Taai, S. H. H., Kanber, H. A., & al-Dulaimi, W. A. M. (2023). The Importance of Using the Internet of Things in Education. *International Journal of Emerging Technologies in Learning*, 18(1), 19–39. 10.3991/ijet.v18i01.35999

An, T. T., & Lee, B. M. (2023). Robust automatic modulation classification in low signal to noise ratio. *IEEE Access : Practical Innovations, Open Solutions*, 11, 7860–7872. 10.1109/ACCESS.2023.3238995

Armi, N., Gharibi, W., Khan, W. Z., Zangoti, H., Rizvi, S., & Wael, C. (2017). Error detection of malicious user attack in cognitive radio networks. *2017 International Conference on Radar, Antenna, Microwave, Electronics, and Telecommunications (ICRAMET)*. IEEE. 10.1109/ICRAMET.2017.8253152

Arshid, K., Jianbiao, Z., Hussain, I., Pathan, M. S., Yaqub, M., Jawad, A., Munir, R., & Ahmad, F. (2022). Energy efficiency in cognitive radio network using cooperative spectrum sensing based on hybrid spectrum handoff. *Egyptian Informatics Journal*, 23(4), 77–88. 10.1016/j.eij.2022.06.008

Awe, O. P., Zhu, Z., & Lambotharan, S. (2013). Eigenvalue and support vector machine techniques for spectrum sensing in cognitive radio networks. *Proceeding of the Conference on Technologies and Applications of Artificial Intelligence*. IEEE. 10.1109/TAAI.2013.52

Awin, F., Abdel-Raheem, E., & Ahmadi, M. (2016). Joint optimal transmission power and sensing time for energy efficient spectrum sensing in cognitive radio system. *IEEE Sensors Journal*, 17(2), 369–376. 10.1109/JSEN.2016.2627884

Awin, F., Abdel-Raheem, E., & Tepe, K. (2018). Blind spectrum sensing approaches for interweaved cognitive radio system: A tutorial and short course. *IEEE Communications Surveys and Tutorials*, 21(1), 238–259. 10.1109/COMST.2018.2863681

Baby, S. M., & James, M. (2016). A Comparative Study on Various Spectrum Sharing Techniques. *Procedia Technology*, 25, 613–620. 10.1016/j.protcy.2016.08.152

Badotra, S., Nagpal, Di., Panda, S. N., Tanwar, S., & Bajaj, S. (2020). IoT-Enabled Healthcare Network with SDN. *ICRITO 2020 - IEEE 8th International Conference on Reliability, Infocom Technologies and Optimization (Trends and Future Directions)*, (pp. 38–42). IEEE. 10.1109/ICRITO48877.2020.9197807

Baggio, G., Bassoli, R., & Granelli, F. (2019). Cognitive Software-Defined Networking Using Fuzzy Cognitive Maps. *IEEE Transactions on Cognitive Communications and Networking*, 5(3), 517–539. 10.1109/TCCN.2019.2920593

Bagustari, B. A., Yazid, S., & Alshehadeh, A. R. (2023). Analyzing fog computing technologies for enhancing m-learning networked environments. *International Journal of Advances in Soft Computing & Its Applications*, 15(1).

Baldo, N., & Zorzi, M. (2008). *Learning and adaptation in cognitive radios using neural networks*. In 5th IEEE CCNC, Las Vegas, NV. 10.1109/ccnc08.2007.229

Baldo, N., & Zorzi, M. (2008). Learning and adaptation in cognitive radios using neural networks. *Proceeding of Conference Record – IAS Annual Meeting (IEEE Industry Applications Society)*. IEEE.

Balieiro, A., Yoshioka, P., Dias, K., Cavalcanti, K., & Cordeiro, C. (2013). Adaptive spectrum sensing for cognitive radio based on multiobjective genetic optimisation. *Electronics Letters*, 49(17), 1099–1101. 10.1049/el.2013.1173

Bany Salameh, H., Samara, M. Q., Elrefae, G. A., & Al-Ajlouni, A. (2023). Profit-maximization spectrum sharing in opportunistic duopoly market under dynamic spectrum pricing and QoS constraints. *Cluster Computing*, 46, 1451–1463. 10.1007/s10586-023-04026-6

Baslam, M., El-Azouzi, R., Sabir, E., & Bouyakhf, E.-H. (2012). New insights from a bounded rationality analysis for strategic price-QoS war. *6th international ICST conference on performance evaluation methodologies and tools*, (pp. 280–289). EUDL. 10.4108/valuetools.2012.250410

Bayhan, S., & Alagoz, F. (2012). Scheduling in centralized cognitive radio networks for energy efficiency. *IEEE Transactions on Vehicular Technology*, 62(2), 582–595. 10.1109/TVT.2012.2225650

Bayhan, S., Zubow, A., Gawlowicz, P., & Wolisz, A. (2019). Smart contracts for spectrum sensing as a service. *IEEE Transactions on Cognitive Communications and Networking*, 5(3), 648–660. 10.1109/TCCN.2019.2936190

Baykasoğlu, A., Ozbakır, L., & Tapkan, P. (2007). Artificial bee colony algorithm and its application to generalized assignment problems. *Swarm intelligence: focus on ant and particle swarm optimization*, 129-156.

Begum, M. B., & Sharmila, K. (2020). Improving performance of Full Duplex Device to Device Communication Underlying Cellular Networks. *International Journal of Innovative Research in Technology*, 7(2), 436–445.

Ben, W., Arfi, S., Gharib, C., & Ben Jabeur, S. (2021). Catboost model and artificial intelligence techniques for corporate failure prediction. *Journal of Tech. Fore.* 0658.doi.org/10.1016/j.techfore.2021.120658

Bhagyashri Wagh, J.V. (2007). Wankhade Sentiment Analysis on Twitter Data usingNaive Bayes. *International Journal of Advanced Research in Computer and Communication Engineering*.

Bharath, C., & Ganesh, J. (2010). A proposal to enhance cross-marketing via social networks by modelling and analyzing market basket like patterns across social network groups. In *proceeding of 2010 IEEE International Workshop on: Business Applications of Social Network Analysis (BASNA)*. IEEE. 10.1109/BASNA.2010.5730301

Bhargav, A., Mathur, R. P., & Bhargav, M. (2014). Market basket analysis using artificial neural network. In *proceedings of International Conference for Convergence for Technology*. IEEE. 10.1109/I2CT.2014.7092091

Bhargavi, D., & Murthy, C. R. (2010). Performance comparison of energy, matched-filter and cyclostationarity-based spectrum sensing. *Signal Processing Advances in Wireless Communications (SPAWC), IEEE Eleventh International Workshop on*, (pp. 1 –5). IEEE. 10.1109/SPAWC.2010.5670882

Bhattacharyya, S., Jha, S., Tharakunnel, K., & Westland, J. C. (2011). Data mining for credit card fraud: A comparative study. *Decision Support Systems*, 50(3), 602–661. 10.1016/j.dss.2010.08.008

Bilal, R., & Khan, B. M. (2019). Software-Defined Networks (SDN): A Survey. B. B. Gupta & Dharma P. Agrawal (eds.) *Handbook of Research on Cloud Computing and Big Data Applications in IoT.* IGI Global. 10.4018/978-1-5225-8407-0.ch023

Biswas, A. A., Mia, M. J., & Majumder, A. (2019). *Forecasting the Number of Road Accidents and Casualties using Random Forest Regression in the Context of Bangladesh*. 2019 10th International Conference on Computing, Communication and Networking Technologies (ICCCNT), Kanpur, India. 10.1109/ICCCNT45670.2019.8944500

Bizanis, N., & Kuipers, F. A. (2016). SDN and Virtualization Solutions for the Internet of Things: A Survey. *IEEE Access : Practical Innovations, Open Solutions*, 4, 5591–5606. 10.1109/ACCESS.2016.2607786

Bkassiny, M. & Jayaweera, S. K. (2012). A Survey on machine-learning techniques in cognitive radios. *IEEE Communications Surveys & Tutorials, 15*, 1136-1159.

Bkassiny, M., Li, Y., & Jayaweera, S. K. (2013). A survey on machine learning techniques in cognitive radios. *IEEE Communications Surveys and Tutorials*, 15(3), 1136–1159. 10.1109/SURV.2012.100412.00017

Blum, C., & Roli, A. (2003, September). Metaheuristics in combinatorial optimization: Overview and conceptual comparison. *ACM Computing Surveys*, 35(3), 268–308. 10.1145/937503.937505

Bobulski, J., & Kubanek, M. (2019). Waste Classification System Using Image Processing and Convolutional Neural Networks. In *International Work Conference on Artificial Neural Networks* (pp. 350-361). Springer. 10.1007/978-3-030-20518-8_30

Borges, P. V. K., & Izquierdo, E. (2010). A probabilistic approach for vision-based fire detection in videos. *IEEE Transactions on Circuits and Systems for Video Technology*, 20(5), 721–731. 10.1109/TCSVT.2010.2045813

BTH. (n.d.). *Applying OSTBC in Cooperative Cognitive Radio Networks - Arkiv EX.* Blekinge Tekniska Högskola. http://www.bth.se/fou/cuppsats.nsf/6753b78eb2944e0ac1256608004f0535/4830a668eb9bd48cc1257735004a7535?OpenDocument

Cabric, D., Mishra, S. M., & Brodersen, R. W. (2004). Implementation issues in spectrum sensing for cognitive radios. *Signals, Systems and Computers:Conference Record of the Thirty-Eighth Asilomar Conference.* IEEE. 10.1109/ACSSC.2004.1399240

Caso, G., Soleimani, H., De Nardis, L., Tosti, A., & Di Benedetto, M. G. (2014). SENSIC: Mobility-aware cluster-based cooperative spectrum sensing for cognitive radio networks. *2014 IEEE International Conference on Ultra-WideBand (ICUWB).* IEEE. 10.1109/ICUWB.2014.6958959

Cattoni, A. F., Ottonello, M., Raffetto, M., & Regazzoni, C. S. (2007). *Neural networks mode classification based on frequency distribution features.* In 2nd Int. Conf. CrownCom, Orlando, FL.

Chamorro-Atalaya, O., Morales-Romero, G., Quispe-And'ıa, A., Caycho-Salas, B., Ramos-Salazar, P., C'aceres-Cayllahua, E., & Auqui-Ramos, R. (2023). Smart environments through the internet of things and its impact on university education. *Systematic Reviews.*

Chen, T.-H., Wu, P.-H., & Chiou, Y.-C. (2004). An early fire-detection method based on image processing. *Proc. Int. Conf. Image Process.* Research Gate.

Chen, G. (2022). Simulation model of AI-assisted cognitive routing algorithm for the dynamic optical network in business. *Optik (Stuttgart)*, 260, 168763. 10.1016/j.ijleo.2022.168763

Cheng, Z., Liu, B., & Huang, J. (2022). Causal Analysis of Road Safety Accidents in Britain Based on a Univariate Decision Tree Method.*2022 International Conference on Data Analytics, Computing and Artificial Intelligence (ICDACAI)*, Zakopane, Poland. 10.1109/ICDACAI57211.2022.00092

Chen, Y., Wang, Z., & Zhang, S. (2023).Optimization Design of Apriori Algorithm Based on Big Data Analysis and Cloud Computing. In *proceedings of 2023 International Conference on Power, Electrical Engineering, Electronics and Control (PEEEC).* IEEE. 10.1109/PEEEC60561.2023.00129

Chen, Z., Xu, Y. Q., Wang, H., & Guo, D. (2021). Federated learning-based cooperative spectrum sensing in cognitive radio. *IEEE Communications Letters*, 26(2), 330–334. 10.1109/LCOMM.2021.3114742

Choo, Y. (2011). An elementary proof of the Jury test for real polynomials. *Automatica*, 47(1), 249–252. 10.1016/j.automatica.2010.10.040

Chu, S. C., & Tsai, P. W. (2007). Computational intelligence based on the behaviour of cats. *International Journal of Innovative Computing, Information, & Control*, 3, 163–173.

Cicio lu, M., Cicio lu, S., & Çalhan, A. (2020). SDN-enabled cognitive radio network architecture. *IET Communications*, 14(18), 3153–3160. 10.1049/iet-com.2019.1346

Cogburn, D. L., & Espinoza-Vasquez, F. K. (2011, February). From networked nominee to networked nation: Examining the impact of web 2.0 and social media on political participation. and civic engagement in the 2008 Obama campaign. *Journal of Political Marketing*, 10(1-2), 189–213. 10.1080/15377857.2011.540224

Columbus, L. (2020). *How E-Commerce's Explosive Growth is Attracting Fraud.* Research Gate.

Cournot, A. (1963). *Researches into the principles of the theory of wealth.* The Macmillan Company.

Cunningham, S. J., & Frank, E. (1999). Market basket analysis of library circulation data. In *Proceeding of 6th International Conference on Neural Information Processing*. Washington: IEEE Computer Society. 10.1109/ICONIP.1999.845702

Daoudi, L., Baslam, M., & Safi, S. (2022). 2022). Dynamics analysis for a duopoly game with bounded rationality in cognitive radio network. In *Networked systems* (Vol. 13464, pp. 236–251). Springer. 10.1007/978-3-031-17436-0_16

Dehghani, M., Kamandi, A., Shabankhah, M., & Moeini, A. (2019). Toward a Distinguishing Approach for Improving the Apriori Algorithm. In *proceedings of 2019 9th International Conference on Computer and Knowledge Engineering (ICCKE)*. IEEE. 10.1109/ICCKE48569.2019.8965206

Devaney, R. L. (2022). *An introduction to chaotic dynamical systems*. CRC Press. 10.1201/9780429280801

Dhar, S. (2018). *Methods for Sentiment Analysis Computer Engineering*. VIVA Institute of Technology, University of Mumbai, India.

Digham, F. F., Alouini, M.-S., & Simon, M. K. (2003). On the energy detection of unknown signals over fading channels. *Communications, ICC '03. IEEE International Conference*. IEEE. 10.1109/ICC.2003.1204119

Dobre, O., Abdi, A., Bar-Ness, Y., & Su, W. (2007, April). Survey of automatic modulation classification techniques: Classical approaches and new trends. *IET Communications*, 1(2), 137–156. 10.1049/iet-com:20050176

Do, C. T., Tran, N. H., Han, Z., Le, L. B., Lee, S., & Hong, C. S. (2014). Optimal Pricing for Duopoly in Cognitive Radio Networks: Cooperate or not Cooperate? *IEEE Transactions on Wireless Communications*, 13(5), 2574–2587. 10.1109/TWC.2014.031914.131363

Economics at the FCC 2019–2020: Spectrum Policy, Universal Service, Inmate Calling Services, and Telehealth. (2020). *Review of Industrial Organization, 57*, 827-858. 10.1007/s11151-020-09791-x

Ejaz, W., & Ibnkahla, M. (2017). Multiband spectrum sensing and resource allocation for IoT in cognitive 5G networks. *IEEE Internet of Things Journal*, 5(1), 150–163. 10.1109/JIOT.2017.2775959

Ejaz, W., Shah, G. A., Hasan, N. U., & Kim, H. S. (2015). Energy and throughput efficient cooperative spectrum sensing in cognitive radio sensor networks. *Transactions on Emerging Telecommunications Technologies*, 26(7), 1019–1030. 10.1002/ett.2803

Elgadi, R., Hilal, A. R., & Basir, O. (2017). A fuzzy logic approach for cooperative spectrum sensing in cognitive radio networks. *Proceeding of IEEE 30th Canadian Conference on Electrical and Computer Engineering (CCECE)*. IEEE. 10.1109/CCECE.2017.7946841

F. C. Commission (2002). *Spectrum policy task force report*. (FCC 02-155). FCC.

Faisal, S. (2022). Deep Learning based small object detection and classification model for Garbage Waste Management in Smart Cities and IoT Environment. *Applied Science*.

FCC. (2015). *Amendment of the commission's rules with regard to commercial operations in the 3550-3650 MHz band*. FCC.

Federal Communication Commission. (2008). *Adopts Rules for Unlicensed Use of Television White Spaces*. FCC. http://hraunfoss.fcc.gov/edocs_public/attachmatch/DOC-286566A1.pdf

Federal Communications Commission. (2002). *Revision of part 15 of the commission's rules regarding ultra-wideband transmission systems*. FCC.

Federal Communications Commission. (2002). *Spectrum policy task force*. ET Docket.

Federal Communications Commission. (2018). FCC online table of frequency allocations. FCC. https://transition.fcc .gov/oet/spectrum/table/fcctable. pdf

Fehske, A., Gaeddert, J., & Reed, J. H. (2005). A new approach to signal classification using spectral correlation and neural networks, in *Proc. 1st IEEE Int. Symp. New Frontiers DySPAN*, (pp. 144–150). IEEE. 10.1109/DYSPAN.2005.1542629

Fernando, X., & Lăzăroiu, G. (2023). Spectrum sensing, clustering algorithms, and energy-harvesting technology for cognitive-radio-based internet-of-things networks. *Sensors (Basel)*, 23(18), 7792. 10.3390/s2318779237765848

Foerster, K., Schmid, S., & Vissicchio, S. (2019). Survey of Consistent Software-Defined Network Updates. *IEEE Communications Surveys and Tutorials*, 21(2), 1435–1461. 10.1109/COMST.2018.2876749

Foggia, P., Saggese, A., & Vento, M. (2015). Real-time fire detection for video-surveillance applications using a combination of experts based on color,shape, and motion. *IEEE Transactions on Circuits and Systems for Video Technology*, 25(9), 1545–1556. 10.1109/TCSVT.2015.2392531

Francois, F., & Gelenbe, E. (2016). Optimizing Secure SDN-Enabled Inter-Data Centre Overlay Networks through Cognitive Routing. *2016 IEEE 24th International Symposium on Modeling, Analysis and Simulation of Computer and Telecommunication Systems (MASCOTS)*, (pp. 283–288). IEEE. https://doi.org/10.1109/MASCOTS.2016.26

Frasch, I. J. (2017). *Algorithmic framework and implementation of spectrum holes detection for cognitive radios.* [Master's thesis, Rochester Institute of Technology].

Gao, Z., Zhu, H., Liu, Y., Li, M., & Cao, Z. (2013, April). Location privacy in database-driven cognitive radio networks: Attacks and countermeasures. In 2013 [IEEE.]. *Proceedings - IEEE INFOCOM*, 2751–2759.

Gardner, W. (1988, August). Signal interception: A unifying theoretical framework for feature detection. *IEEE Transactions on Communications*, 36(8), 897–906. 10.1109/26.3769

Gaur, A., Singh, A., Kumar, A., Kumar, A., & Kapoor, K. (2020). Video flame and smoke based fire detection algorithms: A literature review. *Fire Technology*, 56(5), 1943–1980. 10.1007/s10694-020-00986-y

Gavrilovska, L., Atanasovski, V., Macaluso, I., & DaSilva, L. (2013). L. (2013). Learning and reasoning in cognitive radio networks. *IEEE Communications Surveys and Tutorials*, 15(4), 1761–1777. 10.1109/SURV.2013.030713.00113

Geetha, S., Abhishek, C. S., & Akshayanat, C. S. (2021). Machine vision based fire detection techniques: A survey. *Fire Technology*, 57(2), 591–623. 10.1007/s10694-020-01064-z

Gehlot, A., & Singh, R. (2022). Execution of Market Basket Analysis and Recommendation Systems in Physical Retail Stores to Advance Sales Revenues. In *proceedings of 2022 International Conference on Futuristic Technologies (INCOFT)*. IEEE. 10.1109/INCOFT55651.2022.10094471

Ghosh, C., Cordeiro, C., Agrawal, D. P., & Rao, M. B. (2009). Markov chain existence and hidden Markov models in spectrum sensing. *Proceeding of IEEE International Conference on Pervasive Computing and Communications*, (pp. 1-6). IEEE. 10.1109/PERCOM.2009.4912868

Glassford, A. (2015) *Multiclass Emotion Analysis of Social Media Posts*. Stanford University.

Glover, F. (1986). Future paths for integer programming and links to artificial intelligence. *Computers & Operations Research*, 13(5), 533–549. 10.1016/0305-0548(86)90048-1

Gocs, L. & Johanyak, Z (2023). Catboost algorithm base classifier module for brute force attack prediction. *Tome XXI.* Fascicule 3.

Goldsmith, A., Jafar, S., Maric, I., & Srinivasa, S. (2009, May). Breaking spectrum gridlock with cognitive radios: An information theoretic perspective. *Proceedings of the IEEE*, 97(5), 894–914. 10.1109/JPROC.2009.2015717

Gonzalez, C. R. A., Dietrich, C. B., Kragh, F. E., Sayed, S., Volos, H. I., Gaeddert, J. D., Robert, P. M., & Reed, J. H. (2009, October). Open-source SCA-based core framework and rapid development tools enable software-defined radio education and research. *IEEE Communications Magazine*, 47(10), 48–55. 10.1109/MCOM.2009.5273808

Goswami, N. G., Sharma, P., Arora, A., & Singh, N. (2022). *Traffic improvisation by identifying the accident locations using ML/Data mining approaches.* 2022 4th International Conference on Advances in Computing, Communication Control and Networking (ICAC3N), Greater Noida, India. 10.1109/ICAC3N56670.2022.10074527

Greenis, P. (2009). Smart communications for smart grids, in *Proc. Nanotech Conf. Expo*, Houston, TX, May 3–7, 2009.

Guan, W., Fang, Z., & Yang, C. (2023). IOT Botnet Attack detection model based on DBO- Catboost. *Applied Sciences (Basel, Switzerland)*, 13(12), 7169. 10.3390/app13127169

Haque, M. A., Haque, S., Zeba, S., Kumar, K., Ahmad, S., Rahman, M., & Ahmed, L. (2023). Sustainable and efficient e-learning internet of things system through blockchain technology. *E-Learning and Digital Media*, 20427530231156711.

Hasegawa, M., & Nguyen, T. Ha, Miyamoto, G., Murata, Y. & Kato, S. (2007). Distributed optimization based on neurodynamic for cognitive wireless clouds. *ProceedingIEEE 18th Int. Symp. PIMRC*. IEEE.

Haykin, S. (1999). *Neural Networks: A Comprehensive Foundation*. Prentice-Hall.

Haykin, S. (2005, February). Cognitive radio: Brain-empowered wireless communications. *IEEE Journal on Selected Areas in Communications*, 23(2), 201–220. 10.1109/JSAC.2004.839380

He, A., Gaeddert, J., Tirado, L. M., & Reed, J. H. (2010). A survey of artificial intelligence for cognitive radios. *IEEE Transactions on Vehicular Technology*, 59(4), 1578–1593. 10.1109/TVT.2010.2043968

He, Y., Wang, J., Wu, Z., Hu, L., Xiong, Y., & Fan, W. (2002). Smoke venting and fire safety in an industrial warehouse. *Fire Safety Journal*, 37(2), 191–215. 10.1016/S0379-7112(01)00045-5

Ho, C.-C., & Kuo, T.-H. (2009). Real-time video-based fire smoke detection system. *Advanced Intelligent Mechatronics IEEE/ASME International Conference*, (pp. 1845–1850). IEEE.

Hong, X., Chen, Z., Wang, C. X., Vorobyov, S. A., & Thompson, J. S. (2009). Cognitive radio networks. *IEEE Vehicular Technology Magazine*, 4(4), 76–84. 10.1109/MVT.2009.934672

Hong, X., Wang, J., Wang, C.-X., & Shi, J. (2014). Cognitive radio in 5G: A perspective on energy-spectral efficiency trade-off. *IEEE Communications Magazine*, 52(7), 46–53. 10.1109/MCOM.2014.6852082

Hua, L., Zheng, W., & Gao, S. (2019). Extraction and Analysis of Risk Factors from Chinese Railway Accident Reports. *2019 IEEE Intelligent Transportation Systems Conference (ITSC)*, Auckland, New Zealand10.1109/ITSC.2019.8917094

Hu, N., Yao, Y.-D., & Mitola, J. (2012). Most Active Band (MAB) Attack and Countermeasures in a Cognitive Radio Network. *IEEE Transactions on Wireless Communications*, 11(3), 898–902. 10.1109/TWC.2012.011812.110927

Huo, L., Jiang, D., Qi, S., Song, H., & Miao, L. (2021). An AI-Based Adaptive Cognitive Modeling and Measurement Method of Network Traffic for EIS. *Mobile Networks and Applications*, 26(2), 575–585. 10.1007/s11036-019-01419-z

Hu, S., Yao, Y. D., & Yang, Z. (2014). Cognitive medium access control protocols for secondary users sharing a common channel with time division multiple access primary users. *Wireless Communications and Mobile Computing*, 14(2), 284–296. 10.1002/wcm.2185

Hu, S., Yao, Y. D., & Yang, Z. (2014). MAC protocol identification using support vector machines for cognitive radio networks. *IEEE Wireless Communications*, 21(1), 52–60. 10.1109/MWC.2014.6757897

I., E.E. (2023). Clement JC. GRU-SVM Based Threat Detection in Cognitive Radio Network. *Sensors (Basel)*, 23(3), 1326. 10.3390/s2303132636772366

IEEE. (2023). *Working Group on Wireless Regional Area Networks*. IEEE. https://www.ieee802.org/22/

IEEE10582 (2023). *Standard for Definitions and Concepts for Dynamic Spectrum Access: Terminology Relating to Emerging Wireless Networks, System Functionality, and Spectrum Management*. IEEE. https://standards.ieee.org/ieee/1900.1/10582/

Ismail, S., Khoei, T. T., Marsh, R., & Kaabouch, N. (2021). *A Comparative Study of Machine Learning Models for Cyber-attacks Detection in Wireless Sensor Networks. 2021 IEEE 12th AnnualUbiquitous Computing, Electronics & MobileCommunication Conference*. UEMCON.

Jacob, J., Jose, B. R., & Mathew, J. (2015). A fuzzy approach to decision fusion in cognitive radio. *Procedia Computer Science*, 46, 425–431. 10.1016/j.procs.2015.02.040

Jain, A. K., Mao, J. C., & Mohiuddin, K. M. (1996, March). artificial neural networks: A tutorial. *Computer*, 29(3), 31–44. 10.1109/2.485891

Janjua, H. K., de Miguel, I., Durán Barroso, R. J., Masoumi, M., Hosseini, S., Aguado, J. C., Merayo, N., & Abril, E. J. (2023). A Framework for Next Generation Cloud-Native SDN Cognitive Resource Orchestrator for IoTs (NG2CRO). *Lecture Notes in Networks and Systems, 741*, 399–407. 10.1007/978-3-031-38318-2_39

Janu, D., Singh, K., & Kumar, S. (2022). Machine learning for cooperative spectrum sensing and sharing: A survey. *Transactions on Emerging Telecommunications Technologies*, 33(1), e4352. 10.1002/ett.4352

Jawad, F. H. M. (2018). Software defined networking: An overview. *Journal of Engineering and Applied Sciences, 13*, 10790–10796. https://doi.org/10.36478/jeasci.2018.10790.10796

Jiang, C., Chen, Y., Liu, K. R., & Ren, Y. (2013). Renewal-theoretical dynamic spectrum access in cognitive radio network with unknown primary behavior. *IEEE Journal on Selected Areas in Communications*, 31(3), 406–416. 10.1109/JSAC.2013.130307

Jothiraj, S., & Balu, S. (2019). A novel linear SVM-based compressive collaborative spectrum sensing (CCSS) scheme for IoT cognitive 5G network. *Soft Computing*, 23(18), 8515–8523. 10.1007/s00500-019-04097-x

Jury, E. I. (1963). On the roots of a real polynomial inside the unit circle and a stability criterion for linear discrete systems. *IFAC Proceedings Volumes, 1*, 142-153. https://doi.org/10.1016/S1474-6670(17)69648-4

Kaggle.com. (2022). *Credit Card Fraud Detection*. Kaggle. https://www.kaggle.com/mlg-ulb/creditcardfraud

Kaliyamurthy, N. M., Taterh, S., Shanmugasundaram, S., Saxena, A., Cheikhrouhou, O., & Ben Elhadj, H. (2021). Software-Defined Networking: An Evolving Network Architecture - Programmability and Security Perspective. *Security and Communication Networks*, 2021, 1–7. 10.1155/2021/9971705

Kamruzzaman, M., Alanazi, S., Alruwaili, M., Alshammari, N., Elaiwat, S., AbuZanona, M., Innab, N., Mohammad Elzaghmouri, B., & Ahmed Alanazi, B. (2023). Ai-and iot-assisted sustainable education systems during pandemics, such as covid-19, for smart cities. *Sustainability (Basel)*, 15(10), 8354. 10.3390/su15108354

Kashyap, S., Björnson, E., & Larsson, E. G. (2016). On the feasibility of wireless energy transfer using massive antenna arrays. *IEEE Transactions on Wireless Communications*, 15(5), 3466–3480. 10.1109/TWC.2016.2521739

Kayode, Y., Saheed, A., & Abiodun, S. (2022). *A machine learning based intrusion detection for detecting internet of things network attacks.* Science Direct.

Kelaidonis, D., Somov, A., Foteinos, V., Poulios, G., Stavroulaki, V., Vlacheas, P., & Giaffreda, R. (2012, November). Virtualization and cognitive management of real world objects in the internet of things. In *2012 IEEE International Conference on Green Computing and Communications* (pp. 187-194). IEEE. 10.1109/GreenCom.2012.37

Kennedy, J., & Eberhart, R. (1995). Particle swarm optimization. *Proceeding of the IEEE International Conference on Neural Networks*, (pp. 1942-1948). IEEE. 10.1109/ICNN.1995.488968

Khalid, L., & Anpalagan, A. (2011). Cooperative sensing with correlated local decisions in cognitive radio networks. *IEEE Transactions on Vehicular Technology*, 61(2), 843–849. 10.1109/TVT.2011.2177871

Khan, A. A., Rehmani, M. H., & Rachedi, A. (2017). Cognitive-radio-based internet of things: Applications, architectures, spectrum related functionalities, and future research directions. *IEEE Wireless Communications*, 24(3), 17–25. 10.1109/MWC.2017.1600404

Khan, B. M., & Ali, F. H. (2011). Mobility Adaptive Energy Efficient and Low Latency MAC for Wireless Sensor Networks. *2011 Fifth International Conference on Next Generation Mobile Applications, Services and Technologies*, Cardiff, UK. 10.1109/NGMAST.2011.46

Khan, L. U., Alsenwi, M., Yaqoob, I., Imran, M., Han, Z., & Hong, C. S. (2020). Resource optimized federated learning-enabled cognitive internet of things for smart industries. *IEEE Access : Practical Innovations, Open Solutions*, 8, 168854–168864. 10.1109/ACCESS.2020.3023940

Khanna, A., Rani, P., Sheikh, T. H., Gupta, D., Kansal, V., & Rodrigues, J. J. (2022). Blockchain-based security enhancement and spectrum sensing in cognitive radio network. *Wireless Personal Communications*, 127(3), 1899–1921. 10.1007/s11277-021-08729-0

Khattab, A., Perkins, D., & Bayoumi, M. (2012). *Cognitive radio networks: from theory to practice.* Springer Science & Business Media.

Kim, K., Akbar, I. A., Bae, K. K., Urn, J. S., Spooner, C. M., & Reed, J. H. (2007). Cyclostationary approaches to signal detection and classification in cognitive radio. *Proceeding of the 2nd IEEE international Symposium on New Frontiers in Dynamic Spectrum Access Networks*, (pp. 212-215). IEEE. 10.1109/DYSPAN.2007.35

Kiran, M., Coakley, S., Walkinshaw, N., McMinn, P., & Holcombe, M. (2008). Validation and discovery from computational biology models. *Bio Systems*, 93(1-2), 141–150. 10.1016/j.biosystems.2008.03.01018487010

Kirkpatrick, S., Gelatt, D. C., & Vechhi, M. P. (1983). Optimization by simulated annealing. *Science*, 220(4598), 671–680. 10.1126/science.220.4598.67117813860

Kohlstruck, C., Engel, M., & Gotzhein, R. (2018). *The Selective Clustering Energy Detector for Cognitive Radio Networks – Conceptual Design and Experimental Assessment. 2018 14th International Wireless Communications & Mobile Computing Conference.* IWCMC.

Kreutz, D., Ramos, F. M. V., Verissimo, P. E., Rothenberg, C. E., Azodolmolky, S., & Uhlig, S. (2015). Software-defined networking: A comprehensive survey. *Proceedings of the IEEE*, 103(1), 14–76. 10.1109/JPROC.2014.2371999

Kshirsagar, P. R., Manoharan, H., & Al-Turjman, F. (2020). Design and Testing of Automated Smoke Monitoring Sensors in Vehicles. *IEEE Sensors Journal*, XX, 2020. 10.1109/JSEN.2020.3044604

Kumar, A. (2015). *Emotion Analysis of Twitter using Opinion Mining Dept. of Computer Engineering*. Delhi Technology University.

Kumar, S. D., Soundarapandiyan, K., & Meera, S. (2022).Commiserating Customers Purchasing Pattern Using Market Basket Analysis. *2022 1st International Conference on Computational Science and Technology (ICCST)*. IEEE. 10.1109/ICCST55948.2022.10040320

Kurimoto, T., Urushidani, S., Yamada, H., Yamanaka, K., Nakamura, M., Abe, S., Fukuda, K., Koibuchi, M., Ji, Y., Takakura, H., & Yamada, S. (2016). A fully meshed backbone network for data-intensive sciences and SDN services. *2016 Eighth International Conference on Ubiquitous and Future Networks (ICUFN)*, (pp. 909–911). IEEE. 10.1109/ICUFN.2016.7536928

Kurimoto, T., Urushidani, S., Yamada, H., Yamanaka, K., Nakamura, M., Abe, S., Fukuda, K., Koibuchi, M., Takakura, H., Yamada, S., & Ji, Y. (2017). SINET5: A low-latency and high-bandwidth backbone network for SDN/NFV Era. *2017 IEEE International Conference on Communications (ICC)*, (pp. 1–7). IEEE. 10.1109/ICC.2017.7996843

Lagunas, E., & Najar, M. (2015, July). Spectral feature detection with sub-Nyquist sampling for wideband spectrum sensing. *IEEE Transactions on Wireless Communications*, 14(7), 3978–3990. 10.1109/TWC.2015.2415774

Lai, C.-L., & Yang, J.-C. (2008). Advanced real time fire detection in video surveillance system. *IEEE International Symposium*, (pp. 3542–3545). IEEE.

Lakshmi, S. V. S. S., & Kavilla, S. D. (2018). Machine learning for credit card fraud detection system. *International Journal of Applied Engineering Research: IJAER*, 13(24), 16819–16824.

Lane, W. (2023). *Topic 9: Cognitive Radio Potential for Public Safety*. FCC. http://www.fcc.gov/

Lasaulce, S., Debbah, M., & Altman, E. (2009). Methodologies for analyzing equilibria in wireless games. *IEEE Signal Processing Magazine*, 26(5), 41–52. 10.1109/MSP.2009.933496

Latah, M., & Toker, L. (2019). Artificial intelligence enabled software-defined networking: A comprehensive overview. *IET Networks*, 8(2), 79–99. 10.1049/iet-net.2018.5082

Latif, S. A., Wen, F. B. X., Iwendi, C., Li-Li, F. W., Mohsin, S. M., Han, Z., & Band, S. S. (2022). Ai-empowered, blockchain and sdn integrated security architecture for iot network of cyber physical systems. *Computer Communications*, 181, 274–283. 10.1016/j.comcom.2021.09.029

Lee, C. S., Tiong, A., Tang, W. L., & Yap, K. H. (2019). Data-Driven Market Basket-Pricing and Personalized Health Information Services Using Salesforce's Model-Driven Systems Service Design. In *proceedings of 2019 IEEE International Conference on Industrial Engineering and Engineering Management (IEEM)*. IEEE. 10.1109/IEEM44572.2019.8978835

Lee, W., Kim, M., & Cho, D. (2019). Deep Cooperative Sensing: Cooperative Spectrum Sensing Based on Convolutional Neural Networks. *IEEE Transactions on Vehicular Technology*, 68(3), 3005–3009. 10.1109/TVT.2019.2891291

Lexis/Nexis Solutions. (2020). *True Cost of Fraud Study: E-Commerce/Retail Edition*. LexisNexis Solutions.

Liang, Y. C., Zeng, Y., Peh, E. C., & Hoang, A. T. (2008). Sensing-throughput tradeoff for cognitive radio networks. *IEEE Transactions on Wireless Communications*, 7(4), 1326–1337. 10.1109/TWC.2008.060869

Li, M., Hei, Y., & Qiu, Z. (2017). Optimization of multiband cooperative spectrum sensing with modified artificial bee colony algorithm. *Applied Soft Computing*, 57, 751–759. 10.1016/j.asoc.2017.03.027

Lim, T. (2021). K-Means Clustering-Based Market Basket Analysis: U.K. Online E-Commerce Retailer.[Paper presentation]. *2021 International Conference on Information Technology (ICIT)*, Amman, Jordan. 10.1109/ICIT52682.2021.9491703

Ling-Yu L. (2022). Smart fire-warning materials and sensors: Design principle, performances, and applications. *Mater. Sci. Eng., R, Rep., 150.*

Li, Q., Zhang, Q., & Qin, J. (2016). Beamforming for information and energy cooperation in cognitive non-regenerative two-way relay networks. *IEEE Transactions on Wireless Communications*, 15(8), 5302–5313. 10.1109/TWC.2016.2555913

Li, T., Yuan, J., & Torlak, M. (2018). Network throughput optimization for random access narrowband cognitive radio Internet of Things (NB-CR-IoT). *IEEE Internet of Things Journal*, 5(3), 1436–1448. 10.1109/JIOT.2017.2789217

Liu, F., Zhou, Y., Sun, Z., Du, R., & Sheng, J. (2016). Dynamic attack probability based Spectrum Sensing against Byzantine attack in Cognitive Radio. *2016 2nd IEEE International Conference on Computer and Communications (ICCC)*. IEEE.

Liu, B. (2012, April). Sentiment analysis and opinion mining: Synthesis Lectures on Human Lang. *Tech.*, 5(1), 1–167.

Liu, H., Chen, J., Ding, G., Tsiftsis, T. A., & Rowell, C. (2016, March). Antenna beamforming for energy harvesting in cognitive radio networks. In *2016 IEEE MTT-S International Wireless Symposium (IWS)* (pp. 1-4). IEEE. 10.1109/IEEE-IWS.2016.7585472

Liu, H., Xu, G., Ding, G. Y., & Sun, Y. B. (2014). Human behavior-based particle swarm optimization. *TheScientificWorldJournal*.24883357

Liu, L. L. L. (2024). Current situation and innovative methods of brass music teaching based on network information technology. *Journal of Electrical Systems*, 20(1), 294–312. 10.52783/jes.683

Liu, S., Quan, Z., & Jin, Z. (2023). Implementation of Railway Accident Judgment Criteria Optimization Based on Data Mining and Digital Programming Technology. *2023 Asia-Europe Conference on Electronics, Data Processing and Informatics (ACEDPI)*, Prague, Czech Republic. 10.1109/ACEDPI58926.2023.00043

Li, Y., Zhou, L., Zhu, H., & Sun, L. (2015). Privacy-preserving location proof for securing large-scale database-driven cognitive radio networks. *IEEE Internet of Things Journal*, 3(4), 563–571. 10.1109/JIOT.2015.2481926

Lo, R. (2009). *Moving towards cognitive NetOps*. In JSTeF, San Diego, CA.

Lu, C., Fitzek, F. H. P., & Eggers, P. C. F. (2007, September). A cooperative scheme enabling spatial reuse in wireless networks. *Wireless Communication Magazine*, 15(5), 457–471. 10.1007/978-1-4020-5979-7_23

M.G., , & R.H.R., (2023). Prediction of Road Accidents in the Different States of India using Machine Learning Algorithms. *2023 IEEE International Conference on Integrated Circuits and Communication Systems (ICICACS)*, 1-6. doi: 10.1109/ICICACS57338.2023.10099519

Magdalene, A. H. S., & Thulasimani, L. (2017). Fuzzy Clustering Means (FCM) for Mitigating Spectrum Sensing Data Falsification (SSDF) Attack in Cognitive Radio Networks. *2017 IEEE International Conference on Computational Intelligence and Computing Research (ICCIC)*. IEEE. 10.1109/ICCIC.2017.8524462

Maharjan Sabita, Z. Y., & Stein, G. (2011). Economic Approaches for Cognitive Radio Networks: A Survey. *Wireless Personal Communications*, 57(1), 33–51. 10.1007/s11277-010-0005-9

Maier, M., Chowdhury, M., Rimal, B. P., & Van, D. P. (2016). The tactile internet: Vision, recent progress, and open challenges. *IEEE Communications Magazine*, 54(5), 138–145. 10.1109/MCOM.2016.7470948

Maleh, Y., Qasmaoui, Y., El Gholami, K., Sadqi, Y., & Mounir, S. (2023). A comprehensive survey on sdn security: Threats, mitigations, and future directions. *Journal of Reliable Intelligent Environments*, 9(2), 201–239. 10.1007/s40860-022-00171-8

Mathur, R. (2024). *Pathak- Analyzing Sentiment of Twitter Data using Machine Learning Algorithm*. Academmia.

McHenry, M. (2005). *NSF Spectrum Occupancy Measurements Project Summary*. Shared Spectrum.

McHenry, M. A. (2005). *NSF spectrum occupancy measurements project summary.* (Shared Spectrum Company Report).

Meddeb, R., Jemili, F., Triki, B., & Korbaa, O. (2019). Anomaly-basedbehavioral detection in mobile ad hoc networks. *Procedia Computer Science*, 159, 77–86. 10.1016/j.procs.2019.09.162

Memon, R. A., Qazi, S., & Khan, B. M. (2021). Design and Implementation of a Robust Convolutional Neural Network-Based Traffic Matrix Estimator for Cloud Networks. Wireless Communications and Mobile Computing. 10.1155/2021/1039613

Miao, Y., Lin, J., & Xu, N. (2019). An improved parallel FP-growth algorithm based on Spark and its application. In *proceedings of 2019 Chinese Control Conference (CCC).* IEEE. 10.23919/ChiCC.2019.8866373

Mirjalili, S., & Lewis, A. (2016). The whale optimization algorithm. *Advances in Engineering Software*, 95, 51–67. 10.1016/j.advengsoft.2016.01.008

Mishra, D., & Johansson, H. (2019). Optimal channel estimation for hybrid energy beamforming under phase shifter impairments. *IEEE Transactions on Communications*, 67(6), 4309–4325. 10.1109/TCOMM.2019.2901790

Misra, D., Das, G., Chakrabortty, T., & Das, D. (2018). An IoT-based waste management system monitored by cloud. *Journal of Material Cycles and Waste Management*, 20(3), 1574–1582. 10.1007/s10163-018-0720-y

Mitola, J. (1999). Cognitive radio for flexible mobile multimedia communication, In *Proceedings of the IEEE International Workshop on Mobile Multimedia Communication (MoMuC).* IEEE.

Mitola, J., & Maguire, G. Q. (1999). Cognitive radio: making software radios more personal. *IEEE Personal Communications.* IEEE.

Mitola, J., & Maguire, G. (1999, August). Cognitive radio: Making software radios more personal. *IEEE Personal Communications*, 6(4), 13–18. 10.1109/98.788210

Miz, V., & Hahanov, V. (2014, September). Smart traffic light in terms of the cognitive road traffic management system (CTMS) based on the Internet of Things. In *Proceedings of IEEE east-west design & test symposium (EWDTS 2014)* (pp. 1-5). IEEE.

Mohammadi, F. S., & Kwasinski, A. (2018). *Neural network cognitive engine for autonomous and distributed underlay dynamic spectrum access.* arXiv preprint arXiv:1806.11038.

Moon, B. (2017). Dynamic spectrum access for internet of things service in cognitive radio-enabled LPWANs. *Sensors (Basel)*, 17(12), 2818. 10.3390/s1712281829206215

Muhammad, K., Khan, S., Elhoseny, M., Ahmed, S. H., & Baik, S. W. (2019). Efficient fire detection for uncertain surveillance environment. *IEEE Transactions on Industrial Informatics*, 15(5), 3113–3122. 10.1109/TII.2019.2897594

Muralidharan, L. Rasmussen, D., & Patterson. (2011). *Hope for Haiti: an analysis of facebook and twitter usage during the earthquake relief efforts.* Public Relations Rev.

Musacchio, J., & Walrand, J. (2006). WiFi access point pricing as a dynamic game. *IEEE/ACM Transactions on Networking*, 14(2), 289–301. 10.1109/TNET.2006.872553

Nash, J. (1951). Non-cooperative games. *Annals of Mathematics*, 54(2), 286–295. 10.2307/1969529

Nasser, A., Al Haj Hassan, H., Abou Chaaya, J., Mansour, A., & Yao, K. C. (2021). Spectrum sensing for cognitive radio: Recent advances and future challenge. *Sensors (Basel)*, 21(7), 2408. 10.3390/s2107240833807359

Nderba, W. (2017). *Mint Capstone Project: Design and implementation of QoS aware Software Defined.* Research Gate.

Nedjmedine, O., & Tahar, M. (2022). Analysis of road accident factors using Decision Tree Algorithm: a case of study Algeria. 2022 5th International Symposium on Informatics and its Applications (ISIA), M'sila, Algeria. 10.1109/ISIA55826.2022.9993530

Neel, J. (2007). Synthetic symmetry in cognitive radio networks. *Proc. Forum Tech. Conf. Product Expo. SDR*, Denver.

Neel, J., & Amanna, A. (2009). Introduction to cognitive radio technology for transportation systems. *Proc. SDR Forum Workshop Smart Communication Transport Systems.*, Houston, TX.

Ngai, E. W. T., Hu, Y., Wong, Y. H., Chen, Y., & Sun, X. (2011). The application of data mining techniques in financial fraud detection: A classification framework and an academic review of literature. *Decision Support Systems*, 50(3), 559–569. 10.1016/j.dss.2010.08.006

Nguyen Thanh, N., Ciblat, P., Pham, A. T., & Nguyen, V.-T. (2015). Surveillance Strategies Against Primary User Emulation Attack in Cognitive Radio Networks. *IEEE Transactions on Wireless Communications*, 14(9), 4981–4993. 10.1109/TWC.2015.2430865

Nie, D., Yu, W., Ni, Q., Pervaiz, H., & Min, G. (2023). Cluster control and energy consumption minimization for cooperative prediction based spectrum sensing in cognitive radio networks. *IEEE Transactions on Communications*, 71(9), 5580–5594. 10.1109/TCOMM.2023.3287531

Ning, W., Huang, X., Yang, K., Wu, F., & Leng, S. (2020). Reinforcement learning enabled cooperative spectrum sensing in cognitive radio networks. *Journal of Communications and Networks (Seoul)*, 22(1), 12–22. 10.1109/JCN.2019.000052

Ning, Z., Yu, Y., Song, Q., Peng, Y., & Zhang, B. (2015). Interference aware spectrum sensing mechanisms in cognitive radio networks. *Computers & Electrical Engineering*, 42, 193–206. 10.1016/j.compeleceng.2014.10.011

Niyato, D., & Hossain, E. (2008, December). Spectrum trading in cognitive radio networks: A market-equilibrium-based approach. *IEEE Wireless Communications*, 15(6), 71–80. 10.1109/MWC.2008.4749750

Noh, J. H., & Oh, S. J. (2013). Cognitive radio channel with cooperative multi-antenna secondary systems. *IEEE Journal on Selected Areas in Communications*, 32(3), 539–549. 10.1109/JSAC.2014.140314

Nowak, M., Nowak, S., & Domanska, J. (2019). Cognitive routing for improvement of IoT security. *Proceedings - 2019 IEEE International Conference on Fog Computing*. IEEE. 10.1109/ICFC.2019.00014

Nurelmadina, N., Hasan, M. K., Memon, I., Saeed, R. A., Zainol Ariffin, K. A., Ali, E. S., Mokhtar, R. A., Islam, S., Hossain, E., & Hassan, M. A. (2021). A systematic review on cognitive radio in low power wide area network for industrial IoT applications. *Sustainability (Basel)*, 13(1), 338. 10.3390/su13010338

Obite, F., Usman, A. D., & Okafor, E. (2021). An overview of deep reinforcement learning for spectrum sensing in cognitive radio networks. *Digital Signal Processing*, 113, 103014. 10.1016/j.dsp.2021.103014

Orcay, O., & Ustundag, B. (2008). *Pattern recognition in cognitive communication.* In 23rd ISCIS, Istanbul, Turkey.

OSSIE. (2023). *Wireless @ Virginia Tech.* Open Source SCA Implementation—Embedded. http://wireless.vt.edu/

Ozcan, G., Gursoy, M. C., & Tang, J. (2017). Spectral and energy efficiency in cognitive radio systems with unslotted primary users and sensing uncertainty. *IEEE Transactions on Communications*, 65(10), 4138–4151. 10.1109/TCOMM.2017.2691342

Pal, K. & Campbell, J. A. (1997). An Application of Rule-Based and Case-Based Reasoning within a Single Legal Knowledge Based System. *Data Base for Advances in Information Systems, 28*(4), 48-63.

Pal, K. (1999). *An approach to legal reasoning based on a hybrid decision-support system.* The University College London, Gower Street, London.

Pal, K. (2023). Influence of AI on Cognitive Radio Applications. K. Kumar (Ed.), *Applications of Artificial Intelligence in Wireless Communication Systems*. IGI Global Publication.

Pal, K., & Karakostas, B. (2014). *Multi Agent-Based Service Framework for Supply Chain Management.* In 5th International Conference on Ambient Systems, Networks and Technologies (ANT-2014), Belgium.

Pal, K. (2018). Ontology-Based Web Service Architecture for Retail Supply Chain Management. In *Proceedings of the 9th International Conference on Ambient Systems, Networks and Technologies*, Porto, Portugal. 10.1016/j.procs.2018.04.101

Pal, K. (2022a). Cryptography and Blockchain Solutions for Security Protection of Internet of Things Applications. In Sahoo, B., & Yadav, S. (Eds.), *Information Security Practices of the Internet of Things, 5G, and Next-Generation Wireless Networks*. IGI Global Publication. 10.4018/978-1-6684-3921-0.ch008

Pal, K. (2022b). Security Issues and Solutions for Resource-Constrained IoT Applications Using Lightweight Cryptography. In Verma, S., Vyas, V., & Kaushik, K. (Eds.), *Cybersecurity Issues, Challenges, and Solutions in the Business World*. IGI Global Publication. 10.4018/978-1-6684-5827-3.ch010

Pal, K., & Palmer, O. (2000). A decision-support system for business acquisitions. *Decision Support Systems*, 27(4), 411–429. 10.1016/S0167-9236(99)00083-4

Parra, C., Ponce, C., & Rodrigo, S. F. (2020). *Evaluating the Performance of Explainable Machine Learning Models in Traffic Accidents Prediction in California.* 2020 39th International Conference of the Chilean Computer Science Society (SCCC), Coquimbo, Chile. 10.1109/SCCC51225.2020.9281196

Pati, B. M., Kaneko, M., & Taparugssanagorn, A. (2020). A deep convolutional neural network based transfer learning method for non-cooperative spectrum sensing. *IEEE Access : Practical Innovations, Open Solutions*, 8, 164529–164545. 10.1109/ACCESS.2020.3022513

Pattanayak, S., Venkateswaran, P., & Nandi, R. (2013). Artificial intelligence-based model for channel status prediction: A new spectrum sensing technique for cognitive radio. *International Journal of Communications, Network and Systems Sciences*, 6(3), 139–148. 10.4236/ijcns.2013.63017

Pattnaik, S. S., Bakwad, K. M., Devi, S., Panigrahi, B. K., & Das, S. (2011). Parallel bacterial foraging optimization. *Handbook of Swarm Intelligence, 8*, 487-502.

Pei, Y., Liang, Y. C., Teh, K. C., & Li, K. H. (2011). Energy-efficient design of sequential channel sensing in cognitive radio networks: Optimal sensing strategy, power allocation, and sensing order. *IEEE Journal on Selected Areas in Communications*, 29(8), 1648–1659. 10.1109/JSAC.2011.110914

Peng, M., Li, Y., Zhao, Z., & Wang, C. (2015). System architecture and key technologies for 5G heterogeneous cloud radio access networks. *IEEE Network*, 29(2), 6–14. 10.1109/MNET.2015.7064897

Phillip Ian, W., & John, F. (2006). Facial feature detection using Haar classifiers. *Journal of Computing Sciences in Colleges*, 21, 127–133.

Pincott, J., Tien, P. W., Wei, S., & Calautit, J. K. (2022). Indoor fire detection utilizing computer vision-based strategies. *Journal of Building Engineering*, 61, 105154. 10.1016/j.jobe.2022.105154

Popli, R. R. (2020). A wormhole attack detection in the mobile ad-hoc network using GA and SVM. *International Journal of Engineering Applied Sciences and Technology*, 5(3), 582–588.

Pradana, M. R., Syafrullah, M., Irawan, H., Irawan, J., Chandra, C., & Solichin, A. (2022).Market Basket Analysis Using FP-Growth Algorithm On Retail Sales Data. In *proceedings of 2022 9th International Conference on Electrical Engineering, Computer Science and Informatics (EECSI)*. Research Gate.

Pradhan, P. M., & Panda, G. (2013). Cooperative spectrum sensing in cognitive radio network using multiobjective evolutionary algorithms and fuzzy decision making. *Ad Hoc Networks*, 11(3), 1022–1036. 10.1016/j.adhoc.2012.11.007

Pradhan, P. M., & Panda, G. (2014). Comparative performance analysis of evolutionary algorithm based parameter optimization in cognitive radio engine: A survey. *Ad Hoc Networks*, 17, 129–146. 10.1016/j.adhoc.2014.01.010

Prasad & Tripathi. (n.d.) Wormhole attack detection in adhoc network using machine learning technique. *Communication and Networking Technologies*. IEEE. .10.1109/ICCCNT45670.2019.8944634

Premsankar, G., Di Francesco, M., & Taleb, T. (2018). Edge computing for the Internet of Things: A case study. *IEEE Internet of Things Journal*, 5(2), 1275–1284. 10.1109/JIOT.2018.2805263

Puspitasari, A. A., & Lee, B. M. (2023). A survey on reinforcement learning for reconfigurable intelligent surfaces in wireless communications. *Sensors (Basel)*, 23(5), 2554. 10.3390/s2305255436904758

Qadir, J. (2013). Artificial intelligence based cognitive radio networks routing for cognitive radio networks. *IEEE Communications Surveys and Tutorials*, 2013.

Qineti, Q. (2007). Cognitive Radio Technology a Study for Ofcom—Volume 1, Feb. 12, 2007. [Online]. Available: https://www.ofcom.org.uk/research/technology/research/emer_tech/cograd/

Radoyska, P., & Topalova, I. (2023). A conceptual model for open u-learning platform. *Annals of DAAAM & Proceedings, 34* .

Rajkumari, R., & Marchang, N. (2018). Secure Non-Consensus Based Spectrum Sensing in Non-Centralized Cognitive Radio Networks. *IEEE Sensors Journal*, 18(9), 3883–3890. 10.1109/JSEN.2018.2816595

Ramkumar, B. (2009, June). Automatic modulation classification for cognitive radios using cyclic feature detection. *IEEE Circuits and Systems Magazine*, 9(2), 27–45. 10.1109/MCAS.2008.931739

Ramli, A., & Grace, D. (2010). RF signal Strength based clustering protocols for a self-organizing cognitive radio network. *2010 7th International Symposium on Wireless Communication Systems.*

Ravinder, A., Aakarsha, C., Kohlia, S., & Pratyush Ahujaa, S. G. (2019). *The Impact of Features Extraction on the Sentiment Analysis Noida, India*. Semantic Scholar.

Ray, P. P., & Kumar, N. (2021). SDN/NFV architectures for edge-cloud oriented IoT: A systematic review. *Computer Communications*, 169, 129–153. 10.1016/j.comcom.2021.01.018

Reed, J. H. (2005). *Development of a Cognitive Engine and Analysis of WRAN Cognitive Radio Algorithms—Phase I.* Wireless @ Virginia Tech, Virginia Polytech. Inst. State Univ.

Renault, L., Rodrigues, L., & Pinto, E. L. (2017). HMM models and estimation algorithms for real-time predictive spectrum sensing and cognitive usage. *Proceeding of xxxv simposio brasileiro de telecommunicators Procesamiento de sinais*. Sociedade Brasileria de Telecomunicaoes.

Ren, P., Wang, Y., Du, Q., & Xu, J. (2012). A survey on dynamic spectrum access protocols for distributed cognitive wireless networks. *EURASIP Journal on Wireless Communications and Networking*, 2012(1), 1–21. 10.1186/1687-1499-2012-60

Rondeau, T. W., Le, B., Rieser, C. J., & Bostian, C. W. (2004). *Cognitive radios with genetic algorithms: Intelligent control of software defined radios*. Forum Tech. Conf. SDR, Phoenix, AZ.

Rosen, J. B. (1965). Existence and uniqueness of equilibrium points for concave n-person games. *Econometrica*, 33(3), 520–534. 10.2307/1911749

Roy, P. P., & Muralidhar, D. M. (2015). Hidden Markov Model based channel state prediction in cognitive radio networks. *International Journal of Engineering Research & Technology (Ahmedabad)*, 4(2), 391–394.

Saifuddin, K. M., Reza, K. F., Fahim, M. A. N. I., & Alam, S. S. (2017). Detection of primary user emulation attack in cognitive radio environment. *2017 6th International Conference on Informatics, Electronics and Vision & 2017 7th International Symposium in Computational Medical and Health Technology (ICIEV-ISCMHT)*. IEEE.

Sajjad, M., Nasir, M., Muhammad, K., Khan, S., Jan, Z., Sangaiah, A. K., Elhoseny, M., & Baik, S. W. (2017). Raspberry Pi assisted face recognition framework for enhanced law-enforcement services in smart cities. *Future Generation Computer Systems*. Advance online publication. 10.1016/j.future.2017.11.013

Salahdine, F., & Kaabouch, N. (2020). Security threats, detection, and countermeasures for physical layer in cognitive radio networks: A survey. *Physical Communication*, 39, 101001. 10.1016/j.phycom.2020.101001

Salari, S., & Chan, F. (2023). Maximizing the sum-rate of secondary cognitive radio networks by jointly optimizing beamforming and energy harvesting time. *IEEE Transactions on Vehicular Technology*, 72(6), 8128–8133. 10.1109/TVT.2023.3238347

Salau, H. B., Aibinu, A. M., Onwuka, E. N., Onumanyi, A. J., & Bima, M. E. M. E. (2015). *Applications of artificial Intelligence techniques in cognitive radios: a survey*. Research Gate. https://www.researchgate.net/publication/307935017

Saleem, Y., & Rehmani, M. H. (2014). Primary radio user activity models for cognitive radio networks: A survey. *Journal of Network and Computer Applications*, 43, 1–16. 10.1016/j.jnca.2014.04.001

Saleem, Y., Yau, K.-L. A., Mohamad, H., Ramli, N., Rehmani, M. H., & Ni, Q. (2017). Clustering and Reinforcement-Learning-Based Routing for Cognitive Radio Networks. *IEEE Wireless Communications*, 24(4), 146–151. 10.1109/MWC.2017.1600117

Salman, R. E., & Alzaatreh, A. (2022). Market Basket Analysis of Chicago Road Accidents. In *proceedings of 2022 Advances in Science and Engineering Technology International Conferences (ASET)*. IEEE.10.1109/ASET53988.2022.9734867

Sanjeetha, R. & Raj, A. (2021). Detection and Mitigation of botnet-based DDoS attacks using catboost machine learning algorithm in SDN environment. *International journal of advanced Technology and Engineering Exploration.*

Sari, I. P., Al-Khowarizmi, & Batubara, I. H. (2021). Optimization of the FP-Growth Algorithm in Data Mining Techniques to Get the Electric Power Theft Pattern for the Development of Smart City. In *proceedings of 2021 4th International Conference of Computer and Informatics Engineering (IC2IE)* IEEE.

Sayama, H., Askar, S. S., & Alshamrani, A. M. (2024). Cournot-Bertrand Duopoly Model: Dynamic Analysis Based on a Computed Cost. *Complexity*, 5594918, 1–15. 10.1155/2024/5594918

SDR Forums. (2023). *Software Defined Radio Forum, Cognitive Radio Work Group*. SDR Forums. http://www.sdrforum.org/pages/aboutTheForum/currentProjects.asp

Shah, M. A., Zhang, S., & Maple, C. (2013, September). Cognitive radio networks for Internet of Things: Applications, challenges and future. In *2013 19th International Conference on Automation and Computing* (pp. 1-6). IEEE.

Sharma, P., & Abrol, V. (2013). Optimized cluster head selection & rotation for cooperative spectrum sensing in Cognitive Radio Networks. *2013 Tenth International Conference on Wireless and Optical Communications Networks (WOCN)*. IEEE. 10.1109/WOCN.2013.6616260

Sharma, S., & Nag, A. (2023). Cognitive Software Defined Networking and Network Function Virtualization and Applications. *Future Internet*, 15(2), 78. 10.3390/fi15020078

Sheng, T.J., Islam, M.S., Misran, N., Baharuddin, M.H., Arshad, H., Islam, M.R., Chowdhury, M.E., Rmili, H., & Islam, M.T. (2020). An internet of things based smart waste management system using LoRa and tensorflow deep learning model. *IEEE Access*. IEEE.

Shi, J., Wang, Y., & Zheng, W. (2020). *Correlation Analysis of Causes of Railway Accidents Based on Improved Apriori Algorithm.* 2020 13th International Symposium on Computational Intelligence and Design (ISCID), Hangzhou, China. 10.1109/ISCID51228.2020.00067

Shi, Y., & Eberhart, R. (1998, May). A modified particle swarm optimizer. In *1998 IEEE international conference on evolutionary computation proceedings. IEEE world congress on computational intelligence (Cat. No. 98TH8360)* (pp. 69-73). IEEE. 10.1109/ICEC.1998.699146

Shin, K. G., Kim, H., Min, A. W., & Kumar, A. (2010). Cognitive radios for dynamic spectrum access: From concept to reality. *IEEE Wireless Communications*, 17(6), 64–74. 10.1109/MWC.2010.5675780

Shitharth, S., Yonbawi, S., Manoharan, H., Alahmari, S., Yafoz, A., & Mujlid, H. (2023). Physical Stint Virtual Representation of Biomedical Signals With Wireless Sensors Using Swarm Intelligence Optimization Algorithm. *IEEE Sensors Journal*, 23(4), 3870–3877. 10.1109/JSEN.2022.3233407

Shi, Z., Gao, W., Zhang, S., Liu, J., & Kato, N. (2019). AI-enhanced cooperative spectrum sensing for non-orthogonal multiple access. *IEEE Wireless Communications*, 27(2), 173–179. 10.1109/MNET.001.1900305

Si, J., Harris, S. L., & Yfantis, E. (2018). *A Dynamic ReLU on Neural Network.* 2018 IEEE 13th Dallas Circuits and Systems Conference (DCAS), Dallas, TX, USA. 10.1109/DCAS.2018.8620116

Silva, C., & Saraee, M. (2019). Predicting Road Traffic Accident Severity using Decision Trees and Time-Series Calendar Heatmaps. *2019 IEEE Conference on Sustainable Utilization and Development in Engineering and Technologies (CSUDET)*, Penang, Malaysia. 10.1109/CSUDET47057.2019.9214709

Somov, A., Dupont, C., & Giaffreda, R. (2013, July). Supporting smart-city mobility with cognitive Internet of Things. In *2013 Future Network & Mobile Summit* (pp. 1-10). IEEE.

Spooner, C. M., & Mody, A. N. (2018, June). Wideband cyclostationary signal processing using sparse subsets of narrowband subchannels. *IEEE Transactions on Cognitive Communications and Networking*, 4(2), 162–176. 10.1109/TCCN.2018.2790971

Sridhar, P., Parameswaran, L., & Thangavel, S. K. (2020). An efficient rule based algorithm for fire detection on real time videos. *Journal of Computational and Theoretical Nanoscience*, 17(1), 308–315. 10.1166/jctn.2020.8667

Srivastava, V., & Pandey, R. S. (2021). Machine intelligence approach: To solve load balancing problem with high quality of service performance for multi-controller based Software Defined Network. *Sustainable Computing : Informatics and Systems*, 30, 100511. 10.1016/j.suscom.2021.100511

Sun, G., Liu, G., & Wang, Y. (2014). SDN architecture for cognitive radio networks. *2014 1st International Workshop on Cognitive Cellular Systems*. CCS. 10.1109/CCS.2014.6933795

Sun, Q., Miao, X., Guan, Z., Wang, J., & Gao, D. (2021). Spoofing Attack Detection Using Machine Learning in Cross-Technology Communication. *Security and Communication Networks*, *12*. 10.1155/2021/3314595

Suppala, K., & Rao, N. (2019). Sentiment Analysis Using Naïve Bayes Classifier. *International Journal of Innovative Technology and Exploring Engineering (IJITEE)*.

Swathi, T., & Begum, M. B. (2018). Maximum Connected Load Balancing Cover Tree Algorithm for Wireless Sensor Network. *Journal of Switching Hub*, 3(3), 1–5.

Tabaković, Z. (2011). *A survey of cognitive radio systems*. Semantic Scholar. https://pdfs.semanticscholar.org/19c8/471e97301e408c3e3fe08eb8f5027ef9eed6.pdf

Tahseen, M. (2011). *Analysis of OSTBC in Cooperative Cognitive Radio Networks using 2-hop DF Relaying Protocol - Arkiv EX*. Blekinge Tekniska Högskola.

Thamarai, M. & Vankamamidi, S. (2023). Smart self-power generating Garbage Management System using Deep Learning for Smart Cities. *Microprocessors and Microsystems*. Elsevier.

Thomas, R. (2007). *Cognitive networks*. [Ph.D. dissertation, Virginia Polytech. Inst.]

Tian, Z., Tafesse, Y., & Sadler, B. M. (2012, February). Cyclic feature detection with sub-Nyquist sampling for wideband spectrum sensing. *IEEE Journal of Selected Topics in Signal Processing*, 6(1), 58–69. 10.1109/JSTSP.2011.2181940

Toreyin, B. U., Dedeoğlu, Y., Güdükbay, U., & Çetin, A. E. (2006). Computer vision based method for real-time fire and flame detection. *Pattern Recognition Letters*, 27(1), 49–58. 10.1016/j.patrec.2005.06.015

Tuberquia, M., & Hernandez, C. (2018). New approaches in cognitive radios using evolutionary algorithms. *Iranian Journal of Electrical and Computer Engineering*, 8(3), 1636. 10.11591/ijece.v8i3.pp1636-1646

Turkyilmaz, Y., Senturk, A., & Bayrakdar, M. E. (2023). Employing machine learning based malicious signal detection for cognitive radio networks. *Concurrency and Computation*, 35(2), e7457. 10.1002/cpe.7457

Ukil, A., Braendle, H., & Krippner, P. (2011). Distributed temperature sensing: Review of technology and applications. *IEEE Sensors Journal*, 12(5), 885–892. 10.1109/JSEN.2011.2162060

Ullah, F., Wang, J., Farhan, M., Jabbar, S., Naseer, M. K., & Asif, M. (2020). Lsa based smart assessment methodology for sdn infrastructure in iot environment. *International Journal of Parallel Programming*, 48(2), 162–177. 10.1007/s10766-018-0570-1

Ustundag, B., & Orcay, O. (2008). *Pattern based encoding for cognitive communication*. 3rd International Conference CrownCom, Singapore.

Varatharajana, B., Praveen, E., & Vinoth, E. (2012). Neural network aided enhanced spectrum sensing in cognitive radio. *Procedia Engineering*, 38, 82–88. 10.1016/j.proeng.2012.06.012

Vinod, S., & Bhupendra, S., & Singh. (2013). Face detection by haar cascade classifier with simple and complex backgrounds images using opencv implementation. *International Journal of Advanced Technology in Engineering and Science*, 1, 33–38.

Vyas, M. R., Patel, D. K., & Benitez, M. L. (2017). Artificial neural network based hybrid spectrum sensing scheme for cognitive radio. *Proceeding of IEEE 28th Annual International Symposium on Personal, Indoor, and Mobile Radio Communications (PIMRC)*, (pp. 8-13). IEEE.

Wadhwa, S., Babbar, H., & Rani, S. (2021). A survey on emerging software-defined networking and blockchain in smart health care. *IOP Conference Series. Materials Science and Engineering*, 1022(1), 012056. 10.1088/1757-899X/1022/1/012056

Wang, L., & Delaney, D. T. (2019). QoE oriented cognitive network based on machine learning and SDN. *2019 IEEE 11th International Conference on Communication Software and Networks*. IEEE. 10.1109/ICCSN.2019.8905343

Wang, Y., Zheng, W., Dong, H., & Gao, P. (2020). *Factors correlation mining on railway accidents using association rule learning algorithm.* 2020 IEEE 23rd International Conference on Intelligent Transportation Systems (ITSC), Rhodes, Greece. 10.1109/ITSC45102.2020.9294317

Wang, B., Wu, Y., & Liu, K. J. (2010). Game theory for cognitive radio networks an overview. *Computer Networks*, 54(14), 2537–2561. 10.1016/j.comnet.2010.04.004

Wang, C.-X., You, X., Gao, X., Zhu, X., Li, Z., Zhang, C., Wang, H., Huang, Y., Chen, Y., Haas, H., Thompson, J. S., Larsson, E. G., Renzo, M. D., Tong, W., Zhu, P., Shen, X., Poor, H. V., & Hanzo, L. (2023). On the road to 6g: Visions, requirements, key technologies and testbeds. *IEEE Communications Surveys and Tutorials*, 25(2), 905–974. 10.1109/COMST.2023.3249835

Wang, J. Q., Qu, C., Ran, X., Liu, C., & Chen, B. (2021). *Smart Municipal Waste Management System, Based on Deep Learning and IoT, Waste Management, Elsevier*, 135(November), 20–29.

Wang, N., Liu, Y., Ma, L., Yang, Y., & Wang, H. (2023). Automatic modulation classification based on cnn and multiple kernel maximum mean discrepancy. *Electronics (Basel)*, 12(1).

Wang, R., Tao, M., & Liu, Y. (2012). Optimal linear transceiver designs for cognitive two-way relay networks. *IEEE Transactions on Signal Processing*, 61(4), 992–1005. 10.1109/TSP.2012.2226452

Wang, S., Wang, Y., Coon, J. P., & Doufexi, A. (2011). Energy-efficient spectrum sensing and access for cognitive radio networks. *IEEE Transactions on Vehicular Technology*, 61(2), 906–912. 10.1109/TVT.2011.2180549

Weiss, M. B., Werbach, K., Sicker, D. C., & Bastidas, C. E. C. (2019). On the application of blockchains to spectrum management. *IEEE Transactions on Cognitive Communications and Networking*, 5(2), 193–205. 10.1109/TCCN.2019.2914052

Wei, Z., Tan, W., Elsadany, A. A., & Moroz, I. (2023). Complexity and chaos control in a Cournot duopoly model based on bounded rationality and relative profit maximization. *Nonlinear Dynamics*, 111(18), 17561–17589. 10.1007/s11071-023-08782-3

Wen-Xiu. X., Heng-Nian, Q., & Mei-Li, H. (2010). Market Basket Analysis Based on Text Segmentation and Association Rule Mining. In *proceedings of 2010 First International Conference on Networking and Distributed Computing*. IEEE.

Wilson, T., Wiebe, J., & Hoffmann, P. (2005). *Recognizing contextual polarity in phrase-level sentiment analysis.* Proc. Conf. Human Lang. Tech. and Empirical Methods in Natural Lang. Process., Vancouver, B.C., Canada. 10.3115/1220575.1220619

Wu, C., Chowdhury, K. R., Di Felice, M., & Meleis, W. (2010, May). Spectrum management of cognitive radio using multi-agent reinforcement learning. In *AAMAS* (pp. 1705-1712).

Wu, H., Wu, D., & Zhao, J. (2019). An intelligent fire detection approach through cameras based on computer vision methods. *Process Safety and Environmental Protection*, 127, 245–256. 10.1016/j.psep.2019.05.016

Wulandari, A., Sultan, A., Purbo, Y. S., Bainafila, G. F., Tahyudin, I., & Sholikhatin, S. A. (2022). Analysis of Customer Expense using Market Basket Analysis Model with Hash-Based Algorithm and Association Rules. In *proceedings of 2022 1st International Conference on Smart Technology, Applied Informatics, and Engineering (APICS)*. IEEE. 10.1109/APICS56469.2022.9918742

Wunsch, F., Paisana, F., Rajendran, S., Selim, A., Alvarez, P., Muller, S., Koslowski, S., Van den Bergh, B., & Pollin, S. (2017). DySPAN Spectrum Challenge: Situational Awareness and Opportunistic Spectrum Access Benchmarked. *IEEE Transactions on Cognitive Communications and Networking*, 3(3), 550–562. 10.1109/TCCN.2017.2745682

Wu, Y. J., Hwang, P. C., Hwang, W. S., & Cheng, M. H. (2020). Artificial intelligence enabled routing in software defined networking. *Applied Sciences (Basel, Switzerland)*, 10(18), 6564. 10.3390/app10186564

Xia, W., Wen, Y., Foh, C. H., Niyato, D., & Xie, H. (2015). A Survey on Software-Defined Networking. *IEEE Communications Surveys and Tutorials*, 17(1), 27–51. 10.1109/COMST.2014.2330903

Xia, Z., Zhang, S., Tian, X., & Liu, Y. (2021). Understanding waste sorting behavior and key influencing factors through internet of things: Evidence from college student community. *Resources, Conservation and Recycling*, 174, 105775. https://www.sciencedirect .com/science/article/pii/S0921344921003840. 10.1016/j.resconrec.2021.105775

Xie, L., & Wan, Q. (2017, June). Cyclic feature-based modulation recognition using compressive sensing. *IEEE Wireless Communications Letters*, 6(3), 402–405. 10.1109/LWC.2017.2697853

Xing, Y., Chandramouli, R., & Mangold, S. (2006). Dynamic spectrum access in open spectrum wireless networks. *IEEE Journal on Selected Areas in Communications*, 24(3), 626–637. 10.1109/JSAC.2005.862415

Xu, H., & Huang, G. (2021). *Analysis on Influencing Factors of accidents at national and provincial road intersections based on random forest.* 2021 6th International Conference on Transportation Information and Safety (ICTIS), Wuhan, China. 10.1109/ICTIS54573.2021.9798574

Yadav, Ankita, & Anjum. (2020). Accident Detection Using Deep Learning. 2nd International Conference on Advances in Computing, Communication Control and Networking (ICACCCN), Greater Noida, India.

Yang, A., Cai, Y., & Xu, Y. (2007). A fuzzy collaborative spectrum sensing scheme in cognitive radio. *Proceeding of International Symposium on Intelligent Signal Processing and Communication Systems (ISPACS).* Research Gate.

Yang, C. (2014). *Context-aware learning for sentence-level sentiment analysis with posterior regularization.* Proc. 52nd Annu. Meeting on Assoc. for Computational Linguistics, Baltimore, MD.

Yang, G., Ho, C. K., & Guan, Y. L. (2014). Dynamic resource allocation for multiple-antenna wireless power transfer. *IEEE Transactions on Signal Processing*, 62(14), 3565–3577. 10.1109/TSP.2014.2328980

Yao, N., Liu, D., Chen, B., Hao, X., & Yin, R. (2023). Improved game algorithm for spectrum resource optimization with variable number of channels. *Ad Hoc Networks*, 146, 103190. 10.1016/j.adhoc.2023.103190

Yar, H., Hussain, T., Khan, Z. A., Koundal, D., Lee, M. Y., & Baik, S. W. (2021). Vision sensor-based real-time fire detection in resource-constrained IoT environments. *Computational Intelligence and Neuroscience*, 2021, 1–15. 10.1155/2021/519550834970311

Yazdinejad, A., Parizi, R. M., Dehghantanha, A., Zhang, Q., & Choo, K.-K. R. (2020). An energy-efficient sdn controller architecture for iot networks with blockchain-based security. *IEEE Transactions on Services Computing*, 13(4), 625–638. 10.1109/TSC.2020.2966970

Yongmei, L., & Yong, G. (2009). Application in Market Basket Research Based on FP-Growth Algorithm. In *proceedings of 2009 WRI World Congress on Computer Science and Information Engineering*. IEEE. 10.1109/CSIE.2009.1073

Yu, G., Zheng, W., Wang, L., & Zhang, Z. (2018). *Identification of Significant Factors Contributing to Multi-attribute Railway Accidents Dataset (MARA-D) Using SOM Data Mining.* 2018 21st International Conference on Intelligent Transportation Systems (ITSC), Maui, HI. 10.1109/ITSC.2018.8569336

Yuan, Z., Niyato, D., Li, H., Song, J. B., & Han, Z. (2012). Defeating Primary User Emulation Attacks Using Belief Propagation in Cognitive Radio Networks. *IEEE Journal on Selected Areas in Communications*, 30(10), 1850–1860. 10.1109/JSAC.2012.121102

Yucek T. & Arslan, H. (2009). A survey of spectrum sensing algorithms for cognitive radio applications. *Communications Surveys Tutorials.* IEEE.

Yucek, T., & Arslan, H. (2009). A survey of spectrum sensing algorithms for cognitive radio applications. *IEEE Communications Surveys and Tutorials*, 11(1), 116–130. 10.1109/SURV.2009.090109

Yu, H., & Zikria, Y. B. (2020). Cognitive radio networks for internet of things and wireless sensor networks. *Sensors (Basel)*, 20(18), 5288. 10.3390/s2018528832947832

Zeadally, S. (2024). *An intelligent SDN-IoT enabled intrusion detection system for healthcare systems using a hybrid deep learning and machine learning approach*. IEEE. 10.23919/JCC.ja.2022-0681

Zhang, D., & Zhai, X. (2011). SVM-based spectrum sensing in cognitive radio. *Proceeding of the 7th International Conference on Wireless Communications, Networking and Mobile Computing*. Research Gate.

Zhang, H., & Wang, X. (2011). *A Fuzzy Decision Scheme for Cooperative Spectrum Sensing in Cognitive Radio*. Vehicular Technology Conference (VTC Spring). 10.1109/VETECS.2011.5956116

Zhang, Z., & Xie, X. (2007). Intelligent cognitive radio: Research on learning and evaluation of CR based on neural network, In proceedings *ITI 5th ICICT*, Dhaka, Bangladesh, December 16–18, 2007, 33–37.

Zhang, J., Fan, W., Yao, F.-Q., Zhao, H., & Li, Y. (2010). Cluster-based distributed topology management in Cognitive Radio Ad Hoc networks. *2010 International Conference on Computer Application and System Modeling (ICCASM 2010)*. IEEE.

Zhang, J., Zheng, F. C., Gao, X. Q., & Zhu, H. B. (2014). Sensing-energy efficiency tradeoff for cognitive radio networks. *IET Communications*, 8(18), 3414–3423. 10.1049/iet-com.2014.0400

Zhang, L., Xiao, M., Wu, G., Li, S., & Liang, Y. C. (2016). Energy-efficient cognitive transmission with imperfect spectrum sensing. *IEEE Journal on Selected Areas in Communications*, 34(5), 1320–1335. 10.1109/JSAC.2016.2520166

Zhang, W., Wang, C. X., Chen, D., & Xiong, H. (2015). Energy-spectral efficiency tradeoff in cognitive radio networks. *IEEE Transactions on Vehicular Technology*, 65(4), 2208–2218. 10.1109/TVT.2015.2432073

Zheng, G., Krikidis, I., & Ottersten, B. (2013). Full-duplex cooperative cognitive radio with transmit imperfections. *IEEE Transactions on Wireless Communications*, 12(5), 2498–2511. 10.1109/TWC.2013.032513.121464

Zhou, Q. (2024). Smart library architecture based on internet of things (iot) and software defined networking (sdn). *Heliyon*, 10(3), e25375. 10.1016/j.heliyon.2024.e2537538371976

Zhu, X. L., Liu, Y. A., Weng, W. W., & Yuan, D. M. (2008). *Channel sensing algorithm based on neural networks for cognitive wireless mesh networks*. 4th Int. Conf. WiCOM, Dalian, China.

Zikria, Y. B., Ali, R., Afzal, M. K., & Kim, S. W. (2021). Next-generation internet of things (iot): Opportunities, challenges, and solutions. *Sensors (Basel)*, 21(4), 1174. 10.3390/s2104117433562343

Zormati, M. A., Lakhlef, H., & Ouni, S. (2024). Review and analysis of recent advances in intelligent network softwarization for the internet of things. *Computer Networks*, 241, 110215. 10.1016/j.comnet.2024.110215

Related References

To continue our tradition of advancing academic research, we have compiled a list of recommended IGI Global readings. These references will provide additional information and guidance to further enrich your knowledge and assist you with your own research and future publications.

Abbasnejad, B., Moeinzadeh, S., Ahankoob, A., & Wong, P. S. (2021). The Role of Collaboration in the Implementation of BIM-Enabled Projects. In Underwood, J., & Shelbourn, M. (Eds.), *Handbook of Research on Driving Transformational Change in the Digital Built Environment* (pp. 27–62). IGI Global. https://doi.org/10.4018/978-1-7998-6600-8.ch002

Abdulrahman, K. O., Mahamood, R. M., & Akinlabi, E. T. (2022). Additive Manufacturing (AM): Processing Technique for Lightweight Alloys and Composite Material. In K. Kumar, B. Babu, & J. Davim (Ed.), *Handbook of Research on Advancements in the Processing, Characterization, and Application of Lightweight Materials* (pp. 27-48). IGI Global. https://doi.org/10.4018/978-1-7998-7864-3.ch002

Agrawal, R., Sharma, P., & Saxena, A. (2021). A Diamond Cut Leather Substrate Antenna for BAN (Body Area Network) Application. In Singh, V., Dubey, V., Saxena, A., Tiwari, R., & Sharma, H. (Eds.), *Emerging Materials and Advanced Designs for Wearable Antennas* (pp. 54–59). IGI Global. https://doi.org/10.4018/978-1-7998-7611-3.ch004

Ahmad, F., Al-Ammar, E. A., & Alsaidan, I. (2022). Battery Swapping Station: A Potential Solution to Address the Limitations of EV Charging Infrastructure. In Alam, M., Pillai, R., & Murugesan, N. (Eds.), *Developing Charging Infrastructure and Technologies for Electric Vehicles* (pp. 195–207). IGI Global. 10.4018/978-1-7998-6858-3.ch010

Aikhuele, D. (2018). A Study of Product Development Engineering and Design Reliability Concerns. *International Journal of Applied Industrial Engineering*, 5(1), 79–89. 10.4018/IJAIE.2018010105

Al-Khatri, H., & Al-Atrash, F. (2021). Occupants' Habits and Natural Ventilation in a Hot Arid Climate. In González-Lezcano, R. (Ed.), *Advancements in Sustainable Architecture and Energy Efficiency* (pp. 146–168). IGI Global. https://doi.org/10.4018/978-1-7998-7023-4.ch007

Al-Shebeeb, O. A., Rangaswamy, S., Gopalakrishan, B., & Devaru, D. G. (2017). Evaluation and Indexing of Process Plans Based on Electrical Demand and Energy Consumption. *International Journal of Manufacturing, Materials, and Mechanical Engineering*, 7(3), 1–19. 10.4018/IJMMME.2017070101

Amuda, M. O., Lawal, T. F., & Akinlabi, E. T. (2017). Research Progress on Rheological Behavior of AA7075 Aluminum Alloy During Hot Deformation. *International Journal of Materials Forming and Machining Processes*, 4(1), 53–96. 10.4018/IJMFMP.2017010104

Amuda, M. O., Lawal, T. F., & Mridha, S. (2021). Microstructure and Mechanical Properties of Silicon Carbide-Treated Ferritic Stainless Steel Welds. In Burstein, L. (Ed.), *Handbook of Research on Advancements in Manufacturing, Materials, and Mechanical Engineering* (pp. 395–411). IGI Global. https://doi.org/10.4018/978-1-7998-4939-1.ch019

Anikeev, V., Gasem, K. A., & Fan, M. (2021). Application of Supercritical Technologies in Clean Energy Production: A Review. In Chen, L. (Ed.), *Handbook of Research on Advancements in Supercritical Fluids Applications for Sustainable Energy Systems* (pp. 792–821). IGI Global. https://doi.org/10.4018/978-1-7998-5796-9.ch022

Arafat, M. Y., Saleem, I., & Devi, T. P. (2022). Drivers of EV Charging Infrastructure Entrepreneurship in India. In Alam, M., Pillai, R., & Murugesan, N. (Eds.), *Developing Charging Infrastructure and Technologies for Electric Vehicles* (pp. 208–219). IGI Global. https://doi.org/10.4018/978-1-7998-6858-3.ch011

Araujo, A., & Manninen, H. (2022). Contribution of Project-Based Learning on Social Skills Development: An Industrial Engineer Perspective. In Alves, A., & van Hattum-Janssen, N. (Eds.), *Training Engineering Students for Modern Technological Advancement* (pp. 119–145). IGI Global. https://doi.org/10.4018/978-1-7998-8816-1.ch006

Armutlu, H. (2018). Intelligent Biomedical Engineering Operations by Cloud Computing Technologies. In Kose, U., Guraksin, G., & Deperlioglu, O. (Eds.), *Nature-Inspired Intelligent Techniques for Solving Biomedical Engineering Problems* (pp. 297–317). Hershey, PA: IGI Global. 10.4018/978-1-5225-4769-3.ch015

Atik, M., Sadek, M., & Shahrour, I. (2017). Single-Run Adaptive Pushover Procedure for Shear Wall Structures. In Plevris, V., Kremmyda, G., & Fahjan, Y. (Eds.), *Performance-Based Seismic Design of Concrete Structures and Infrastructures* (pp. 59–83). Hershey, PA: IGI Global. 10.4018/978-1-5225-2089-4.ch003

Attia, H. (2021). Smart Power Microgrid Impact on Sustainable Building. In González-Lezcano, R. (Ed.), *Advancements in Sustainable Architecture and Energy Efficiency* (pp. 169–194). IGI Global. https://doi.org/10.4018/978-1-7998-7023-4.ch008

Aydin, A., Akyol, E., Gungor, M., Kaya, A., & Tasdelen, S. (2018). Geophysical Surveys in Engineering Geology Investigations With Field Examples. In Ceryan, N. (Ed.), *Handbook of Research on Trends and Digital Advances in Engineering Geology* (pp. 257–280). Hershey, PA: IGI Global. 10.4018/978-1-5225-2709-1.ch007

Ayoobkhan, M. U. D., Y., A., J., Easwaran, B., & R., T. (2021). Smart Connected Digital Products and IoT Platform With the Digital Twin. In P. Vasant, G. Weber, & W. Punurai (Ed.), *Research Advancements in Smart Technology, Optimization, and Renewable Energy* (pp. 330-350). IGI Global. https://doi.org/10.4018/978-1-7998-3970-5.ch016

Baeza Moyano, D., & González Lezcano, R. A. (2021). The Importance of Light in Our Lives: Towards New Lighting in Schools. In González-Lezcano, R. (Ed.), *Advancements in Sustainable Architecture and Energy Efficiency* (pp. 239–256). IGI Global. https://doi.org/10.4018/978-1-7998-7023-4.ch011

Bagdadee, A. H. (2021). A Brief Assessment of the Energy Sector of Bangladesh. *International Journal of Energy Optimization and Engineering*, 10(1), 36–55. 10.4018/IJEOE.2021010103

Baklezos, A. T., & Hadjigeorgiou, N. G. (2021). Magnetic Sensors for Space Applications and Magnetic Cleanliness Considerations. In Nikolopoulos, C. (Ed.), *Recent Trends on Electromagnetic Environmental Effects for Aeronautics and Space Applications* (pp. 147–185). IGI Global. https://doi.org/10.4018/978-1-7998-4879-0.ch006

Bas, T. G. (2017). Nutraceutical Industry with the Collaboration of Biotechnology and Nutrigenomics Engineering: The Significance of Intellectual Property in the Entrepreneurship and Scientific Research Ecosystems. In Bas, T., & Zhao, J. (Eds.), *Comparative Approaches to Biotechnology Development and Use in Developed and Emerging Nations* (pp. 1–17). Hershey, PA: IGI Global. 10.4018/978-1-5225-1040-6.ch001

Bazeer Ahamed, B., & Periakaruppan, S. (2021). Taxonomy of Influence Maximization Techniques in Unknown Social Networks. In P. Vasant, G. Weber, & W. Punurai (Eds.), *Research Advancements in Smart Technology, Optimization, and Renewable Energy* (pp. 351-363). IGI Global. https://doi.org/10.4018/978-1-7998-3970-5.ch017

Beale, R., & André, J. (2017). *Design Solutions and Innovations in Temporary Structures*. Hershey, PA: IGI Global. 10.4018/978-1-5225-2199-0

Behnam, B. (2017). Simulating Post-Earthquake Fire Loading in Conventional RC Structures. In Samui, P., Chakraborty, S., & Kim, D. (Eds.), *Modeling and Simulation Techniques in Structural Engineering* (pp. 425–444). Hershey, PA: IGI Global. 10.4018/978-1-5225-0588-4.ch015

Ben Hamida, I., Salah, S. B., Msahli, F., & Mimouni, M. F. (2018). Distribution Network Reconfiguration Using SPEA2 for Power Loss Minimization and Reliability Improvement. *International Journal of Energy Optimization and Engineering*, 7(1), 50–65. 10.4018/IJEOE.2018010103

Bentarzi, H. (2021). Fault Tree-Based Root Cause Analysis Used to Study Mal-Operation of a Protective Relay in a Smart Grid. In Recioui, A., & Bentarzi, H. (Eds.), *Optimizing and Measuring Smart Grid Operation and Control* (pp. 289–308). IGI Global. https://doi.org/10.4018/978-1-7998-4027-5.ch012

Beysens, D. A., Garrabos, Y., & Zappoli, B. (2021). Thermal Effects in Near-Critical Fluids: Piston Effect and Related Phenomena. In Chen, L. (Ed.), *Handbook of Research on Advancements in Supercritical Fluids Applications for Sustainable Energy Systems* (pp. 1–31). IGI Global. https://doi.org/10.4018/978-1-7998-5796-9.ch001

Bhaskar, S. V., & Kudal, H. N. (2017). Effect of TiCN and AlCrN Coating on Tribological Behaviour of Plasma-nitrided AISI 4140 Steel. *International Journal of Surface Engineering and Interdisciplinary Materials Science*, 5(2), 1–17. 10.4018/IJSEIMS.2017070101

Bhuyan, D. (2018). Designing of a Twin Tube Shock Absorber: A Study in Reverse Engineering. In Kumar, K., & Davim, J. (Eds.), *Design and Optimization of Mechanical Engineering Products* (pp. 83–104). Hershey, PA: IGI Global. 10.4018/978-1-5225-3401-3.ch005

Blumberg, G. (2021). Blockchains for Use in Construction and Engineering Projects. In Underwood, J., & Shelbourn, M. (Eds.), *Handbook of Research on Driving Transformational Change in the Digital Built Environment* (pp. 179–208). IGI Global. https://doi.org/10.4018/978-1-7998-6600-8.ch008

Bolboaca, A. M. (2021). Considerations Regarding the Use of Fuel Cells in Combined Heat and Power for Stationary Applications. In Badea, G., Felseghi, R., & A chilean, I. (Eds.), *Hydrogen Fuel Cell Technology for Stationary Applications* (pp. 239–275). IGI Global. https://doi.org/10.4018/978-1-7998-4945-2.ch010

Burstein, L. (2021). Simulation Tool for Cable Design. In Burstein, L. (Ed.), *Handbook of Research on Advancements in Manufacturing, Materials, and Mechanical Engineering* (pp. 54–74). IGI Global. https://doi.org/10.4018/978-1-7998-4939-1.ch003

Calderon, F. A., Giolo, E. G., Frau, C. D., Rengel, M. G., Rodriguez, H., Tornello, M., & Gallucci, R. (2018). Seismic Microzonation and Site Effects Detection Through Microtremors Measures: A Review. In Ceryan, N. (Ed.), *Handbook of Research on Trends and Digital Advances in Engineering Geology* (pp. 326–349). Hershey, PA: IGI Global. 10.4018/978-1-5225-2709-1.ch009

Ceryan, N., & Can, N. K. (2018). Prediction of The Uniaxial Compressive Strength of Rocks Materials. In Ceryan, N. (Ed.), *Handbook of Research on Trends and Digital Advances in Engineering Geology* (pp. 31–96). Hershey, PA: IGI Global. 10.4018/978-1-5225-2709-1.ch002

Ceryan, S. (2018). Weathering Indices Used in Evaluation of the Weathering State of Rock Material. In Ceryan, N. (Ed.), *Handbook of Research on Trends and Digital Advances in Engineering Geology* (pp. 132–186). Hershey, PA: IGI Global. 10.4018/978-1-5225-2709-1.ch004

Chen, H., Padilla, R. V., & Besarati, S. (2017). Supercritical Fluids and Their Applications in Power Generation. In Chen, L., & Iwamoto, Y. (Eds.), *Advanced Applications of Supercritical Fluids in Energy Systems* (pp. 369–402). Hershey, PA: IGI Global. 10.4018/978-1-5225-2047-4.ch012

Chen, H., Padilla, R. V., & Besarati, S. (2021). Supercritical Fluids and Their Applications in Power Generation. In Chen, L. (Ed.), *Handbook of Research on Advancements in Supercritical Fluids Applications for Sustainable Energy Systems* (pp. 566–599). IGI Global. https://doi.org/10.4018/978-1-7998-5796-9.ch016

Chen, L. (2017). Principles, Experiments, and Numerical Studies of Supercritical Fluid Natural Circulation System. In Chen, L., & Iwamoto, Y. (Eds.), *Advanced Applications of Supercritical Fluids in Energy Systems* (pp. 136–187). Hershey, PA: IGI Global. 10.4018/978-1-5225-2047-4.ch005

Chen, L. (2021). Principles, Experiments, and Numerical Studies of Supercritical Fluid Natural Circulation System. In Chen, L. (Ed.), *Handbook of Research on Advancements in Supercritical Fluids Applications for Sustainable Energy Systems* (pp. 219–269). IGI Global. https://doi.org/10.4018/978-1-7998-5796-9.ch007

Chiba, Y., Marif, Y., Henini, N., & Tlemcani, A. (2021). Modeling of Magnetic Refrigeration Device by Using Artificial Neural Networks Approach. *International Journal of Energy Optimization and Engineering*, 10(4), 68–76. https://doi.org/10.4018/IJEOE.2021100105

Clementi, F., Di Sciascio, G., Di Sciascio, S., & Lenci, S. (2017). Influence of the Shear-Bending Interaction on the Global Capacity of Reinforced Concrete Frames: A Brief Overview of the New Perspectives. In Plevris, V., Kremmyda, G., & Fahjan, Y. (Eds.), *Performance-Based Seismic Design of Concrete Structures and Infrastructures* (pp. 84–111). Hershey, PA: IGI Global. 10.4018/978-1-5225-2089-4.ch004

Codinhoto, R., Fialho, B. C., Pinti, L., & Fabricio, M. M. (2021). BIM and IoT for Facilities Management: Understanding Key Maintenance Issues. In Underwood, J., & Shelbourn, M. (Eds.), *Handbook of Research on Driving Transformational Change in the Digital Built Environment* (pp. 209–231). IGI Global. 10.4018/978-1-7998-6600-8.ch009

Cortés-Polo, D., Calle-Cancho, J., Carmona-Murillo, J., & González-Sánchez, J. (2017). Future Trends in Mobile-Fixed Integration for Next Generation Networks: Classification and Analysis. *International Journal of Vehicular Telematics and Infotainment Systems*, 1(1), 33–53. 10.4018/IJVTIS.2017010103

Costa, H. G., Sheremetieff, F. H., & Araújo, E. A. (2022). Influence of Game-Based Methods in Developing Engineering Competences. In Alves, A., & van Hattum-Janssen, N. (Eds.), *Training Engineering Students for Modern Technological Advancement* (pp. 69–88). IGI Global. https://doi.org/10.4018/978-1-7998-8816-1.ch004

Cui, X., Zeng, S., Li, Z., Zheng, Q., Yu, X., & Han, B. (2018). Advanced Composites for Civil Engineering Infrastructures. In Kumar, K., & Davim, J. (Eds.), *Composites and Advanced Materials for Industrial Applications* (pp. 212–248). Hershey, PA: IGI Global. 10.4018/978-1-5225-5216-1.ch010

Dalgıç, S., & Kuşku, İ. (2018). Geological and Geotechnical Investigations in Tunneling. In Ceryan, N. (Ed.), *Handbook of Research on Trends and Digital Advances in Engineering Geology* (pp. 482–529). Hershey, PA: IGI Global. 10.4018/978-1-5225-2709-1.ch014

Dang, C., & Hihara, E. (2021). Study on Cooling Heat Transfer of Supercritical Carbon Dioxide Applied to Transcritical Carbon Dioxide Heat Pump. In Chen, L. (Ed.), *Handbook of Research on Advancements in Supercritical Fluids Applications for Sustainable Energy Systems* (pp. 451–493). IGI Global. https://doi.org/10.4018/978-1-7998-5796-9.ch013

Daus, Y., Kharchenko, V., & Yudaev, I. (2021). Research of Solar Energy Potential of Photovoltaic Installations on Enclosing Structures of Buildings. *International Journal of Energy Optimization and Engineering*, 10(4), 18–34. https://doi.org/10.4018/IJEOE.2021100102

Daus, Y., Kharchenko, V., & Yudaev, I. (2021). Optimizing Layout of Distributed Generation Sources of Power Supply System of Agricultural Object. *International Journal of Energy Optimization and Engineering*, 10(3), 70–84. https://doi.org/10.4018/IJEOE.2021070104

de la Varga, D., Soto, M., Arias, C. A., van Oirschot, D., Kilian, R., Pascual, A., & Álvarez, J. A. (2017). Constructed Wetlands for Industrial Wastewater Treatment and Removal of Nutrients. In Val del Río, Á., Campos Gómez, J., & Mosquera Corral, A. (Eds.), *Technologies for the Treatment and Recovery of Nutrients from Industrial Wastewater* (pp. 202–230). Hershey, PA: IGI Global. 10.4018/978-1-5225-1037-6.ch008

Deb, S., Ammar, E. A., AlRajhi, H., Alsaidan, I., & Shariff, S. M. (2022). V2G Pilot Projects: Review and Lessons Learnt. In Alam, M., Pillai, R., & Murugesan, N. (Eds.), *Developing Charging Infrastructure and Technologies for Electric Vehicles* (pp. 252–267). IGI Global. https://doi.org/10.4018/978-1-7998-6858-3.ch014

Dekhandji, F. Z., & Rais, M. C. (2021). A Comparative Study of Power Quality Monitoring Using Various Techniques. In Recioui, A., & Bentarzi, H. (Eds.), *Optimizing and Measuring Smart Grid Operation and Control* (pp. 259–288). IGI Global. https://doi.org/10.4018/978-1-7998-4027-5.ch011

Deperlioglu, O. (2018). Intelligent Techniques Inspired by Nature and Used in Biomedical Engineering. In Kose, U., Guraksin, G., & Deperlioglu, O. (Eds.), *Nature-Inspired Intelligent Techniques for Solving Biomedical Engineering Problems* (pp. 51–77). Hershey, PA: IGI Global. 10.4018/978-1-5225-4769-3.ch003

Dhurpate, P. R., & Tang, H. (2021). Quantitative Analysis of the Impact of Inter-Line Conveyor Capacity for Throughput of Manufacturing Systems. *International Journal of Manufacturing, Materials, and Mechanical Engineering*, 11(1), 1–17. https://doi.org/10.4018/IJMMME.2021010101

Dinkar, S., & Deep, K. (2021). A Survey of Recent Variants and Applications of Antlion Optimizer. *International Journal of Energy Optimization and Engineering*, 10(2), 48–73. 10.4018/IJEOE.2021040103

Dixit, A. (2018). Application of Silica-Gel-Reinforced Aluminium Composite on the Piston of Internal Combustion Engine: Comparative Study of Silica-Gel-Reinforced Aluminium Composite Piston With Aluminium Alloy Piston. In Kumar, K., & Davim, J. (Eds.), *Composites and Advanced Materials for Industrial Applications* (pp. 63–98). Hershey, PA: IGI Global. 10.4018/978-1-5225-5216-1.ch004

Drabecki, M. P., & Kułak, K. B. (2021). Global Pandemics on European Electrical Energy Markets: Lessons Learned From the COVID-19 Outbreak. *International Journal of Energy Optimization and Engineering*, 10(3), 24–46. https://doi.org/10.4018/IJEOE.2021070102

Dutta, M. M. (2021). Nanomaterials for Food and Agriculture. In Bhat, M., Wani, I., & Ashraf, S. (Eds.), *Applications of Nanomaterials in Agriculture, Food Science, and Medicine* (pp. 75–97). IGI Global. 10.4018/978-1-7998-5563-7.ch004

Dutta, M. M., & Goswami, M. (2021). Coating Materials: Nano-Materials. In Roy, S., & Bose, G. (Eds.), *Advanced Surface Coating Techniques for Modern Industrial Applications* (pp. 1–30). IGI Global. 10.4018/978-1-7998-4870-7.ch001

Elsayed, A. M., Dakkama, H. J., Mahmoud, S., Al-Dadah, R., & Kaialy, W. (2017). Sustainable Cooling Research Using Activated Carbon Adsorbents and Their Environmental Impact. In Kobayashi, T. (Ed.), *Applied Environmental Materials Science for Sustainability* (pp. 186–221). Hershey, PA: IGI Global. 10.4018/978-1-5225-1971-3.ch009

Ercanoglu, M., & Sonmez, H. (2018). General Trends and New Perspectives on Landslide Mapping and Assessment Methods. In Ceryan, N. (Ed.), *Handbook of Research on Trends and Digital Advances in Engineering Geology* (pp. 350–379). Hershey, PA: IGI Global. 10.4018/978-1-5225-2709-1.ch010

Faroz, S. A., Pujari, N. N., Rastogi, R., & Ghosh, S. (2017). Risk Analysis of Structural Engineering Systems Using Bayesian Inference. In Samui, P., Chakraborty, S., & Kim, D. (Eds.), *Modeling and Simulation Techniques in Structural Engineering* (pp. 390–424). Hershey, PA: IGI Global. 10.4018/978-1-5225-0588-4.ch014

Fekik, A., Hamida, M. L., Denoun, H., Azar, A. T., Kamal, N. A., Vaidyanathan, S., Bousbaine, A., & Benamrouche, N. (2022). Multilevel Inverter for Hybrid Fuel Cell/PV Energy Conversion System. In Fekik, A., & Benamrouche, N. (Eds.), *Modeling and Control of Static Converters for Hybrid Storage Systems* (pp. 233–270). IGI Global. https://doi.org/10.4018/978-1-7998-7447-8.ch009

Fekik, A., Hamida, M. L., Houassine, H., Azar, A. T., Kamal, N. A., Denoun, H., Vaidyanathan, S., & Sambas, A. (2022). Power Quality Improvement for Grid-Connected Photovoltaic Panels Using Direct Power Control. In Fekik, A., & Benamrouche, N. (Eds.), *Modeling and Control of Static Converters for Hybrid Storage Systems* (pp. 107–142). IGI Global. https://doi.org/10.4018/978-1-7998-7447-8.ch005

Fernando, P. R., Hamigah, T., Disne, S., Wickramasingha, G. G., & Sutharshan, A. (2018). The Evaluation of Engineering Properties of Low Cost Concrete Blocks by Partial Doping of Sand with Sawdust: Low Cost Sawdust Concrete Block. *International Journal of Strategic Engineering*, 1(2), 26–42. 10.4018/IJoSE.2018070103

Ferro, G., Minciardi, R., Parodi, L., & Robba, M. (2022). Optimal Charging Management of Microgrid-Integrated Electric Vehicles. In Alam, M., Pillai, R., & Murugesan, N. (Eds.), *Developing Charging Infrastructure and Technologies for Electric Vehicles* (pp. 133–155). IGI Global. https://doi.org/10.4018/978-1-7998-6858-3.ch007

Flumerfelt, S., & Green, C. (2022). Graduate Lean Leadership Education: A Case Study of a Program. In Alves, A., & van Hattum-Janssen, N. (Eds.), *Training Engineering Students for Modern Technological Advancement* (pp. 202–224). IGI Global. https://doi.org/10.4018/978-1-7998-8816-1.ch010

Galli, B. J. (2021). Implications of Economic Decision Making to the Project Manager. *International Journal of Strategic Engineering*, 4(1), 19–32. https://doi.org/10.4018/IJoSE.2021010102

Gento, A. M., Pimentel, C., & Pascual, J. A. (2022). Teaching Circular Economy and Lean Management in a Learning Factory. In Alves, A., & van Hattum-Janssen, N. (Eds.), *Training Engineering Students for Modern Technological Advancement* (pp. 183–201). IGI Global. https://doi.org/10.4018/978-1-7998-8816-1.ch009

Ghosh, S., Mitra, S., Ghosh, S., & Chakraborty, S. (2017). Seismic Reliability Analysis in the Framework of Metamodelling Based Monte Carlo Simulation. In Samui, P., Chakraborty, S., & Kim, D. (Eds.), *Modeling and Simulation Techniques in Structural Engineering* (pp. 192–208). Hershey, PA: IGI Global. 10.4018/978-1-5225-0588-4.ch006

Gil, M., & Otero, B. (2017). Learning Engineering Skills through Creativity and Collaboration: A Game-Based Proposal. In Alexandre Peixoto de Queirós, R., & Pinto, M. (Eds.), *Gamification-Based E-Learning Strategies for Computer Programming Education* (pp. 14–29). Hershey, PA: IGI Global. 10.4018/978-1-5225-1034-5.ch002

Gill, J., Ayre, M., & Mills, J. (2017). Revisioning the Engineering Profession: How to Make It Happen! In Gray, M., & Thomas, K. (Eds.), *Strategies for Increasing Diversity in Engineering Majors and Careers* (pp. 156–175). Hershey, PA: IGI Global. 10.4018/978-1-5225-2212-6.ch008

Godzhaev, Z., Senkevich, S., Kuzmin, V., & Melikov, I. (2021). Use of the Neural Network Controller of Sprung Mass to Reduce Vibrations From Road Irregularities. In P. Vasant, G. Weber, & W. Punurai (Ed.), *Research Advancements in Smart Technology, Optimization, and Renewable Energy* (pp. 69-87). IGI Global. https://doi.org/10.4018/978-1-7998-3970-5.ch005

Gomes de Gusmão, C. M. (2022). Digital Competencies and Transformation in Higher Education: Upskilling With Extension Actions. In Alves, A., & van Hattum-Janssen, N. (Eds.), *Training Engineering Students for Modern Technological Advancement* (pp. 313–328). IGI Global. https://doi.org/10.4018/978-1-7998-8816-1.ch015A

Goyal, N., Ram, M., & Kumar, P. (2017). Welding Process under Fault Coverage Approach for Reliability and MTTF. In Ram, M., & Davim, J. (Eds.), *Mathematical Concepts and Applications in Mechanical Engineering and Mechatronics* (pp. 222–245). Hershey, PA: IGI Global. 10.4018/978-1-5225-1639-2.ch011

Gray, M., & Lundy, C. (2017). Engineering Study Abroad: High Impact Strategy for Increasing Access. In Gray, M., & Thomas, K. (Eds.), *Strategies for Increasing Diversity in Engineering Majors and Careers* (pp. 42–59). Hershey, PA: IGI Global. 10.4018/978-1-5225-2212-6.ch003

Güler, O., & Varol, T. (2021). Fabrication of Functionally Graded Metal and Ceramic Powders Synthesized by Electroless Deposition. In Roy, S., & Bose, G. (Eds.), *Advanced Surface Coating Techniques for Modern Industrial Applications* (pp. 150–187). IGI Global. https://doi.org/10.4018/978-1-7998-4870-7.ch007

Guraksin, G. E. (2018). Internet of Things and Nature-Inspired Intelligent Techniques for the Future of Biomedical Engineering. In Kose, U., Guraksin, G., & Deperlioglu, O. (Eds.), *Nature-Inspired Intelligent Techniques for Solving Biomedical Engineering Problems* (pp. 263–282). Hershey, PA: IGI Global. 10.4018/978-1-5225-4769-3.ch013

Hamida, M. L., Fekik, A., Denoun, H., Ardjal, A., & Bokhtache, A. A. (2022). Flying Capacitor Inverter Integration in a Renewable Energy System. In Fekik, A., & Benamrouche, N. (Eds.), *Modeling and Control of Static Converters for Hybrid Storage Systems* (pp. 287–306). IGI Global. https://doi.org/10.4018/978-1-7998-7447-8.ch011

Hasegawa, N., & Takahashi, Y. (2021). Control of Soap Bubble Ejection Robot Using Facial Expressions. *International Journal of Manufacturing, Materials, and Mechanical Engineering*, 11(2), 1–16. https://doi.org/10.4018/IJMMME.2021040101

Hejazi, T., & Akbari, L. (2017). A Multiresponse Optimization Model for Statistical Design of Processes with Discrete Variables. In Ram, M., & Davim, J. (Eds.), *Mathematical Concepts and Applications in Mechanical Engineering and Mechatronics* (pp. 17–37). Hershey, PA: IGI Global. 10.4018/978-1-5225-1639-2.ch002

Hejazi, T., & Hejazi, A. (2017). Monte Carlo Simulation for Reliability-Based Design of Automotive Complex Subsystems. In Ram, M., & Davim, J. (Eds.), *Mathematical Concepts and Applications in Mechanical Engineering and Mechatronics* (pp. 177–200). Hershey, PA: IGI Global. 10.4018/978-1-5225-1639-2.ch009

Hejazi, T., & Poursabbagh, H. (2017). Reliability Analysis of Engineering Systems: An Accelerated Life Testing for Boiler Tubes. In Ram, M., & Davim, J. (Eds.), *Mathematical Concepts and Applications in Mechanical Engineering and Mechatronics* (pp. 154–176). Hershey, PA: IGI Global. 10.4018/978-1-5225-1639-2.ch008

Henao, J., Poblano-Salas, C. A., Vargas, F., Giraldo-Betancur, A. L., Corona-Castuera, J., & Sotelo-Mazón, O. (2021). Principles and Applications of Thermal Spray Coatings. In Roy, S., & Bose, G. (Eds.), *Advanced Surface Coating Techniques for Modern Industrial Applications* (pp. 31–70). IGI Global. https://doi.org/10.4018/978-1-7998-4870-7.ch002

Henao, J., & Sotelo, O. (2018). Surface Engineering at High Temperature: Thermal Cycling and Corrosion Resistance. In Pakseresht, A. (Ed.), *Production, Properties, and Applications of High Temperature Coatings* (pp. 131–159). Hershey, PA: IGI Global. 10.4018/978-1-5225-4194-3.ch006

Hrnčič, M. K., Cör, D., & Knez, Ž. (2021). Supercritical Fluids as a Tool for Green Energy and Chemicals. In Chen, L. (Ed.), *Handbook of Research on Advancements in Supercritical Fluids Applications for Sustainable Energy Systems* (pp. 761–791). IGI Global. 10.4018/978-1-7998-5796-9.ch021

Ibrahim, O., Erdem, S., & Gurbuz, E. (2021). Studying Physical and Chemical Properties of Graphene Oxide and Reduced Graphene Oxide and Their Applications in Sustainable Building Materials. In González-Lezcano, R. (Ed.), *Advancements in Sustainable Architecture and Energy Efficiency* (pp. 221–238). IGI Global. https://doi.org/10.4018/978-1-7998-7023-4.ch010

Ihianle, I. K., Islam, S., Naeem, U., & Ebenuwa, S. H. (2021). Exploiting Patterns of Object Use for Human Activity Recognition. In Nwajana, A., & Ihianle, I. (Eds.), *Handbook of Research on 5G Networks and Advancements in Computing, Electronics, and Electrical Engineering* (pp. 382–401). IGI Global. https://doi.org/10.4018/978-1-7998-6992-4.ch015

Ijemaru, G. K., Ngharamike, E. T., Oleka, E. U., & Nwajana, A. O. (2021). An Energy-Efficient Model for Opportunistic Data Collection in IoV-Enabled SC Waste Management. In Nwajana, A., & Ihianle, I. (Eds.), *Handbook of Research on 5G Networks and Advancements in Computing, Electronics, and Electrical Engineering* (pp. 1–19). IGI Global. https://doi.org/10.4018/978-1-7998-6992-4.ch001

Ilori, O. O., Adetan, D. A., & Umoru, L. E. (2017). Effect of Cutting Parameters on the Surface Residual Stress of Face-Milled Pearlitic Ductile Iron. *International Journal of Materials Forming and Machining Processes*, 4(1), 38–52. 10.4018/IJMFMP.2017010103

Imam, M. H., Tasadduq, I. A., Ahmad, A., Aldosari, F., & Khan, H. (2017). Automated Generation of Course Improvement Plans Using Expert System. *International Journal of Quality Assurance in Engineering and Technology Education*, 6(1), 1–12. 10.4018/IJQAETE.2017010101

Injeti, S. K., & Kumar, T. V. (2018). A WDO Framework for Optimal Deployment of DGs and DSCs in a Radial Distribution System Under Daily Load Pattern to Improve Techno-Economic Benefits. *International Journal of Energy Optimization and Engineering*, 7(2), 1–38. 10.4018/IJEOE.2018040101

Ishii, N., Anami, K., & Knisely, C. W. (2018). *Dynamic Stability of Hydraulic Gates and Engineering for Flood Prevention*. Hershey, PA: IGI Global. 10.4018/978-1-5225-3079-4

Iwamoto, Y., & Yamaguchi, H. (2021). Application of Supercritical Carbon Dioxide for Solar Water Heater. In Chen, L. (Ed.), *Handbook of Research on Advancements in Supercritical Fluids Applications for Sustainable Energy Systems* (pp. 370–387). IGI Global. https://doi.org/10.4018/978-1-7998-5796-9.ch010

Jayapalan, S. (2018). A Review of Chemical Treatments on Natural Fibers-Based Hybrid Composites for Engineering Applications. In Kumar, K., & Davim, J. (Eds.), *Composites and Advanced Materials for Industrial Applications* (pp. 16–37). Hershey, PA: IGI Global. 10.4018/978-1-5225-5216-1.ch002

Kapetanakis, T. N., Vardiambasis, I. O., Ioannidou, M. P., & Konstantaras, A. I. (2021). Modeling Antenna Radiation Using Artificial Intelligence Techniques: The Case of a Circular Loop Antenna. In Nikolopoulos, C. (Ed.), *Recent Trends on Electromagnetic Environmental Effects for Aeronautics and Space Applications* (pp. 186–225). IGI Global. https://doi.org/10.4018/978-1-7998-4879-0.ch007

Karkalos, N. E., Markopoulos, A. P., & Dossis, M. F. (2017). Optimal Model Parameters of Inverse Kinematics Solution of a 3R Robotic Manipulator Using ANN Models. *International Journal of Manufacturing, Materials, and Mechanical Engineering*, 7(3), 20–40. 10.4018/IJMMME.2017070102

Kelly, M., Costello, M., Nicholson, G., & O'Connor, J. (2021). The Evolving Integration of BIM Into Built Environment Programmes in a Higher Education Institute. In Underwood, J., & Shelbourn, M. (Eds.), *Handbook of Research on Driving Transformational Change in the Digital Built Environment* (pp. 294–326). IGI Global. https://doi.org/10.4018/978-1-7998-6600-8.ch012

Kesimal, A., Karaman, K., Cihangir, F., & Ercikdi, B. (2018). Excavatability Assessment of Rock Masses for Geotechnical Studies. In Ceryan, N. (Ed.), *Handbook of Research on Trends and Digital Advances in Engineering Geology* (pp. 231–256). Hershey, PA: IGI Global. 10.4018/978-1-5225-2709-1.ch006

Knoflacher, H. (2017). The Role of Engineers and Their Tools in the Transport Sector after Paradigm Change: From Assumptions and Extrapolations to Science. In Knoflacher, H., & Ocalir-Akunal, E. (Eds.), *Engineering Tools and Solutions for Sustainable Transportation Planning* (pp. 1–29). Hershey, PA: IGI Global. 10.4018/978-1-5225-2116-7.ch001

Kose, U. (2018). Towards an Intelligent Biomedical Engineering With Nature-Inspired Artificial Intelligence Techniques. In Kose, U., Guraksin, G., & Deperlioglu, O. (Eds.), *Nature-Inspired Intelligent Techniques for Solving Biomedical Engineering Problems* (pp. 1–26). Hershey, PA: IGI Global. 10.4018/978-1-5225-4769-3.ch001

Kostić, S. (2018). A Review on Enhanced Stability Analyses of Soil Slopes Using Statistical Design. In Ceryan, N. (Ed.), *Handbook of Research on Trends and Digital Advances in Engineering Geology* (pp. 446–481). Hershey, PA: IGI Global. 10.4018/978-1-5225-2709-1.ch013

Kumar, A., Patil, P. P., & Prajapati, Y. K. (2018). *Advanced Numerical Simulations in Mechanical Engineering*. Hershey, PA: IGI Global. 10.4018/978-1-5225-3722-9

Kumar, G. R., Rajyalakshmi, G., & Manupati, V. K. (2017). Surface Micro Patterning of Aluminium Reinforced Composite through Laser Peening. *International Journal of Manufacturing, Materials, and Mechanical Engineering*, 7(4), 15–27. 10.4018/IJMMME.2017100102

Kumar, N., Basu, D. N., & Chen, L. (2021). Effect of Flow Acceleration and Buoyancy on Thermal-hydraulics of sCO2 in Mini/Micro-Channel. In Chen, L. (Ed.), *Handbook of Research on Advancements in Supercritical Fluids Applications for Sustainable Energy Systems* (pp. 161–182). IGI Global. 10.4018/978-1-7998-5796-9.ch005

Kumari, N., & Kumar, K. (2018). Fabrication of Orthotic Calipers With Epoxy-Based Green Composite. In Kumar, K., & Davim, J. (Eds.), *Composites and Advanced Materials for Industrial Applications* (pp. 157–176). Hershey, PA: IGI Global. 10.4018/978-1-5225-5216-1.ch008

Kuppusamy, R. R. (2018). Development of Aerospace Composite Structures Through Vacuum-Enhanced Resin Transfer Moulding Technology (VERTMTy): Vacuum-Enhanced Resin Transfer Moulding. In Kumar, K., & Davim, J. (Eds.), *Composites and Advanced Materials for Industrial Applications* (pp. 99–111). Hershey, PA: IGI Global. 10.4018/978-1-5225-5216-1.ch005

Kurganov, V. A., Zeigarnik, Y. A., & Maslakova, I. V. (2021). Normal and Deteriorated Heat Transfer Under Heating Turbulent Supercritical Pressure Coolants Flows in Round Tubes. In Chen, L. (Ed.), *Handbook of Research on Advancements in Supercritical Fluids Applications for Sustainable Energy Systems* (pp. 494–532). IGI Global. https://doi.org/10.4018/978-1-7998-5796-9.ch014

Li, H., & Zhang, Y. (2021). Heat Transfer and Fluid Flow Modeling for Supercritical Fluids in Advanced Energy Systems. In Chen, L. (Ed.), *Handbook of Research on Advancements in Supercritical Fluids Applications for Sustainable Energy Systems* (pp. 388–422). IGI Global. https://doi.org/10.4018/978-1-7998-5796-9.ch011

Loy, J., Howell, S., & Cooper, R. (2017). Engineering Teams: Supporting Diversity in Engineering Education. In Gray, M., & Thomas, K. (Eds.), *Strategies for Increasing Diversity in Engineering Majors and Careers* (pp. 106–129). Hershey, PA: IGI Global. 10.4018/978-1-5225-2212-6.ch006

Macher, G., Armengaud, E., Kreiner, C., Brenner, E., Schmittner, C., Ma, Z., & Krammer, M. (2018). Integration of Security in the Development Lifecycle of Dependable Automotive CPS. In Druml, N., Genser, A., Krieg, A., Menghin, M., & Hoeller, A. (Eds.), *Solutions for Cyber-Physical Systems Ubiquity* (pp. 383–423). Hershey, PA: IGI Global. 10.4018/978-1-5225-2845-6.ch015

Madhu, M. N., Singh, J. G., Mohan, V., & Ongsakul, W. (2021). Transmission Risk Optimization in Interconnected Systems: Risk-Adjusted Available Transfer Capability. In P. Vasant, G. Weber, & W. Punurai (Ed.), *Research Advancements in Smart Technology, Optimization, and Renewable Energy* (pp. 183-199). IGI Global. https://doi.org/10.4018/978-1-7998-3970-5.ch010

Mahendramani, G., & Lakshmana Swamy, N. (2018). Effect of Weld Groove Area on Distortion of Butt Welded Joints in Submerged Arc Welding. *International Journal of Manufacturing, Materials, and Mechanical Engineering*, 8(2), 33–44. 10.4018/IJMMME.2018040103

Makropoulos, G., Koumaras, H., Setaki, F., Filis, K., Lutz, T., Montowtt, P., Tomaszewski, L., Dybiec, P., & Järvet, T. (2021). 5G and Unmanned Aerial Vehicles (UAVs) Use Cases: Analysis of the Ecosystem, Architecture, and Applications. In Nwajana, A., & Ihianle, I. (Eds.), *Handbook of Research on 5G Networks and Advancements in Computing, Electronics, and Electrical Engineering* (pp. 36–69). IGI Global. https://doi.org/10.4018/978-1-7998-6992-4.ch003

Meric, E. M., Erdem, S., & Gurbuz, E. (2021). Application of Phase Change Materials in Construction Materials for Thermal Energy Storage Systems in Buildings. In González-Lezcano, R. (Ed.), *Advancements in Sustainable Architecture and Energy Efficiency* (pp. 1–20). IGI Global. https://doi.org/10.4018/978-1-7998-7023-4.ch001

Mihret, E. T., & Yitayih, K. A. (2021). Operation of VANET Communications: The Convergence of UAV System With LTE/4G and WAVE Technologies. *International Journal of Smart Vehicles and Smart Transportation*, 4(1), 29–51. https://doi.org/10.4018/IJSVST.2021010103

Mir, M. A., Bhat, B. A., Sheikh, B. A., Rather, G. A., Mehraj, S., & Mir, W. R. (2021). Nanomedicine in Human Health Therapeutics and Drug Delivery: Nanobiotechnology and Nanobiomedicine. In Bhat, M., Wani, I., & Ashraf, S. (Eds.), *Applications of Nanomaterials in Agriculture, Food Science, and Medicine* (pp. 229–251). IGI Global. 10.4018/978-1-7998-5563-7.ch013

Mohammadzadeh, S., & Kim, Y. (2017). Nonlinear System Identification of Smart Buildings. In Samui, P., Chakraborty, S., & Kim, D. (Eds.), *Modeling and Simulation Techniques in Structural Engineering* (pp. 328–347). Hershey, PA: IGI Global. 10.4018/978-1-5225-0588-4.ch011

Molina, G. J., Aktaruzzaman, F., Soloiu, V., & Rahman, M. (2017). Design and Testing of a Jet-Impingement Instrument to Study Surface-Modification Effects by Nanofluids. *International Journal of Surface Engineering and Interdisciplinary Materials Science*, 5(2), 43–61. 10.4018/IJSEIMS.2017070104

Moreno-Rangel, A., & Carrillo, G. (2021). Energy-Efficient Homes: A Heaven for Respiratory Illnesses. In González-Lezcano, R. (Ed.), *Advancements in Sustainable Architecture and Energy Efficiency* (pp. 49–71). IGI Global. https://doi.org/10.4018/978-1-7998-7023-4.ch003

Msomi, V., & Jantjies, B. T. (2021). Correlative Analysis Between Tensile Properties and Tool Rotational Speeds of Friction Stir Welded Similar Aluminium Alloy Joints. *International Journal of Surface Engineering and Interdisciplinary Materials Science*, 9(2), 58–78. https://doi.org/10.4018/IJSEIMS.2021070104

Muigai, M. N., Mwema, F. M., Akinlabi, E. T., & Obiko, J. O. (2021). Surface Engineering of Materials Through Weld-Based Technologies: An Overview. In Roy, S., & Bose, G. (Eds.), *Advanced Surface Coating Techniques for Modern Industrial Applications* (pp. 247–260). IGI Global. 10.4018/978-1-7998-4870-7.ch011

Mukherjee, A., Saeed, R. A., Dutta, S., & Naskar, M. K. (2017). Fault Tracking Framework for Software-Defined Networking (SDN). In Singhal, C., & De, S. (Eds.), *Resource Allocation in Next-Generation Broadband Wireless Access Networks* (pp. 247–272). Hershey, PA: IGI Global. 10.4018/978-1-5225-2023-8.ch011

Mukhopadhyay, A., Barman, T. K., & Sahoo, P. (2018). Electroless Nickel Coatings for High Temperature Applications. In Kumar, K., & Davim, J. (Eds.), *Composites and Advanced Materials for Industrial Applications* (pp. 297–331). Hershey, PA: IGI Global. 10.4018/978-1-5225-5216-1.ch013

Mwema, F. M., & Wambua, J. M. (2022). Machining of Poly Methyl Methacrylate (PMMA) and Other Olymeric Materials: A Review. In Kumar, K., Babu, B., & Davim, J. (Eds.), *Handbook of Research on Advancements in the Processing, Characterization, and Application of Lightweight Materials* (pp. 363–379). IGI Global. https://doi.org/10.4018/978-1-7998-7864-3.ch016

Mykhailyshyn, R., Savkiv, V., Boyko, I., Prada, E., & Virgala, I. (2021). Substantiation of Parameters of Friction Elements of Bernoulli Grippers With a Cylindrical Nozzle. *International Journal of Manufacturing, Materials, and Mechanical Engineering*, 11(2), 17–39. https://doi.org/10.4018/IJMMME.2021040102

Náprstek, J., & Fischer, C. (2017). Dynamic Stability and Post-Critical Processes of Slender Auto-Parametric Systems. In Plevris, V., Kremmyda, G., & Fahjan, Y. (Eds.), *Performance-Based Seismic Design of Concrete Structures and Infrastructures* (pp. 128–171). Hershey, PA: IGI Global. 10.4018/978-1-5225-2089-4.ch006

Nautiyal, L., Shivach, P., & Ram, M. (2018). Optimal Designs by Means of Genetic Algorithms. In Ram, M., & Davim, J. (Eds.), *Soft Computing Techniques and Applications in Mechanical Engineering* (pp. 151–161). Hershey, PA: IGI Global. 10.4018/978-1-5225-3035-0.ch007

Nazir, R. (2017). Advanced Nanomaterials for Water Engineering and Treatment: Nano-Metal Oxides and Their Nanocomposites. In Saleh, T. (Ed.), *Advanced Nanomaterials for Water Engineering, Treatment, and Hydraulics* (pp. 84–126). Hershey, PA: IGI Global. 10.4018/978-1-5225-2136-5.ch005

Nikolopoulos, C. D. (2021). Recent Advances on Measuring and Modeling ELF-Radiated Emissions for Space Applications. In Nikolopoulos, C. (Ed.), *Recent Trends on Electromagnetic Environmental Effects for Aeronautics and Space Applications* (pp. 1–38). IGI Global. https://doi.org/10.4018/978-1-7998-4879-0.ch001

Nogueira, A. F., Ribeiro, J. C., Fernández de Vega, F., & Zenha-Rela, M. A. (2018). Evolutionary Approaches to Test Data Generation for Object-Oriented Software: Overview of Techniques and Tools. In M. Khosrow-Pour, D.B.A. (Ed.), *Incorporating Nature-Inspired Paradigms in Computational Applications* (pp. 162-194). Hershey, PA: IGI Global. https://doi.org/10.4018/978-1-5225-5020-4.ch006

Nwajana, A. O., Obi, E. R., Ijemaru, G. K., Oleka, E. U., & Anthony, D. C. (2021). Fundamentals of RF/Microwave Bandpass Filter Design. In Nwajana, A., & Ihianle, I. (Eds.), *Handbook of Research on 5G Networks and Advancements in Computing, Electronics, and Electrical Engineering* (pp. 149–164). IGI Global. https://doi.org/10.4018/978-1-7998-6992-4.ch005

Ogbodo, E. A. (2021). Comparative Study of Transmission Line Junction vs. Asynchronously Coupled Junction Diplexers. In Nwajana, A., & Ihianle, I. (Eds.), *Handbook of Research on 5G Networks and Advancements in Computing, Electronics, and Electrical Engineering* (pp. 326–336). IGI Global. https://doi.org/10.4018/978-1-7998-6992-4.ch013

Orosa, J. A., Vergara, D., Fraguela, F., & Masdías-Bonome, A. (2021). Statistical Understanding and Optimization of Building Energy Consumption and Climate Change Consequences. In González-Lezcano, R. (Ed.), *Advancements in Sustainable Architecture and Energy Efficiency* (pp. 195–220). IGI Global. https://doi.org/10.4018/978-1-7998-7023-4.ch009

Osho, M. B. (2018). Industrial Enzyme Technology: Potential Applications. In Bharati, S., & Chaurasia, P. (Eds.), *Research Advancements in Pharmaceutical, Nutritional, and Industrial Enzymology* (pp. 375–394). Hershey, PA: IGI Global. 10.4018/978-1-5225-5237-6.ch017

Ouadi, A., & Zitouni, A. (2021). Phasor Measurement Improvement Using Digital Filter in a Smart Grid. In Recioui, A., & Bentarzi, H. (Eds.), *Optimizing and Measuring Smart Grid Operation and Control* (pp. 100–117). IGI Global. https://doi.org/10.4018/978-1-7998-4027-5.ch005

Padmaja, P., & Marutheswar, G. (2017). Certain Investigation on Secured Data Transmission in Wireless Sensor Networks. *International Journal of Mobile Computing and Multimedia Communications*, 8(1), 48–61. 10.4018/IJMCMC.2017010104

Palmer, S., & Hall, W. (2017). An Evaluation of Group Work in First-Year Engineering Design Education. In Tucker, R. (Ed.), *Collaboration and Student Engagement in Design Education* (pp. 145–168). Hershey, PA: IGI Global. 10.4018/978-1-5225-0726-0.ch007

Panchenko, V. (2021). Prospects for Energy Supply of the Arctic Zone Objects of Russia Using Frost-Resistant Solar Modules. In P. Vasant, G. Weber, & W. Punurai (Eds.), *Research Advancements in Smart Technology, Optimization, and Renewable Energy* (pp. 149-169). IGI Global. https://doi.org/10.4018/978-1-7998-3970-5.ch008

Panchenko, V. (2021). Photovoltaic Thermal Module With Paraboloid Type Solar Concentrators. *International Journal of Energy Optimization and Engineering*, 10(2), 1–23. https://doi.org/10.4018/IJEOE.2021040101

Pandey, K., & Datta, S. (2021). Dry Machining of Inconel 825 Superalloys: Performance of Tool Inserts (Carbide, Cermet, and SiAlON). *International Journal of Manufacturing, Materials, and Mechanical Engineering*, 11(4), 26–39. 10.4018/IJMMME.2021100102

Panneer, R. (2017). Effect of Composition of Fibers on Properties of Hybrid Composites. *International Journal of Manufacturing, Materials, and Mechanical Engineering*, 7(4), 28–43. 10.4018/IJMMME.2017100103

Pany, C. (2021). Estimation of Correct Long-Seam Mismatch Using FEA to Compare the Measured Strain in a Non-Destructive Testing of a Pressurant Tank: A Reverse Problem. *International Journal of Smart Vehicles and Smart Transportation*, 4(1), 16–28. 10.4018/IJSVST.2021010102

Paul, S., & Roy, P. (2018). Optimal Design of Power System Stabilizer Using a Novel Evolutionary Algorithm. *International Journal of Energy Optimization and Engineering*, 7(3), 24–46. 10.4018/IJEOE.2018070102

Paul, S., & Roy, P. K. (2021). Oppositional Differential Search Algorithm for the Optimal Tuning of Both Single Input and Dual Input Power System Stabilizer. In P. Vasant, G. Weber, & W. Punurai (Eds.), *Research Advancements in Smart Technology, Optimization, and Renewable Energy* (pp. 256-282). IGI Global. https://doi.org/10.4018/978-1-7998-3970-5.ch013

Pavaloiu, A. (2018). Artificial Intelligence Ethics in Biomedical-Engineering-Oriented Problems. In Kose, U., Guraksin, G., & Deperlioglu, O. (Eds.), *Nature-Inspired Intelligent Techniques for Solving Biomedical Engineering Problems* (pp. 219–231). Hershey, PA: IGI Global. 10.4018/978-1-5225-4769-3.ch010

Pioro, I., Mahdi, M., & Popov, R. (2017). Application of Supercritical Pressures in Power Engineering. In Chen, L., & Iwamoto, Y. (Eds.), *Advanced Applications of Supercritical Fluids in Energy Systems* (pp. 404–457). Hershey, PA: IGI Global. 10.4018/978-1-5225-2047-4.ch013

Plaksina, T., & Gildin, E. (2017). Rigorous Integrated Evolutionary Workflow for Optimal Exploitation of Unconventional Gas Assets. *International Journal of Energy Optimization and Engineering*, 6(1), 101–122. 10.4018/IJEOE.2017010106

Popat, J., Kakadiya, H., Tak, L., Singh, N. K., Majeed, M. A., & Mahajan, V. (2021). Reliability of Smart Grid Including Cyber Impact: A Case Study. In Singh, R., Singh, A., Dwivedi, A., & Nagabhushan, P. (Eds.), *Computational Methodologies for Electrical and Electronics Engineers* (pp. 163–174). IGI Global. https://doi.org/10.4018/978-1-7998-3327-7.ch013

Quiza, R., La Fé-Perdomo, I., Rivas, M., & Ramtahalsing, V. (2021). Triple Bottom Line-Focused Optimization of Oblique Turning Processes Based on Hybrid Modeling: A Study Case on AISI 1045 Steel Turning. In Burstein, L. (Ed.), *Handbook of Research on Advancements in Manufacturing, Materials, and Mechanical Engineering* (pp. 215–241). IGI Global. https://doi.org/10.4018/978-1-7998-4939-1.ch010

Rahmani, M. K. (2022). Blockchain Technology: Principles and Algorithms. In Khan, S., Syed, M., Hammad, R., & Bushager, A. (Eds.), *Blockchain Technology and Computational Excellence for Society 5.0* (pp. 16–27). IGI Global. https://doi.org/10.4018/978-1-7998-8382-1.ch002

Ramdani, N., & Azibi, M. (2018). Polymer Composite Materials for Microelectronics Packaging Applications: Composites for Microelectronics Packaging. In Kumar, K., & Davim, J. (Eds.), *Composites and Advanced Materials for Industrial Applications* (pp. 177–211). Hershey, PA: IGI Global. 10.4018/978-1-5225-5216-1.ch009

Ramesh, M., Garg, R., & Subrahmanyam, G. V. (2017). Investigation of Influence of Quenching and Annealing on the Plane Fracture Toughness and Brittle to Ductile Transition Temperature of the Zinc Coated Structural Steel Materials. *International Journal of Surface Engineering and Interdisciplinary Materials Science*, 5(2), 33–42. 10.4018/IJSEIMS.2017070103

Robinson, J., & Beneroso, D. (2022). Project-Based Learning in Chemical Engineering: Curriculum and Assessment, Culture and Learning Spaces. In Alves, A., & van Hattum-Janssen, N. (Eds.), *Training Engineering Students for Modern Technological Advancement* (pp. 1–19). IGI Global. https://doi.org/10.4018/978-1-7998-8816-1.ch001

Rondon, B. (2021). Experimental Characterization of Admittance Meter With Crude Oil Emulsions. *International Journal of Electronics, Communications, and Measurement Engineering*, 10(2), 51–59. https://doi.org/10.4018/IJECME.2021070104

Rudolf, S., Biryuk, V. V., & Volov, V. (2018). Vortex Effect, Vortex Power: Technology of Vortex Power Engineering. In Kharchenko, V., & Vasant, P. (Eds.), *Handbook of Research on Renewable Energy and Electric Resources for Sustainable Rural Development* (pp. 500–533). Hershey, PA: IGI Global. 10.4018/978-1-5225-3867-7.ch021

Sah, A., Bhadula, S. J., Dumka, A., & Rawat, S. (2018). A Software Engineering Perspective for Development of Enterprise Applications. In Elçi, A. (Ed.), *Handbook of Research on Contemporary Perspectives on Web-Based Systems* (pp. 1–23). Hershey, PA: IGI Global. 10.4018/978-1-5225-5384-7.ch001

Sahli, Y., Zitouni, B., & Hocine, B. M. (2021). Three-Dimensional Numerical Study of Overheating of Two Intermediate Temperature P-AS-SOFC Geometrical Configurations. In Badea, G., Felseghi, R., & A chilean, I. (Eds.), *Hydrogen Fuel Cell Technology for Stationary Applications* (pp. 186–222). IGI Global. https://doi.org/10.4018/978-1-7998-4945-2.ch008

Sahoo, P., & Roy, S. (2017). Tribological Behavior of Electroless Ni-P, Ni-P-W and Ni-P-Cu Coatings: A Comparison. *International Journal of Surface Engineering and Interdisciplinary Materials Science*, 5(1), 1–15. 10.4018/IJSEIMS.2017010101

Sahoo, S. (2018). Laminated Composite Hypar Shells as Roofing Units: Static and Dynamic Behavior. In Kumar, K., & Davim, J. (Eds.), *Composites and Advanced Materials for Industrial Applications* (pp. 249–269). Hershey, PA: IGI Global. 10.4018/978-1-5225-5216-1.ch011

Sahu, H., & Hungyo, M. (2018). Introduction to SDN and NFV. In Dumka, A. (Ed.), *Innovations in Software-Defined Networking and Network Functions Virtualization* (pp. 1–25). Hershey, PA: IGI Global. 10.4018/978-1-5225-3640-6.ch001

Salem, A. M., & Shmelova, T. (2018). Intelligent Expert Decision Support Systems: Methodologies, Applications, and Challenges. In Shmelova, T., Sikirda, Y., Rizun, N., Salem, A., & Kovalyov, Y. (Eds.), *Socio-Technical Decision Support in Air Navigation Systems: Emerging Research and Opportunities* (pp. 215–242). Hershey, PA: IGI Global. 10.4018/978-1-5225-3108-1.ch007

Samal, M. (2017). FE Analysis and Experimental Investigation of Cracked and Un-Cracked Thin-Walled Tubular Components to Evaluate Mechanical and Fracture Properties. In Samui, P., Chakraborty, S., & Kim, D. (Eds.), *Modeling and Simulation Techniques in Structural Engineering* (pp. 266–293). Hershey, PA: IGI Global. 10.4018/978-1-5225-0588-4.ch009

Samal, M., & Balakrishnan, K. (2017). Experiments on a Ring Tension Setup and FE Analysis to Evaluate Transverse Mechanical Properties of Tubular Components. In Samui, P., Chakraborty, S., & Kim, D. (Eds.), *Modeling and Simulation Techniques in Structural Engineering* (pp. 91–115). Hershey, PA: IGI Global. 10.4018/978-1-5225-0588-4.ch004

Samarasinghe, D. A., & Wood, E. (2021). Innovative Digital Technologies. In Underwood, J., & Shelbourn, M. (Eds.), *Handbook of Research on Driving Transformational Change in the Digital Built Environment* (pp. 142–163). IGI Global. https://doi.org/10.4018/978-1-7998-6600-8.ch006

Sawant, S. (2018). Deep Learning and Biomedical Engineering. In Kose, U., Guraksin, G., & Deperlioglu, O. (Eds.), *Nature-Inspired Intelligent Techniques for Solving Biomedical Engineering Problems* (pp. 283–296). Hershey, PA: IGI Global. 10.4018/978-1-5225-4769-3.ch014

Schulenberg, T. (2021). Energy Conversion Using the Supercritical Steam Cycle. In Chen, L. (Ed.), *Handbook of Research on Advancements in Supercritical Fluids Applications for Sustainable Energy Systems* (pp. 659–681). IGI Global. 10.4018/978-1-7998-5796-9.ch018

Sezgin, H., & Berkalp, O. B. (2018). Textile-Reinforced Composites for the Automotive Industry. In Kumar, K., & Davim, J. (Eds.), *Composites and Advanced Materials for Industrial Applications* (pp. 129–156). Hershey, PA: IGI Global. 10.4018/978-1-5225-5216-1.ch007

Shaaban, A. A., & Shehata, O. M. (2021). Combining Response Surface Method and Metaheuristic Algorithms for Optimizing SPIF Process. *International Journal of Manufacturing, Materials, and Mechanical Engineering*, 11(4), 1–25. https://doi.org/10.4018/IJMMME.2021100101

Shafaati Shemami, M., & Sefid, M. (2022). Implementation and Demonstration of Electric Vehicle-to-Home (V2H) Application: A Case Study. In Alam, M., Pillai, R., & Murugesan, N. (Eds.), *Developing Charging Infrastructure and Technologies for Electric Vehicles* (pp. 268–293). IGI Global. https://doi.org/10.4018/978-1-7998-6858-3.ch015

Shah, M. Z., Gazder, U., Bhatti, M. S., & Hussain, M. (2018). Comparative Performance Evaluation of Effects of Modifier in Asphaltic Concrete Mix. *International Journal of Strategic Engineering*, 1(2), 13–25. 10.4018/IJoSE.2018070102

Sharma, N., & Kumar, K. (2018). Fabrication of Porous NiTi Alloy Using Organic Binders. In Kumar, K., & Davim, J. (Eds.), *Composites and Advanced Materials for Industrial Applications* (pp. 38–62). Hershey, PA: IGI Global. 10.4018/978-1-5225-5216-1.ch003

Shivach, P., Nautiyal, L., & Ram, M. (2018). Applying Multi-Objective Optimization Algorithms to Mechanical Engineering. In Ram, M., & Davim, J. (Eds.), *Soft Computing Techniques and Applications in Mechanical Engineering* (pp. 287–301). Hershey, PA: IGI Global. 10.4018/978-1-5225-3035-0.ch014

Shmelova, T. (2018). Stochastic Methods for Estimation and Problem Solving in Engineering: Stochastic Methods of Decision Making in Aviation. In Kadry, S. (Ed.), *Stochastic Methods for Estimation and Problem Solving in Engineering* (pp. 139–160). Hershey, PA: IGI Global. 10.4018/978-1-5225-5045-7.ch006

Siero González, L. R., & Romo Vázquez, A. (2017). Didactic Sequences Teaching Mathematics for Engineers With Focus on Differential Equations. In Ramírez-Montoya, M. (Ed.), *Handbook of Research on Driving STEM Learning With Educational Technologies* (pp. 129–151). Hershey, PA: IGI Global. 10.4018/978-1-5225-2026-9.ch007

Sim, M. S., You, K. Y., Esa, F., & Chan, Y. L. (2021). Nanostructured Electromagnetic Metamaterials for Sensing Applications. In Bhat, M., Wani, I., & Ashraf, S. (Eds.), *Applications of Nanomaterials in Agriculture, Food Science, and Medicine* (pp. 141–164). IGI Global. https://doi.org/10.4018/978-1-7998-5563-7.ch009

Singh, R., & Dutta, S. (2018). Visible Light Active Nanocomposites for Photocatalytic Applications. In Kumar, K., & Davim, J. (Eds.), *Composites and Advanced Materials for Industrial Applications* (pp. 270–296). Hershey, PA: IGI Global. 10.4018/978-1-5225-5216-1.ch012

Skripov, P. V., Yampol'skiy, A. D., & Rutin, S. B. (2021). High-Power Heat Transfer in Supercritical Fluids: Microscale Times and Sizes. In Chen, L. (Ed.), *Handbook of Research on Advancements in Supercritical Fluids Applications for Sustainable Energy Systems* (pp. 424–450). IGI Global. https://doi.org/10.4018/978-1-7998-5796-9.ch012

Sözbilir, H., Özkaymak, Ç., Uzel, B., & Sümer, Ö. (2018). Criteria for Surface Rupture Microzonation of Active Faults for Earthquake Hazards in Urban Areas. In Ceryan, N. (Ed.), *Handbook of Research on Trends and Digital Advances in Engineering Geology* (pp. 187–230). Hershey, PA: IGI Global. 10.4018/978-1-5225-2709-1.ch005

Stanciu, I. (2018). Stochastic Methods in Microsystems Engineering. In Kadry, S. (Ed.), *Stochastic Methods for Estimation and Problem Solving in Engineering* (pp. 161–176). Hershey, PA: IGI Global. 10.4018/978-1-5225-5045-7.ch007

Strebkov, D., Nekrasov, A., Trubnikov, V., & Nekrasov, A. (2018). Single-Wire Resonant Electric Power Systems for Renewable-Based Electric Grid. In Kharchenko, V., & Vasant, P. (Eds.), *Handbook of Research on Renewable Energy and Electric Resources for Sustainable Rural Development* (pp. 449–474). Hershey, PA: IGI Global. 10.4018/978-1-5225-3867-7.ch019

Sukhyy, K., Belyanovskaya, E., & Sukhyy, M. (2021). *Basic Principles for Substantiation of Working Pair Choice*. IGI Global. 10.4018/978-1-7998-4432-7.ch002

Suri, M. S., & Kaliyaperumal, D. (2022). Extension of Aspiration Level Model for Optimal Planning of Fast Charging Stations. In Fekik, A., & Benamrouche, N. (Eds.), *Modeling and Control of Static Converters for Hybrid Storage Systems* (pp. 91–106). IGI Global. https://doi.org/10.4018/978-1-7998-7447-8.ch004

Tallet, E., Gledson, B., Rogage, K., Thompson, A., & Wiggett, D. (2021). Digitally-Enabled Design Management. In Underwood, J., & Shelbourn, M. (Eds.), *Handbook of Research on Driving Transformational Change in the Digital Built Environment* (pp. 63–89). IGI Global. https://doi.org/10.4018/978-1-7998-6600-8.ch003

Terki, A., & Boubertakh, H. (2021). A New Hybrid Binary-Real Coded Cuckoo Search and Tabu Search Algorithm for Solving the Unit-Commitment Problem. *International Journal of Energy Optimization and Engineering*, 10(2), 104–119. https://doi.org/10.4018/IJEOE.2021040105

Tüdeş, Ş., Kumlu, K. B., & Ceryan, S. (2018). Integration Between Urban Planning and Natural Hazards For Resilient City. In Ceryan, N. (Ed.), *Handbook of Research on Trends and Digital Advances in Engineering Geology* (pp. 591–630). Hershey, PA: IGI Global. 10.4018/978-1-5225-2709-1.ch017

Ulamis, K. (2018). Soil Liquefaction Assessment by Anisotropic Cyclic Triaxial Test. In Ceryan, N. (Ed.), *Handbook of Research on Trends and Digital Advances in Engineering Geology* (pp. 631–664). Hershey, PA: IGI Global. 10.4018/978-1-5225-2709-1.ch018

Valente, M., & Milani, G. (2017). Seismic Assessment and Retrofitting of an Under-Designed RC Frame Through a Displacement-Based Approach. In Plevris, V., Kremmyda, G., & Fahjan, Y. (Eds.), *Performance-Based Seismic Design of Concrete Structures and Infrastructures* (pp. 36–58). Hershey, PA: IGI Global. 10.4018/978-1-5225-2089-4.ch002

Vargas-Bernal, R. (2021). Advances in Electromagnetic Environmental Shielding for Aeronautics and Space Applications. In Nikolopoulos, C. (Ed.), *Recent Trends on Electromagnetic Environmental Effects for Aeronautics and Space Applications* (pp. 80–96). IGI Global. https://doi.org/10.4018/978-1-7998-4879-0.ch003

Vasant, P. (2018). A General Medical Diagnosis System Formed by Artificial Neural Networks and Swarm Intelligence Techniques. In Kose, U., Guraksin, G., & Deperlioglu, O. (Eds.), *Nature-Inspired Intelligent Techniques for Solving Biomedical Engineering Problems* (pp. 130–145). Hershey, PA: IGI Global. 10.4018/978-1-5225-4769-3.ch006

Verner, C. M., & Sarwar, D. (2021). Avoiding Project Failure and Achieving Project Success in NHS IT System Projects in the United Kingdom. *International Journal of Strategic Engineering*, 4(1), 33–54. https://doi.org/10.4018/IJoSE.2021010103

Verrollot, J., Tolonen, A., Harkonen, J., & Haapasalo, H. J. (2018). Challenges and Enablers for Rapid Product Development. *International Journal of Applied Industrial Engineering*, 5(1), 25–49. 10.4018/IJAIE.2018010102

Wan, A. C., Zulu, S. L., & Khosrow-Shahi, F. (2021). Industry Views on BIM for Site Safety in Hong Kong. In Underwood, J., & Shelbourn, M. (Eds.), *Handbook of Research on Driving Transformational Change in the Digital Built Environment* (pp. 120–140). IGI Global. https://doi.org/10.4018/978-1-7998-6600-8.ch005

Yardimci, A. G., & Karpuz, C. (2018). Fuzzy Rock Mass Rating: Soft-Computing-Aided Preliminary Stability Analysis of Weak Rock Slopes. In Ceryan, N. (Ed.), *Handbook of Research on Trends and Digital Advances in Engineering Geology* (pp. 97–131). Hershey, PA: IGI Global. 10.4018/978-1-5225-2709-1.ch003

You, K. Y. (2021). Development Electronic Design Automation for RF/Microwave Antenna Using MATLAB GUI. In Nwajana, A., & Ihianle, I. (Eds.), *Handbook of Research on 5G Networks and Advancements in Computing, Electronics, and Electrical Engineering* (pp. 70–148). IGI Global. https://doi.org/10.4018/978-1-7998-6992-4.ch004

Yousefi, Y., Gratton, P., & Sarwar, D. (2021). Investigating the Opportunities to Improve the Thermal Performance of a Case Study Building in London. *International Journal of Strategic Engineering*, 4(1), 1–18. https://doi.org/10.4018/IJoSE.2021010101

Zindani, D., & Kumar, K. (2018). Industrial Applications of Polymer Composite Materials. In Kumar, K., & Davim, J. (Eds.), *Composites and Advanced Materials for Industrial Applications* (pp. 1–15). Hershey, PA: IGI Global. 10.4018/978-1-5225-5216-1.ch001

Zindani, D., Maity, S. R., & Bhowmik, S. (2018). A Decision-Making Approach for Material Selection of Polymeric Composite Bumper Beam. In Kumar, K., & Davim, J. (Eds.), *Composites and Advanced Materials for Industrial Applications* (pp. 112–128). Hershey, PA: IGI Global. 10.4018/978-1-5225-5216-1.ch006

About the Contributors

S. Kannadhasan is working as an Associate Professor in the department of Electronics and Communication Engineering in Study World College of Engineering, Tamilnadu, India. He is Completed the Ph.D in the field of Smart Antenna for Anna University in 2022. He is Twelve years of teaching and research experience. He obtained his B.E in ECE from Sethu Institute of Technology, Kariapatti in 2009 and M.E in Communication Systems from Velammal College of Engineering and Technology, Madurai in 2013.

R. Dhaya has 16 years experience in teaching and research in the field of Computer Science and Engineering. She published more than 80 research papers in peer reviewed international Journals. She was the recipient of IEI Young women Engineer award. Her areas of interests are wireless sensor networks, embedded systems, Machine Learning, Communication Systems.

R. Nagarajan received his B.E. in Electrical and Electronics Engineering from Madurai Kamarajar University, Madurai, India, in 1997. He received his M.E. in Power Electronics and Drives from Anna University, Chennai, India, in 2008. He received his Ph.D in Electrical Engineering from Anna University, Chennai, India, in 2014. He has worked in the industry as an Electrical Engineer. He is currently working as Professor of Electrical and Electronics Engineering at Gnanamani College of Technology, Namakkal, and Tamilnadu, India. His current research interest includes Power Electronics, Power System, Soft Computing Techniques and Renewable Energy Sources. He is published more than ninety research papers in various referred international journal.

Alagar Karthick working as Associate professor in the Electrical and Electronics Engineering department in KPR Institute of Engineering and Technology, Coimbatore, Tamilnadu, India. He has Published more than 100 International Journal and also reviewer for various journals such as Solar Energy, Fuel, Journal of cleaner Production, Heliyon, Building services Engineering research and Technology. Hereceived his Doctor of Philosophy in the field of Building Integrated Photovoltaic (BIPV) from Anna University, Chennai in 2018. He received his Master degree from Energy Engineering and bachelor degree in Electrical and Electronics Engineering. He has received Best Paper Award for his research articles in Biomass conversion. His research area includes Solar Photovoltaic, Bioenergy, zero energy buildings, Energy with Artificial Intelligence, Machine learning and Deep learning algorithms.

***

Arifa Ahmed received the B.Tech. degree in Electronics and Communication Engineering from North-Eastern Hill University (NEHU), Shillong, in 2011 and the M.Tech. degree from the same department, North Eastern Regional Institute of Science and Technology (NERIST), Itanagar, in 2014. She completed her Ph.D. degree in Electronics and Communication Engineering at National Institute of Technology (NIT) Silchar, India in 2022. She is currently working as an Assistant Professor in the Department of Electronics and Communication Engineering, Madanapalle Institute of Technology & Science, Andhra Pradesh, India. Her research interests include wireless communications, wireless networking, cognitive radio, and optimization.

|Aarif Ali - Contributing Author|Electrical Engineering student

C. Annadurai is the Associate Professor in the Department of ECE at SSN College of Engineering, Kalavakkam, and Chennai, India. He received B.E degree and M.E degree in 1991 and 2002, respectively, from Bharathiar University, Coimbatore India, and Ph.D from Anna University, Chennai, India in 2016. He is Life time member of ISTE and IEI. His research interests include several aspects of wireless communications such as MIMO, Cooperative communication, Machine Learning, Deep learning and Embedded Design.

Bhavanishankar Yamini has completed her Bachelor of Engineering in Computer Science and Engineering from Mailam Engineering College, in the year 2003. She pursued her Master of Technology in Information Technology from Sathyabama University, Chennai in the year 2007. She was awarded the Doctor of Philosophy in Computer Science and Engineering from Sathyabama Institute of Science and Technology, Chennai in the year 2020. She has published papers in various International and National Conferences and Journals. She is currently working at SRM Institute of Science and Technology, College of Engineering and Technology as Associate Professor in the Department of Networking and Communications. Her areas of interests include Network Security, Cyber Forensics, Image Processing, and Information Retrieval system, Machine Learning, Deep Learning and Cloud Computing.

Krishna Lal Baishnab received the B.E. degree from Regional Engineering College (currently, NIT Silchar) in 1995, the M.Tech. degree from IIT Kharagpur in 2004, and the Ph.D. degree from NIT Silchar, India, in 2014, where he is currently working as an Associate Professor with the Department of Electronics and Communication Engineering. His research interests include wireless systems, analog VLSI, optimization, and RF design.

Sameera Banu M is an active researcher who is currently pursuing her Bachelor of Engineering degree in Computer Science and Engineering at University College of Engineering Villupuram, India. She is doing her research work on Artificial Intelligence, Machine learning and Data Science projects.

Rabia Bilal completed her MPhil from the University of Sussex UK; Through out her research career, she is actively involved in design and developing hardware and software bassed systems. Her fields of expertise include AI; ML; Algorithms; Computer Vision; CT; Biomedical Systems; Image Processing

Gajalakshmi Duraikannu is an active researcher in the field of Interfacing Chemistry with Electronics for sensing applications, Machine Learning for scientific solutions, Synthesizing Plant derived Nano Materials for optoelectronic applications, Modelling Molecules for optoelectronic applications. she received her Ph.D. degree from the Anna University Chennai, Tamil Nadu, India in 2017. She is currently working as an Assistant Professor in the Department of Chemistry, University College of Engineering Villupuram, Tamil Nadu, India. Her current research is mainly focused on Interfacing Chemicals with microcontrollers for sensing application, Converting chemicals into sound signal for digitalization etc. She has published more than 50 papers in international journals and conferences. She has granted 1 patent and filed 3 patents and a trademark.

Nelson Iruthayanathan received the B.E., M.E., and Ph.D. degrees from the University of Madras, Chennai, India, in 1995, College of Engineering, Guindy and Anna University, Chennai in 1995, 1998 and 2018 respectively. His research field includes MIMO, multi-carrier communications in underwater communication and applications of machine learning algorithms.

Faheem Khuhawar is currently working as an Associate Professor at Department of Telecommunication Engineering, MUET. In 2009, he was awarded HEC's Open Merit Scholarship to pursue PhD studies in Italy and received the degree in Electronics and Communications Engineering from Politecnico di Torino, Italy in 2014. During his Ph.D., he was awarded 1000 Euro Grant by 25th International Teletraffic Congress (ITC), held at Shanghai, China to present his paper. In 2018, he received the award of "engineer of the year 2017" from IEEEP, PK. In 2019, he received the award of "excellent tutor" from Huawei Technologies Co., Ltd. within Middle East region. He has numerous national and international research publications to his credit. He holds various certifications and has been conducting various training workshops and tutorials in the domain of communication networks. He is also member of professional bodies such as IEEE Communication Society. Presently, he is focusing his research interests in energy efficient tcp/ip networks. In past, he has worked on Neural Networks and its application in data networks.

M. Nalini is an Associate Professor in the Department of Computer Science and Engineering, S.A.Engineering College, Chennai, India. She received her B.E. in Computer Science and Engineering from Anna University, Chennai in 2010 and M.Tech. in Computer Science and Engineering from B.S.Abdur Rahman Crescent Institute of Science & Technology, Chennai, India in 2012. She was awarded Ph.D. in Computer Science and Engineering from St. Peter's Institute of Higher Education and Research, Chennai, India in 2018. She has 10 years teaching experience. Her research interests include Data Mining, Big Data Analytics and Networking. She has published many articles in reputed Journals

Muhammad Imran Majid is Head of Electrical Engineering Department and Associate Professor at Institute of Business Management. He also served as Visiting Associate Professor University of Warwick a Commonwealth Ph.D. in Electronic Engineering and application for Science and Technology Development, a Commonwealth Fellow, and a Chartered Engineer registered with IET, UK. Currently, he heads the electrical engineering department at the Institute of Business Management, Karachi Pakistan. He has worked at University of Toronto, Surrey Satellite Technology Limited and Telenor Pakistan (pvt) Ltd. His research interests include network information theory, wireless communication, advanced network prototyping and humanitarian aspects of engineering

Deepak Mishra (Member, IEEE) received the B.Tech. degree in electronics and communication engineering from Guru Gobind Singh Indraprastha University, New Delhi, India, in 2012, and the Ph.D. degree in electrical engineering from the Indian Institutes of Technology Delhi, New Delhi, in 2017. He has been a Senior Lecturer with the School of Electrical Engineering and Telecommunications, University of New South Wales Sydney, Australia. Before that, he was a Postdoctoral Researcher with the Department of Electrical Engineering (ISY), Linköping University, Linköping, Sweden, from August 2017 to July 2019. He has also been a Visiting Researcher with the Northeastern University, Boston, MA, USA, University of Rochester, Rochester, NY, USA, Huawei Technologies, Paris, France, and Southwest Jiaotong University, Chengdu, China. His current research interests include energy harvesting cooperative communication networks, massive MIMO, backscattering, physical layer security, as well as signal processing and energy optimization schemes for the uninterrupted operation of wireless networks. He was a recipient of the IBM Ph.D. Fellowship Award in 2016, the Raman Charpak Fellowship Award in 2017, and the Endeavour Research Fellowship Award in 2018. He was also a Technical Program Committee Member for the IEEE International Conference on Communications from 2018 to 2022, IEEE Global Communications Conference (GLOBECOM) in 2021, IEEE Vehicular Technology Conference (VTC Spring) in 2021, and International Workshop on Wirelessly Powered Systems and Networks (WPSN) in 2019 and 2021. He was selected as an Exemplary Reviewer of IEEE TRANSACTIONS ON WIRELESS COMMUNICATIONS in 2017 and 2018, IEEE WIRELESS COMMUNICATIONS LETTERS in 2019, and IEEE TRANSACTIONS ON COMMUNICATIONS in 2019 and 2020.

Shabana Muhammad is working as a researcher & content writer. She has MS degree in Engineering Management and Bachelor's in Electrical Engineering. Her research interests include engineering project management, wireless sensor network, underwater acoustic communication, energy conservation, and applying engineering knowledge using AI/ML and Next Generation Networks.

Kamalendu Pal is currently associated with the Department of Computer Science, School of Science and Technology at City, University of London. He has a diverse academic background, having received a BSc (Hons) degree in Physics from Calcutta University, India, followed by a Postgraduate Diploma in Computer Science from Pune, India. He has also worked as a research assistant for the government of India's two research and development organizations, under the Council of Scientific and Industrial Research (CSIR). Later, Kamalendu graduated with his MSc in Software Systems Technology from the University of Sheffield, a Postgraduate Diploma in Artificial Intelligence from Kingston University, an MPhil in Computer Science from University College London, an MBA from the University of Hull, and a diploma in Higher Education teaching practice from City, University of London, United Kingdom. Kamalendu is an accomplished researcher, having published over a hundred international research articles (including book chapters) in the scientific community. His research papers have been published in prestigious journals such as the ACM SIGMIS Database, Expert Systems with Applications, Decision Support Systems, and international conferences. His research interests include various fields of computer science - such as knowledge-based systems, decision support systems, teaching and learning practices, blockchain technology, software engineering, service-oriented computing, sensor network simulation, and supply chain management. Kamalendu is also on the editorial board of an international computer science journal. He is a member of the British Computer Society, the Institution of Engineering and Technology, and the IEEE Computer Society.

Poornima Pandian is working as an Assistant Professor in the Department of Electronics and Communication Engineering, Sri Sai Ram Engineering College, Chennai. Her research areas includes cognitive radio network, machine learning techniques and optimisation techniques.

Arjun Paramarthalingam is an active researcher in the field of Computer Vision, Image Processing, and IoT. He received Ph.D. degree from the Anna University Chennai, Tamilnadu, India in 2017. He is currently working as an Assistant Professor in the Department of Computer Science and Engineering, University College of Engineering Villpuram, Tamilnadu, India. His current research is mainly on image processing, computer vision, pattern recognition, etc. He has published more than 25 papers in International journals and conferences

Ganesh Prasad (Member, IEEE) received the M.Tech. degree in communication systems from the Department of Electronics Engineering, IIT-BHU, Varanasi, and the Ph.D. degree from NIT Silchar. He joined NIT Raipur as an Assistant Professor, and after working briefly there, he joined NIT Silchar as an Assistant Professor with the Department of Electronics and Communication Engineering. He is also working on the project granted by CCBT, MeitY, Government of India, 2020. He has published his research works in many reputed conferences and journals in IEEE Communication Society along with various book chapters in Springer and IET. His research interests include cooperative 5G and beyond communications, underwater IoT networks, free-space optical communications, green communications, OFDM, NOMA, MIMO, classical convex optimization, and optimizations using machine learning. He is also a potential reviewer of many journals, such as IEEE Transactions on Wireless Communications, IEEE Transactions on Vehicular Technology, and IEEE Systems Journal. Since the last eight years, he is an active member of the IEEE Communication Society and has chaired many technical sessions in flagship IEEE Conferences, such as IEEE PIMRC and IEEE CCNC.

Samyuktha S R is an active scholar who is currently pursuing her Bachelor of Engineering degree in the department of Computer Science and Engineering at University College of Engineering Villupuram. She is doing her research work in Artificial Intelligence and Machine Learning related projects.

Said Safi received the B.Sc. degree from Cadi Ayyad University, Marrakech, Morocco in 1995, M.Sc. and the doctorate degree from Chouaib Doukkali University and Cadi Ayyad Univerrsity, in 1997 and 2002, respectively. He has been a Professor of information theory and telecommunication systems at the National School for applied Sciences, Tangier, Morocco, from 2003 to 2005. Since 2006, he is a Professor of applied mathematics and programming at Polydisciplinary Faculty, Sultan Moulay Slimane University, Beni Mellal, Morocco. In 2008 he received the Ph.D. degree in Telecommunication and Informatics from the Cadi Ayyad University. In 2015 he received the degree of Professor in Sciences at Sultan Moulay Slimane University. His general interests span the areas of communications and signal processing, estimation, time-series analysis, and system identification subjects on which he has published more than 70 journal papers and more than 80 conference papers. Current research topics focus on transmitter and receiver diversity techniques for single- and multi-user fading communication channels, and wide-band wireless communication systems.

Faisal K. Shaikh is working as a Professor at Department of Telecommunication Engg., Mehran University of Engineering & Technology (MUET), Jamshoro. He received Ph.D. in Computer Science from the Technische Universität Darmstadt, Germany. His research areas include Internet of Things (IoT), Wireless Sensor Networks, Vehicular Adhoc Networks, Smart Homes and Cities, Body Area Networks, and Underwater Sensor Networks. He is a founder of IoT Research Laboratory at MUET. He has published more than 125 refereed journal, conference, and book chapters. He is a senior member of IEEE, IEEE COMSOC, IEEE Computer Society and life member of PEC.

Tariq Rahim Soomro, professor of Computer Science, and Acting Rector at Institute of Business Management, has received BSc (Hons) and M.Sc degrees in Computer Science from University of Sindh, Jamshoro, Pakistan and his PhD in Computer Applications from Zhejiang University, Hangzhou, China and became the first Pakistani to receive all three degrees in the field of Computer Science. He has more than 29 years of extensive and diverse experience as an administrator, computer programmer, researcher, and teacher. As an administrator, He served as Coordinator, Head of Department, Head of Faculty, Dean of Faculty, Head of Academic Affairs and having wide experience in accreditation related matters, including ABET USA, NCEAC & HEC Pakistan, KHDA United Arab Emirates (UAE) and Ministry of Higher Education and Scientific Research (MoHESR), UAE. He has published over 100 peer-reviewed papers. He has been a senior member of IEEE since 2005 and IEEE Member since 2000. He is currently serving as Chair IEEE Karachi Section (2024-25) and IEEE Computer Society R10 Southern Area Coordinator of Computer Society along with Member FRC IEEE Educational Activities Committee.

Priyanga Subbiah received the B.Tech degree in Information Technology from New Prince Shri Bhavani College of Engineering and Technology, Chennai, India, in 2015 and the M.E. degree in Computer Science and Engineering from Thiagarajar College of Engineering, Madurai, in 2020. She is currently pursuing the Ph.D. degree in Computer Science and Engineering at SRM Institute of Science and Technology, Chengalpattu, Tamil Nadu, India. From 2022, she was a Research Scholar with the Computer Science and Engineering. His research interest includes the processing of images and detection of disease in the plant using Artificial Intelligence.

Index

R

S

T

W

www.ingramcontent.com/pod-product-compliance
Lightning Source LLC
LaVergne TN
LVHW080525300926
849162LV00007B/1538